Instabilities and Nonequilibrium Structures IV

Mathematics and Its Applications

Volume 267

Instabilities and Nonequilibrium Structures IV

edited by

E. Tirapegui
Facultad de Ciencias Físicas y Matemáticas,
Universidad de Chile,
Santiago, Chile

and

W. Zeller
Instituto de Física,
Universidad Católica de Valparaíso,
Valparaíso, Chile

SPRINGER-SCIENCE+BUSINESS MEDIA, B.V.

A C.I.P. Catalogue record for this book is available from the Library of Congress.

ISBN 978-94-010-4842-2 ISBN 978-94-011-1906-1 (eBook)
DOI 10.1007/978-94-011-1906-1

Printed on acid-free paper

TABLE OF CONTENTS

FOREWORD

We present here a selection of the lectures given at the Fourth International Workshop on Instabilities and Nonequilibrium Structures in Valparaíso, Chile, in December 1991. The Workshop was organized by Facultad de Ciencias Físicas y Matemáticas of Universidad de Chile and by Instituto de Física of Universidad Católica de Valparaíso where it took place.

This periodic meeting takes places every two years in Chile and aims to contribute to the efforts in Latin America towards the development of scientific research. This development is certainly a necessary condition for progress in our countries and we thank our lectures for their warm collaboration to fulfill this need. We are also very much indebted to the Chilean Academy of Sciences for sponsoring officially this Workshop.

We thank CONICYT (Comisión Nacional de Investigación Científica y Tecnológica de Chile) and his President Professor E. d'Etigny for his valuable help and also Fundación Andes of Chile. The efforts of M.D. d'Ollone, Scientific Advisor at the French Ambassy in Santiago, have been essential for our success and we acknowledge here the generous support of the Scientific Cooperation Program of France for Chile.

The organizers are especially grateful to Mr. J.A. Guzmán and Mr. B. Philippi, President and Director General of CHILGENER S.A., for their great interest and generous financial support to our Workshop which initiates a new style of collaboration between the chilean private sector and the scientific activity. We also thank Banco de Crédito e Inversiones for their support. Ms. Alda Bertoni deserves a special mention for her remarkable work during the realization of the Workshop. We are grateful to Professor Michiel Hazewinkel for including this book in his series and to Dr. David Larner of Kluwer for his continuous interest and support to this project.

E. Tirapegui
W. Zeller

LIST OF SPONSORS OF THE WORKSHOP

- Academia Chilena de Ciencias
- Facultad de Ciencias Físicas y Matemáticas de la U. de Chile
- Instituto de Física de la Universidad Católica de Valparaíso
- CONICYT (Chile)
- Ministere Francais des Affaires Etrangeres
- International Centre for Theoretical Physics (Trieste)
- DFG (Germany)
- FNRS (Belgium)
- Fundación Andes (Chile)
- Sociedad Chilena de Física
- Departamento Técnico de Investigación y de Relaciones Internacionales de la Universidad de Chile
- CHILGENER S.A.
- Banco de Crédito e Inversiones

PREFACE

We have classified the articles presented here in two Sections according to their general content.

In Part I we have included papers which deal with statistical mechanics, mathematical aspects of dynamical systems and sthochastic effects in nonequilibrium systems.

Part II is devoted mainly to instabilities and self-organization in extended nonequilibrium systems. The study of partial differential equations by numerical and analytic methods plays a great role here and many works are related to this subject. Most recent developments in this fascinating and rapidly growing area are discussed.

PART I

STATISTICAL MECHANICS AND RELATED TOPICS

NONEQUILIBRIUM POTENTIALS FOR PERIOD DOUBLING

R. Graham and A. Hamm
Fachbereich Physik, Universität Gesamthochschule Essen
D4300 Essen 1
Germany

ABSTRACT. In this lecture we consider the influence of weak stochastic perturbations on period doubling using nonequilibrium potentials, a concept which is explained in section 1 and formulated for the case of maps in section 2. In section 3 nonequilibrium potentials are considered for the family of quadratic maps (a) at the Feigenbaum 'attractor' with Gaussian noise, (b) for more general non-Gaussian noise, and (c) for the case of a strange repeller. Our discussion will be informal. A more detailed account of this and related material can be found in our papers [1-3] and in the reviews [4, 5], where further references to related work are also given.

1. Concept of a nonequilibrium potential

This concept arises in two different ways:

1. Let a set of dynamical equations of the general form $\dot{q}^\nu = K^\nu(q)$ be given where the $q^\nu(\nu = 1, \dots, n)$ is a set of dynamical variables and the $K^\nu(q)$ describe their instantaneous rate of change. Furthermore, let a set of transport coefficients $Q^{\nu\mu} = Q^{\mu\nu}$ be given, possibly depending on the q. The matrix $Q^{\nu\mu}$ is assumed to be positive definite. It is then possible, in principle, to cast $K^\nu(q)$ into a form familiar from thermodynamics by writing

$$K^\nu(q) = -\frac{1}{2}Q^{\nu\mu}(q)\frac{\partial\phi(q)}{\partial q^\mu} + r^\nu(q) \tag{1.1}$$

with an appropriately chosen nonequilibrium potential $\phi(q)$ and functions $r^\nu(q)$, $(\nu = 1, \dots, n)$ satisfying the constraint

$$r^\nu(q)\frac{\partial\phi(q)}{\partial q^\nu} = 0 \tag{1.2}$$

Thus within the metric provided by the transport matrix, eqs. (1.1), (1.2) split the drift in two parts, one being the gradient of a potential, the other being orthogonal to the gradient. In thermodynamics the gradient part would be irreversible and dissipative, the orthogonal part would be reversible and conservative. In special cases it may happen that $K^\nu(q)$ admits a $\phi(q)$ such that $r^\nu(q) = 0$ in eq. (1.1), in which case eq. (1.2) is trivially satisfied. The dynamics described by $K^\nu(q)$ is then a gradient dynamics i.e. K^ν has a potential $K^\nu(q) = -\frac{1}{2}Q^{\nu\mu}\partial\phi/\partial q^\mu$. Eqs. (1.1), (1.2) are, of course, more general and can be applied also in cases where K^ν has no potential. Eliminating $r^\nu(q)$ we obtain from eqs. (1.1), (1.2) a local condition on ϕ

$$K^\nu\frac{\partial\phi}{\partial q^\nu} + \frac{1}{2}Q^{\nu\mu}\frac{\partial\phi}{\partial q^\nu}\frac{\partial\phi}{\partial q^\mu} = 0 \tag{1.3}$$

which has the form of a Hamilton-Jacobi equation.

2. Let a noisy dynamical system be given by the stochastic differential equations (in the sense of Ito)

E. Tirapegui and W. Zeller (eds.), Instabilities and Nonequilibrium Structures IV, 5–13.

$$\dot{q}^\nu = K^\nu(q) + \sqrt{\eta}\, g_i^\nu(q)\xi_i(t) \tag{1.4}$$

where the K^ν are the same as before and the $\xi_i(t)$ $(i = 1 \ldots n)$ are independent Gaussian white noise sources of unit strength. η is a formal small parameter. The noise-coupling functions $g^{\nu i}(q)$ can define a matrix $Q^{\nu\mu}$ via

$$Q^{\nu\mu} = \sum_i g_i^\nu g_i^\mu \tag{1.5}$$

Let us now assume that the stochastic process has a stationary probability density $W_\eta(q)$ whose dependence on the strength of the noise we indicate explicitely. For small noise-intensity $W_\eta(q)$ assumes a form again reminiscent of thermodynamics, namely, for $\eta \to 0$

$$W_\eta(q) \simeq Z(q) \exp\left(-\frac{\phi(q)}{\eta}\right) \tag{1.6}$$

$\phi(q)$ is again the nonequilibrium potential. In fact, eq. (1.3) can be derived from eqs. (1.4)-(1.6) via the Fokker Planck equation, which is stochastically equivalent to (1.4). An important property following from (1.1), (1,2) with $\dot{q}^\nu = K^\nu$ is

$$\frac{d\phi}{dt} = -\frac{1}{2} Q^{\nu\mu} \frac{\partial\phi}{\partial q^\nu}\frac{\partial\phi}{\partial q^\mu} \leq 0 \tag{1.7}$$

If $\phi(q)$ is bounded from below, eq. (1.7) implies that ϕ is a Lyapunov function of $\dot{q}^\nu = K^\nu$ which is minimal in attractors of the system. The Hamilton Jacobi equation (1.3) implies that ϕ satisfies the variational principle related to such equations. In the simplest case where the system has only a single attractor A it takes the form

$$\phi(\bar{q}) = \inf S[q] \tag{1.8}$$

with the action functional

$$S[q] = \int_A^{\bar{q}} d\tau\, \frac{1}{2} Q_{\nu\mu}(q)\, (\dot{q}^\nu - K^\nu(q))(\dot{q}^\mu - K^\mu(q)) \tag{1.9}$$

where $Q_{\nu\mu}$ is the matrix inverse to $Q^{\nu\mu}$ and the paths $q(\tau)$ begin on A and end in $\bar{q}$. The time τ serves as an integration parameter. The infimum is taken over all paths starting in the attractor where ϕ is taken to vanish and ending in the point q. A canonical formalism can be set up for the problem with a Lagrangean given by the integrand of (1.9), a canonical momentum $p_\nu = Q_{\nu\mu}(\dot{q}^\mu - K^\mu)$, a Hamiltonian defined as usual, and a corresponding canonical form of the action

$$\phi(q) = \inf \left\{ \int_A^q p_\nu(q) dq^\nu \right\} \tag{1.10}$$

Here the integral is performed along solutions of the canonical equations of motion

$$\dot{q}^\nu = \frac{\partial H}{\partial p_\nu}, \quad \dot{p}_\nu = -\frac{\partial H}{\partial q^\nu} \tag{1.11}$$

with

$$H = \frac{1}{2} Q^{\nu\mu} p_\nu p_\mu + K^\nu p_\nu \tag{1.12}$$

2. Nonequilibrium potentials for maps

Maps occur e.g. for dynamical systems in discrete time. For noisy maps nonequilibrium potentials may again be introduced (see e.g. ref. [6] besides [1]-[3]).

Let us consider, e.g., a noisy map of the form

$$q_{n+1} = f(q_n) + \sqrt{\eta}\, g(q_n)\xi_n \tag{2.1}$$

with δ-correlated Gaussian noise ξ_n. Supposing the process has a stationary probability density $W_\eta(q)$ we have in general, for $\eta \to 0$

$$W_\eta(q) \simeq Z(q) \exp\left(-\frac{\phi(q)}{\eta}\right) \tag{2.2}$$

where $\phi(\bar{q})$ satisfies (for $g(q) \neq 0$)

$$\begin{aligned} \phi(\bar{q}) &= \inf S[q] \\ S[q] &= \frac{1}{2}\sum_{j=0}^{N-1} \left(q_{j+1} - f(q_j)\right)^2 / g^2(q_j). \end{aligned} \tag{2.3}$$

The infimum is taken over all sequences $q_j (0 \leq j \leq N)$ starting in the attractor (assuming for this discussion that there is only one) and ending in the point $q_N = \bar{q}$. For simplicity and concreteness we shall assume from now on that $g(q) = 1$. Eq. (1.7) has an analog

$$\phi(f(q_n)) \leq \phi(q_n) \tag{2.4}$$

and an analog of the canonical equations of motion also exists in the form (if $f'(q_n) \neq 0$)

$$\begin{aligned} q_{n+1} &= f(q_n) + p_{n+1} \\ p_{n+1} &= p_n / f'(q_n). \end{aligned} \tag{2.5}$$

If these equations are solved for the boundary conditions $q_0 = x$, $q_N = y$ then the analog of eq. (1.10) yields a function $V_N(x, y)$ via

$$V_N(x,y) = \frac{1}{2}\sum_{n=0}^{N-1} p_{n+1}^2 \tag{2.6}$$

which can be used to obtain $\phi(y)$ from the relation

$$\phi(y) = \inf_{(x)} \left(V_N(x,y) + \phi(x)\right). \tag{2.7}$$

As was shown by Reimann and Talkner [6] eqs. (2.6), (2.7) can be used even for $N = 1$ to obtain a rapidly converging numerical scheme by which $\phi(y)$ can be determined recursively, for $l \to \infty$, from

$$\begin{aligned} \phi_{l+1}(y) &= \inf_{(x)} \left(V_1(x,y) + \phi_l(x)\right) \\ V_1(x,y) &= \frac{1}{2}(y - f(x))^2 \end{aligned} \tag{2.8}$$

Another convenient method, which we have used to obtain figs. 1, 2 below, consists in solving eqs. (2.5) starting from a series of points close to the attractor and identifying $\phi(q)$ with the envelope from below of the functions

$$\tilde{\phi}_N(q) = \frac{1}{2} \sum_{n=0}^{N-1} p_{n+1}^2, \tag{2.9}$$

generated by these solutions for large N

$$\phi(q) = \inf \tilde{\phi}_N(q) \tag{2.10}$$

3. The family of quadratic maps

Let us now apply the concepts introduced in the preceding sections to the family of quadratic maps

$$x_{n+1} = f_\mu(x_n) = 1 - \mu x_n^2. \tag{3.1}$$

We shall consider the nonequilibrium potential at the Feigenbaum point $\mu = \mu_\infty = 1.4011\ldots$ in subsection 3.1 and inside the period-3 window with a strange repeller in subsection 3.3. In subsection 3.2 we discuss a generalization of the theory for weak non-Gaussian noise, again considering the case $\mu = \mu_\infty$.

3.1. NONEQUILIBRIUM POTENTIAL AT THE LIMIT OF INFINITE PERIOD DOUBLING

Fig. 1 gives a logarithmic plot of the nonequilibrium potential for $\mu = \mu_\infty$, determined by the method described in the preceding section. The following features are noteworthy: $\phi(x)$ is minimal (and zero) in all points of the 2^∞-cycle, which form a Cantor set indicated by the bars in fig. 1. In each gap of the Cantor set $\phi(x)$ has a sharp maximum at the unstable periodic point in this gap. $\phi(x)$ has the same value in all periodic points of a common cycle, i.e. $\phi(x) = \phi^{(n+1)}$ if x is a point of the unstable 2^n-cycle. Increasing n by 1 the height $\phi^{(n+1)}$ of the maxima decreases by a factor λ_n^{-1}

$$\frac{\phi^{(n)}}{\phi^{(n+1)}} = \lambda_n \tag{3.2}$$

which becomes independent of n, $\lambda_n \to \lambda$ for n large enough. From the figure one may infer $\lambda \approx 44$. Using the thermodynamic formalism for fractals it was shown in [2] that for Gaussian noise $\lambda = \kappa^2$ where κ is the noise-scaling constant introduced in refs. [7,8]. Some details are given in the following section. It is easy to understand this result in a qualitative way as follows: If $\phi^{(n)}$ is the potential maximum in the gap occupied by the unstable 2^{n-1}-cycle, then the iterated map $f^2 = f(f(x))$ must give a maximum

$$\bar{\phi}^{(n+1)} = \alpha^{-2}\phi^{(n)} \tag{3.3}$$

where α has its usual meaning as the rescaling parameter of the x-coordinate in the doubling transformation. Eq. (2.11) follows from the fact that the gap described by f^2 is smaller by a factor α^{-1} and ϕ increases from zero quadratically if x varies from the attractor into the gap. However, the noise in the noise-perturbed map f^2 is larger (in amplitude) by a factor $\alpha^{-1}\kappa$ compared to the noise in f [7] and therefore

$$\phi^{(n+1)} \sim \frac{\bar{\phi}^{(n+1)}}{(\alpha^{-1}\kappa)^2} = \kappa^{-2}\phi^{(n)}. \tag{3.4}$$

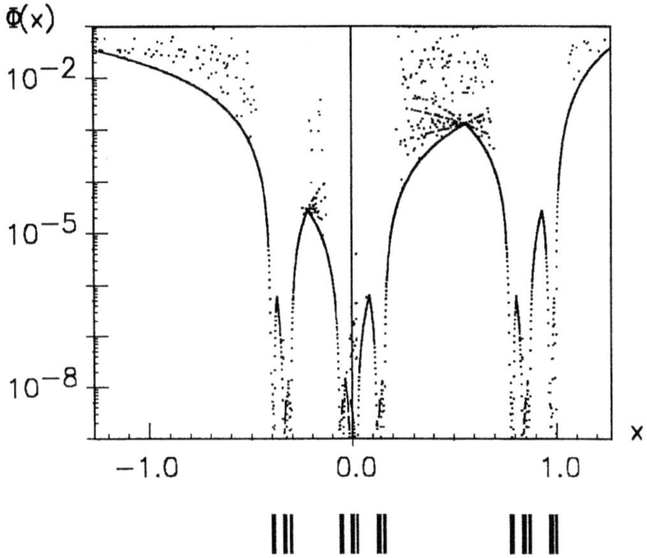

Figure 1: The nonequilibrium potential (logarithmic scale) of the logistic map with $\mu = 1.4011\ldots$. The scaling of the potential maxima at the unstable periodic orbits is characterized by κ.

This reasoning gives the correct result in the present case but it is somewhat misleading in its simplicity because the limits of validity are hidden. For instance it was often believed that the noise scaling with the same scaling constant applies universally also for non-Gaussian noise. This turns out not to be the case as will be discussed next.

3.2. NON-GAUSSIAN NOISE

In computer simulations of noisy maps localized noise is frequently used, whose amplitude is equidistributed over a small interval of size σ around zero. A continuous family of noise-distribution functions

$$W_r(\xi) \sim \exp\left(-\frac{1}{r}\left(\frac{|\xi|}{\sigma}\right)^r\right) \tag{3.5}$$

can be defined, with the Gaussian noise and the equidistributed amplitude limited noise as special cases $r = 2$ and $r \to \infty$, respectively.

Let us consider again the case $\mu = \mu_\infty$ of the quadratic maps, and let $F(x)$ be the function with quadratic maximum satisfying the Feigenbaum-Cvitanovich equation

$$F(x) = -\alpha F(F(\alpha^{-1}x)),\ \alpha = 2.503\ldots \tag{3.6}$$

Consider now the potential maxima in the gaps $\Delta_0^{(n)} = [-\alpha^{-n},\ \alpha^{-n}]$ of generation n near the origin. We shall now state a shadowing formula, which has been proved in [2]:

If an N-sequence minimizing the action

$$\begin{aligned} S_{N,r}\left[(q_i)_{0\le i\le N}\right] &= \sum_{j=0}^{N-1} \frac{1}{r}|q_{j+1} - F(q_j)|^r \\ V_{N,r}(x,y) &= \min S_{N,r} \end{aligned} \tag{3.7}$$

from $q_0 = x$ to $q_N = y$ is approximated to order ϵ by the unperturbed orbit of x generated by F, then we have the order ϵ estimate

$$V_{N,r}(x,y) = (1+0(\epsilon))\frac{1}{r}|y - F^N(x)|^r \left(\sum_{j=1}^{N}\prod_{k=j}^{N-1} |F'(F^k(x)|^{\frac{r}{r-1}}\right)^{1-r}. \tag{3.8}$$

In order to apply this formula, let us take $x = 0$ at the critical point of the map F and consider a 2^n-sequence of points close to the unperturbed orbit of x ending in the point of the unstable 2^{n-1}-cycle inside the gap. In this case $\phi(x) = 0$ and $V_{N,r}(x,y)$ with $N = 2^n$ gives an estimate of the maximal value $\phi_r^{(n)}$ of ϕ on the unstable 2^{n-1}-cycle.

The last bracket in eq. (3.8), which we denote by $(\mathcal{D}_r^{(n)})^{(1-r)}$, may be rewritten by using, for n large,

$$\mathcal{D}_r^{(n)} \approx |\Delta_{2^n}^{(n)}|^{\frac{r}{r-1}} \sum_{j=1}^{2^n} |\Delta_j^{(n)}|^{\frac{r}{1-r}} \tag{3.9}$$

where

$$\Delta_l^{(n)} = F^l(\Delta_0^{(n)});\quad l \le 2^n \tag{3.10}$$

Here we have employed the approximation

$$\Delta_l^{(n)} \simeq \sum_{k=0}^{l-1} \left|F'(F^k(0))\right| \Delta_0^{(n)} \tag{3.11}$$

The 'free energy' of the Cantor set is defined by the asymptotic relation

$$\sum_j |\Delta_j^{(n)}|^\beta \simeq e^{-n\beta\mathcal{F}(\beta)} \tag{3.12}$$

Here we need $\mathcal{F}$ for $\beta = \frac{r}{1-r}$ and obtain

$$\lim_{n\to\infty} \frac{1}{n} \ln \phi_r^{(n)} = -r\mathcal{F}\left(\frac{r}{1-r}\right). \tag{3.13}$$

For $r = 2$ we find

$$\frac{\phi^{(n)}}{\phi^{(n+1)}} = \exp(2\mathcal{F}(-2)) \approx (6.619)^2 \tag{3.14}$$

i.e. the result discussed in the preceding section. For equidistributed amplitude-limited noise, $r \to \infty$, it is appropriate to consider the scaling of $(\phi_r^{(n)})^{1/r}$. We obtain for $r \to \infty$

$$\left(\frac{\phi_r^{(n)}}{\phi_r^{(n+1)}}\right)^{1/r} \to \exp(\mathcal{F}(-1)) \simeq 8.490 \qquad (3.15)$$

in agreement with the scaling observed in numerical calculations with amplitude-limited noise in ref. [9].

3.3. STRANGE REPELLERS

Let us now consider the case of a strange repeller which occurs e.g. in the period-3 window of the quadratic map. Let $\{x_1, x_2, x_3\}$ denote, in increasing order, the stable and $\{z_1, z_2, z_3\}$ the nearby unstable period-3 orbit. The unstable invariant Cantor set existing in the period-3 window can be generated by defining the intervals

$$I_1 = [z_1, z_2], J_2 = (-z_2, z_2), I_2 = [z_2, z_3]$$

and repeatedly removing the preimages of J_2 from I_1, I_2. It can be shown that there is no path from the attractor (i.e. the attracting 3-cycle) to any $y \in I_1 \cup I_2$ with smaller action (3.7) than a minimizing path which goes to y via one of the boundary points z_i of $I_1 \cup I_2$, therefore $\phi(y) \geq \phi(z_i)$. On the other hand, the construction of the Cantor set implies that each of its gaps has an iterated preimage arbitrarily close to a z_i, hence $\phi(y) \leq \phi(z_i)$ for all $y \in I_1 \cup J_2 \cup I_2$ and $\phi(y) = \phi(z_i) = \text{const}$ for all $y \in I_1 \cup I_2$. The nonequilibrium potential is constant on the full interval $I_1 \cup I_2$ despite of the fact that the Lebesgue measure of the strange repeller vanishes. The situation is shown in fig. 2, which also shows the degenerate minima of ϕ on the stable 3-cycle.

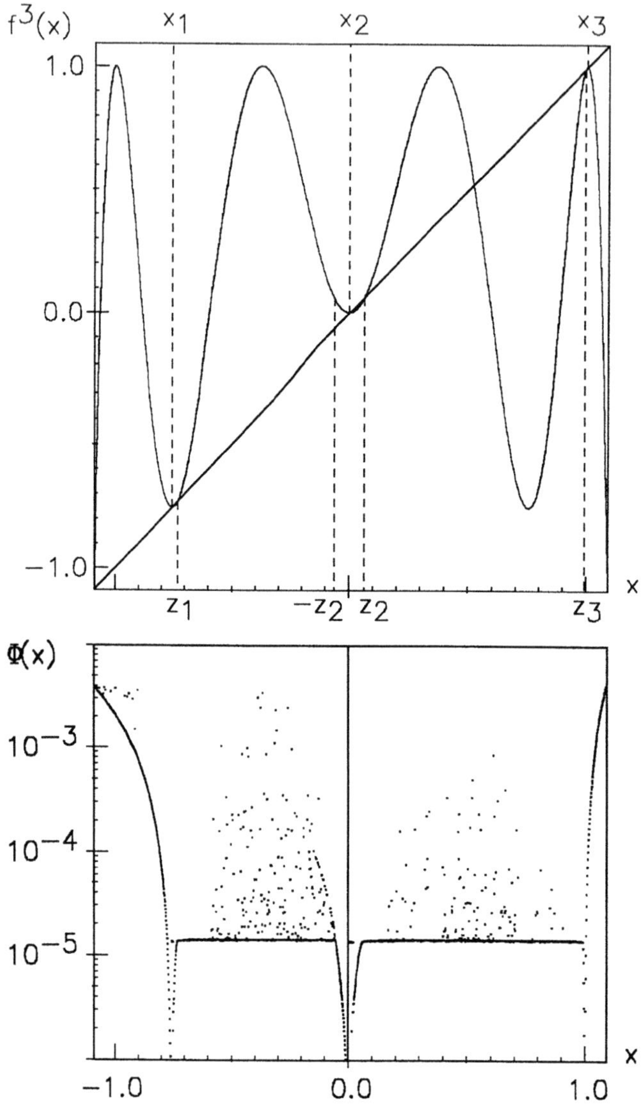

Figure 2: The third iterate (upper part) and the nonequilibrium potential (lower part) of the logistic map with $\mu = 1.7548$.

References

[1] R. Graham, A. Hamm and T. Tél, *Phys. Rev. Lett.* **66**, 3089 (1991).

[2] A. Hamm and R. Graham, *J. Stat. Phys.* **66**, 689 (1992).

[3] A. Hamm and R. Graham, 'Scaling for Small Random Perturbations of Golden Critical Circle Maps', preprint (Essen, 1992).

[4] R. Graham and A. Hamm, in *From Phase Transitions to Chaos* eds. G. Györgyi, I. Kondor, L. Sasvári, and T. Tél, (World Scientific, Singapore 1992).

[5] R. Graham and A. Hamm, in *Evolution of Dynamical Structures in Complex Systems* eds. A. Wunderlin and R. Friedrich, (Springer, Berlin 1992).

[6] P. Reimann and P. Talkner, *Phys. Rev.* **A44**, 6348 (1991).

[7] J. Crutchfield, M. Nauenberg and J. Rudnick, *Phys. Rev. Lett.* **46**, 933 (1981).

[8] B. I. Shraiman, C. E. Wayne, and P. C. Martin, *Phys. Rev. Lett.* **46**, 935 (1981).

[9] G. Mayer-Kress and H. Haken, *J. Stat. Phys.* **26**, 149 (1981).

ON SOLITON INSTABILITIES IN 1+1 DIMENSIONAL INTEGRABLE SYSTEMS

F. Lambert, J. Springael and R. Willox*.
Vrije Universiteit Brussel,
Dienst Theoretische Natuurkunde,
Pleinlaan 2, B-1050 Brussel.

*Research Assistant, National Fund for Scientific Research, Belgium.

ABSTRACT. One space dimensional soliton resonances are investigated for a family of integrable equations which are related by *translation* to the Sawada-Kotera hierarchy. These resonances are found to induce a structural instability responsible for the decay of solitons with amplitude belonging to a definite interval.

1. Introduction

The simplest solitons are sech-squared solitary wave solutions of integrable nonlinear partial differential equations (NLPDE's) in 1+1 dimensions, such as the KdV equation, the Sawada-Kotera (SK) equation and the Boussinesq equations[1)]. The fact that these equations admit exact N-soliton solutions, $N = 2, 3, ...$, which break up into N separated sech-squared pulses at large positive and negative times, has lead to the familiar picture of the soliton as a particularly stable excitation which retains its shape and velocity after collision with any number of other such excitations.

An additional feature of multisoliton interaction is its "two-ness", according to which the effect (phaseshift) on each soliton of a multiple collision is equal to a sum of two-soliton effects, as if each soliton were always interacting with only one other soliton at a time. An interesting exception are the s.c. "resonant soliton triads" (such triads are a characteristic of soliton interaction in two space dimensions[2,3)]) the first 1+1 dimensional examples of which were observed for the "good" Boussinesq (gB) equation[4,5)] (also called nonlinear string equation[6)]):

$$u_{tt} - u_{xx} + u_{xxxx} + 3(u^2)_{xx} = 0. \tag{1}$$

The discovery of these triads and of the corresponding resonant decay of large gB-solitons (with amplitude above a critical value) into two smaller solitons, was an indication that the integrability of a 1+1 dimensional soliton system does *not* guarantee the stability of its solitons.

The structural instability of large gB-solitons, confirmed by the observed instability of similar lattice excitations of a nonlinear atomic chain modelled by equ.(1) in the continuum limit[7)], was first regarded as an exceptional phenomenon due to a peculiarity of the gB-dispersion according to which small gB-solitons move faster

E. Tirapegui and W. Zeller (eds.), Instabilities and Nonequilibrium Structures IV, 15–25.

than larger ones[4]. In fact, it is easy to verify that a necessary "resonance condition" cannot be satisfied with KdV (SK) dispersion or higher-order KdV (SK) dispersion, nor with any simple dispersion mode characterized by a phase velocity $v(k)$ which is a continuous and monotonically increasing function of k.
However, with the discovery of other examples of one space dimensional soliton resonan-ces[8,9] it was soon realized that the possible splitting of an initial soliton profile is a more general ingredient of sech-squared soliton dynamics, which depends less on the dispersion law (linear terms of the equation) than on the particular form of the two-soliton coupling (linear plus quadratic terms). Thus, the existence of a critical amplitude for the gB-solitons can be predicted at the two-soliton level by the existence of an elliptic boundary of the regularity domain of the gB two-soliton solutions (in the two-soliton parameter plane) on which the two-soliton coupling factor is unbounded and on which the resonance condition is satisfied (with non trivial values of the soliton parameters). The unstable gB-solitons correspond to those sech-squared pulses which appear *near resonance* as "intermediate solitary waves" (formed by a temporary merger of two incoming solitons), and which are therefore bound to decay[10].
As soliton equations are known to appear in infinite families (hierarchies) with a characteristic two-soliton coupling it is interesting to investigate whether the resonance phenomenon and the corresponding soliton instabilities can be traced throughout a family. In this paper we discuss soliton resonances for integrable evolution equations, of first order in time, which belong to a hierarchy related to the Sawada-Kotera hierarchy[11]. These equations can be obtained with a direct binary operator method[12] which allows a systematic construction of NLPDE's (of increasing order) with two-soliton solutions:

$$u_2(x,t) = 2\partial_x^2 \ln\, f_2(\theta_1, \theta_2; A_{12})$$

$$f_2 = 1 + \exp\theta_1 + \exp\theta_2 + A_{12}\exp(\theta_1+\theta_2), \quad \theta_i = -k_i x + \omega_i t + \tau_i \tag{2}$$

with a prescribed two-soliton coupling factor A_{12}. The procedure is based on the introduction of multidimensional bilinear forms (involving auxiliary time variables) followed by a dimensional reduction with the help of lower order equations which were obtained at a previous stage.
The present results, predict the existence of a definite interval of soliton instability (resonance window) for all equations in the hierarchy.

2. Key-properties of the binary operators

We first review some basic properties of Hirota's binary operators which are defined by their action on a pair of functions $f, g : \mathbb{R}^n \to \mathbb{R}$

$$D^{p_1}_{x_1}\dots D^{p_n}_{x_n} f.g = \left(\partial_{x_1}-\partial_{x'_1}\right)^{p_1}\dots\left(\partial_{x_n}-\partial_{x'_n}\right)^{p_n} f(x_1,\dots x_n)g(x'_1,\dots x'_n)\Big|_{x'_i=x_i} \qquad (3)$$

The following two properties related to the action of D-operators on exponentials, follow straight away from this definition:

PROPERTY 1.

If $\Theta = \sum_{i=1}^{n}\alpha_i x_i + \tau$ and $\Theta' = \sum_{i=1}^{n}\beta_i x_i + \tau'$ with $\alpha_i, \beta_i, \tau, \tau' =$ constant (4)

then: $$D^{p_1}_{x_1}\dots D^{p_n}_{x_n}\exp\Theta.\exp\Theta' = \left[\prod_{i=1}^{n}(\alpha_i-\beta_i)^{p_i}\right]\exp(\Theta+\Theta'). \qquad (5)$$

PROPERTY 2.

Let $(p_1,\dots p_n;m)$ be a set of integers with $p_1+\dots+p_n = 2m.$

If $q_{p_1x_1,\dots p_nx_n} = \partial^{p_1}_{x_1}\dots\partial^{p_n}_{x_n}q(x_1,\dots x_n)$, then:

$$f^{-2}D^{p_1}_{x_1}\dots D^{p_n}_{x_n}f.f\Big|_{f=\exp q/2} = P_{p_1x_1,\dots p_nx_n}(q), \qquad (6)$$

where the r.h.s. of equ.(6) denotes that "partitional polynomial" in even order partial derivatives of q which contains as many different terms of degree p $(1\le p\le m)$ of the form:

$$C(r_{ij})\prod_{i=1}^{p}q_{r_{i1}x_1,\dots r_{in}x_n},\quad \sum_{i=1}^{p}r_{ij}=p_j,\quad \sum_{j=1}^{n}r_{ij}=2m_i,$$
$$m_i = \text{ integer},\quad r_{ij} = \text{ integer or zero}, \qquad (7)$$

as there are different ways of dividing $2m$ elements, of which p_i are of type x_i, among p indistinguishable non-empty boxes with the condition that each box must contain an even number of elements, each coefficient $C(r_{ij})$ being equal to the combinatorial weight of the corresponding partition (number of such partitions if all elements can be distinguished).
This property is related to the familiar partitional property of the ordinary n^{th} order derivative acting on the exponential of $q(x)$:

$$\exp(-q)\partial^n_x\exp(q) \equiv \sum\left[c_1!\dots c_n!(2!)^{c_2}\dots(n!)^{c_n}\right]^{-1} n!q_x^{c_1}q_{2x}^{c_2}\dots q_{nx}^{c_n} \qquad (8)$$

where the sum is taken over all partitions of n which are written as follows:

$$n = c_1 + 2c_2 + \ldots + nc_n, \quad c_i = \text{number of parts equal to } i.$$

3. The fifth-order translated SK equation

An interesting equation to start with is a fifth-order equation for $u(x,t)$, which mixes SK-dispersion with KdV-dispersion, and which was discussed by Hirota and Ito[8]. The integrability of this equation - we shall call it $\widetilde{SK}_5$ - follows from the fact that it is simply related to the fifth-order SK equation:

$$SK_5(u) \equiv u_t + u_{5x} + 15(uu_{2x} + u^3)_x = 0, \tag{9}$$

by the transformation:

$$u \longrightarrow u + c, \qquad c = \text{constant}. \tag{10}$$

$$\widetilde{SK}_5(u) \equiv u_t + u_{5x} + 15cu_{3x} + 45c^2u_x + 15(uu_{2x} + 3cu^2 + u^3)_x = 0, \tag{11}$$

We notice that this *translation* opens the way to resonances as it transforms the monotonic SK_5-phase velocity $v(k) = k^4$ into a non-monotonic one (when $c < 0$):

$$\widetilde{v}(k) = k^4 + 15ck^2 + 45c^2. \tag{12}$$

Another point to be made is that the above translation preserves the *bilinearizability* of the equation: both SK_5 and $\widetilde{SK}_5$ can be derived from a two-dimensional bilinear equation which guarantees the existence of two-soliton solutions $u_2(x,t)$ which vanish as $x \to \pm\infty$.
Indeed, an easy way to derive the SK_5 equation is by looking for an NLPDE with linear part:

$$L_5(u) = (\partial_t + \partial_x^5)u(x,t) \tag{13}$$

and with two-soliton solutions of type (2) with $\theta_i = -k_i x + k_i^5 t + \tau_i$. It follows from the structure of f_2 and from property 1 that:

$$\begin{aligned} D_x(D_t + D_x^5)f_2 \cdot f_2 = {} & 10k_1k_2\{(k_1+k_2)^2(k_1^2 + k_1k_2 + k_2^2)A_{12} \\ & - (k_1-k_2)^2(k_1^2 - k_1k_2 + k_2^2)\}\exp(\theta_1+\theta_2), \end{aligned} \tag{14}$$

indicating that f_2 solves the bilinear equation:

$$\text{Bil } SK_5(f) \equiv F_6(D_x, D_t) f \cdot f = 0, \qquad F_6(D_x, D_t) = D_x(D_t + D_x^5), \tag{15}$$

iff

$$A_{12} = A_{12}^{SK} \equiv \frac{(k_1 - k_2)^2 (k_1^2 - k_1 k_2 + k_2^2)}{(k_1 + k_2)^2 (k_1^2 + k_1 k_2 + k_2^2)}. \tag{16}$$

On account of property 2 it is clear that $q_2(\theta_1, \theta_2; A_{12}^{SK}) = 2\ln f_2(\theta_1, \theta_2; A_{12}^{SK})$ solves the "primary" SK_5 equation:

$$PSK_5(q) \equiv f^{-2} \text{Bil } SK_5(f)\big|_{f=\exp(q/2)}$$

$$\equiv P_{x,t}(q) + P_{6x}(q) \equiv q_{x,t} + q_{6x} + 15(q_{2x} q_{4x} + q_{2x}^3) = 0, \tag{17}$$

and, after differentiation with respect to x, that $u_2(\theta_1, \theta_2; A_{12}^{SK}) = \partial_x^2 q_2(\theta_1, \theta_2, A_{12}^{SK})$ solves the equation (9).
It is easy to see that the $S\widetilde{K}_5$ equation (11) can similarly be derived from a primary equation:

$$PS\widetilde{K}_5(q) \equiv P_{x,t}(q) + P_{6x}(q) + 15cP_{4x}(q) + 45c^2 P_{2x}(q)$$

$$\equiv f^{-2}\left[D_x D_t + D_x^6 + 15cD_x^4 + 45c^2 D_x^2\right] f \cdot f\big|_{f=\exp(q/2)} = 0 \tag{18}$$

which corresponds to the translated bilinear equation:

$$\text{Bil } S\widetilde{K}_5(f) \equiv \widetilde{F}_6(D_x, D_t) f \cdot f = 0,$$

$$\widetilde{F}_6(D_x, D_t) = D_x(D_t + D_x^5 + 15cD_x^3 + 45c^2 D_x) \tag{19}$$

Substitution of $f_2(\widetilde{\theta}_1, \widetilde{\theta}_2; A_{12}^{S\widetilde{K}})$ with:

$$\widetilde{\theta}_i = -k_i x + \widetilde{\omega}_i^{(5)} t + \tau_i, \qquad \widetilde{\omega}_i^{(5)} = \widetilde{\omega}^{(5)}(k_i) = k_i^5 + 15ck_i^3 + 45c^2 k_i \tag{20}$$

and with

$$A_{12}^{S\widetilde{K}} = -\frac{\widetilde{F}(-k_1 + k_2, \widetilde{\omega}_1^{(5)} - \widetilde{\omega}_2^{(5)})}{\widetilde{F}(-k_1 - k_2, \widetilde{\omega}_1^{(5)} + \widetilde{\omega}_2^{(5)})} = \frac{(k_1 - k_2)^2 (k_1^2 - k_1 k_2 + k_2^2 + 9c)}{(k_1 + k_2)^2 (k_1^2 + k_1 k_2 + k_2^2 + 9c)}, \tag{21}$$

into this equation shows that the $S\widetilde{K}_5$ equation does still admit two-soliton solutions $u_2(\widetilde{\theta}_1, \widetilde{\theta}_2; A_{12}^{S\widetilde{K}})$ in addition to the translated multisoliton solutions of equ.(9) which obey non-zero boundary conditions.

4. Soliton resonances for $S\widetilde{K}_5$

Inspection of $A_{12}^{S\widetilde{K}}$ shows the existence, for $c < 0$, of resonant triads involving sech-squared solitons with parameters (k_1, k_2) on the ellipse:

$$k_1^2 + k_1 k_2 + k_2^2 - 9|c| = 0, \tag{22}$$

where $A_{12}^{S\widetilde{K}}$ is unbounded, and where the resonance condition

$$\widetilde{\omega}^{(5)}(k_1 + k_2) = \widetilde{\omega}^{(5)}(k_1) + \widetilde{\omega}^{(5)}(k_2) \tag{23}$$

is fulfilled:

$$\widetilde{\omega}^{(5)}(k_1 + k_2) - \widetilde{\omega}^{(5)}(k_1) - \widetilde{\omega}^{(5)}(k_2) = 5k_1 k_2 (k_1 + k_2)(k_1^2 + k_1 k_2 + k_2^2 - 9|c|) = 0 \tag{24}$$

The analogy with the gB-case suggests that these resonances should lead to a structural instability of some of the $S\widetilde{K}_5$-solitons. In order to identify these unstable solitons it is instructive to consider the behaviour near resonance of solutions:

$$\widetilde{u}_2 = u_2(\widetilde{\theta}_1, \widetilde{\theta}_2; A_{12}^{S\widetilde{K}})\Big|_{\tau_1 = -\ln\, A_{12}^{S\widetilde{K}},\ \tau_2 = 0} \tag{25}$$

Let us approach the elliptic boundary (22) from neighbouring points inside the ellipse, say with $0 < k_1 < \sqrt{3}\sqrt{|c|} < k_2 < 3\sqrt{|c|}$, where $0 < A_{12}^{S\widetilde{K}} < \infty$ and where the soliton velocities $v_i = \widetilde{\omega}_i^{(5)}/k_i$ obey the ordering $v_1 > v_2$. The solution $\widetilde{u}_2 = 2\partial_x^2 \ln\, \widetilde{f}_2$ describes a regular two-soliton collision process as indicated by the asymptotic analysis of:

$$\begin{aligned}\widetilde{f}_2 =& 1 + \frac{1}{A_{12}^{S\widetilde{K}}} \exp(-k_1 x + \widetilde{\omega}_1^{(5)} t) + \exp(-k_2 x + \widetilde{\omega}_2^{(5)} t) \\ & + \exp\left[-(k_1 + k_2)x + (\widetilde{\omega}_1^{(5)} + \widetilde{\omega}_2^{(5)})t\right]\end{aligned} \tag{26}$$

as $t \to \pm\infty$.

Following soliton "j", $j = 1, 2$, by setting $x = v_j t + \xi$ one sees that $\widetilde{f}_2(\xi, t)$ is asymptotically dominated at large $|t|$ by two "competing" exponential terms with the same t-dependence, and such that:

$$\widetilde{u}_2(\xi,t) \underset{t\to-\infty}{\longrightarrow} 2\partial_\xi^2 \ln\left[1+\exp(-k_j\xi)\right] = \frac{1}{2}k_j^2 \text{sech}^2\left(\frac{k_j\xi}{2}\right)$$

$$\widetilde{u}_2(\xi,t) \underset{t\to+\infty}{\longrightarrow} \frac{1}{2}k_j^2 \text{sech}^2\left[\frac{k_j\xi-(-)^j \ln A_{12}^{\widetilde{SK}}}{2}\right] \tag{27}$$

It is important to emphasize that, as (k_1, k_2) approaches the ellipse (22), the asymptotic solitons "1" and "2" are not the only sech^2-shaped pulses which can emerge from $\widetilde{u}_2$. Indeed, setting $x = v_R t + \xi$ with

$$v_R = \frac{\widetilde{\omega}_1^{(5)} + \widetilde{\omega}_2^{(5)}}{k_1+k_2} = k_1^4 - k_1^3 k_2 + k_1^2 k_2^2 - k_1 k_2^3 + k_2^4 - 15|c|(k_1^2 - k_1 k_2 + k_2^2) + 45c^2, \tag{28}$$

it is clear that:

$$k_1(v_1 - v_R) = -k_2(v_2 - v_R) = a_R, a_R = k_1 k_2(k_1 - k_2)(k_1^2 + k_2^2 - 15|c|) > 0, \tag{29}$$

and that the expression

$$\widetilde{f}_2(\xi,t) = 1 + \frac{1}{A_{12}^{\widetilde{SK}}}\exp(a_R t - k_1\xi) + \exp(-a_R t - k_2\xi) + \exp\left[-(k_1+k_2)\xi\right], \tag{30}$$

is dominated near $\xi = 0$ by its t-independent terms at *times* t which are such that:

$$1 \ll \exp(a_R t) \ll A_{12}^{\widetilde{SK}}. \tag{31}$$

By taking (k_1, k_2) close enough to the ellips one can make $A_{12}^{\widetilde{SK}}$ large enough so as to get $\widetilde{f}_2(\xi,t)$ as close as one wishes to the one-soliton expression $1+\exp\left[-(k_1+k_2)\xi\right]$ during a significant period of time (of the order of $\frac{1}{a_R}\ln A_{12}^{\widetilde{SK}}$). During this time $\widetilde{u}_2(x,t)$ will be almost identical with the travelling solitary wave profile of the one-soliton solution:

$$\widetilde{u}_{\text{sol}} = \frac{1}{2}k_R^2 \text{sech}^2\left[\frac{k_R(x - v_R t)}{2}\right], \qquad k_R = k_1 + k_2. \tag{32}$$

This solitary peak cannot survive asymptotically (as $t \to \pm\infty$) as long as $A_{12}^{\widetilde{SK}}$ remains finite. However, when (k_1, k_2) reaches the ellipse ($k_1 = \frac{1}{2}\left(-k_2 + \sqrt{36|c| - 3k_2^2}\right)$, $\sqrt{3|c|} < k_2 < 3\sqrt{|c|}$) it follows from formula (30) that the intermediate solitary

wave survives, as $t \to +\infty$, as an outgoing soliton: the solution degenerates into a "resonant soliton triad" describing the merger of two incoming solitons (with parameters k_1, v_1 and k_2, v_2) into one large outgoing soliton (with parameters $k_R = k_1 + k_2$ and v_R).
Thus, by taking (k_1, k_2) near the elliptic boundary $(k_1^2 + k_1 k_2 + k_2^2 - 9|c| = 0$, $0 < k_1 < \sqrt{3|c|})$ it is found that the $S\widetilde{K}_5$ equation has regular two-soliton solutions $\widetilde{u}_2$ which describe a particle-like "resonance process" in which two incoming solitons collide and fuse to form a larger intermediate solitary wave which, eventually, decays into the same outgoing solitons. The amplitude of the intermediate sech-squared pulse is determined by $k_R = k_1 + k_2$, with $3\sqrt{|c|} < k_R < 2\sqrt{3|c|}$. Its shape remains practically unchanged during its lifetime except for continuous but unperceivable changes in the details of its profile. The structural instability of the corresponding $S\widetilde{K}_5$-solitons, with amplitude ranging from $\frac{9}{2}|c|$ to $6|c|$, is linked to the fact that their sech-squared profiles can be approximated with an arbitrary accuracy by intermediate solitary waves which are bound to decay after a finite period of time.

5. The seventh-order translated SK equation

It is natural to regard $S\widetilde{K}_5$ as belonging to an integrable "translated SK-hierarchy" which comprises the SK-hierarchy (at $c = 0$). The next equation of this $S\widetilde{K}$ family should be of seventh order and should correspond (through translation) to the seventh-order equation of the SK-hierarchy. We shall now see that both equations can be derived from similar three-dimensional bilinear systems.
We look first for an equation with linear part $L_7(u) = (\partial_t + \partial_x^7)u(x,t)$ and with two-soliton solutions of type (2) with $\theta_i = -k_i x + k_i^7 t + \tau_i$ and $A_{12} = A_{12}^{SK}$.
It then follows from property 1 that:

$$
\begin{aligned}
D_x(D_t + D_x^7) f_2 \cdot f_2 = {} & 14 k_1 k_2 \left\{ (k_1 + k_2)^2 (k_1^2 + k_1 k_2 + k_2^2)^2 A_{12}^{SK} \right. \\
& \left. -(k_1 - k_2)^2 (k_1^2 - k_1 k_2 + k_2^2)^2 \right\} \exp(\theta_1 + \theta_2) \\
= {} & 28 k_1^2 k_2^2 (k_1 - k_2)^2 (k_1^2 - k_1 k_2 + k_2^2) \exp(\theta_1 + \theta_2) \qquad (33)
\end{aligned}
$$

An appropriate "counter-term" can easily be found in terms of an auxiliary t-variable, say t_5, which corresponds to SK_5 dispersion. Thus, setting $\theta_i = -k_i x + k_i^7 t + k_i^5 t_5 + \tau_i'$ (by choosing $\tau_i = k_i^5 t_5 + \tau_i'$) and keeping $A_{12} = A_{12}^{SK}$ it is found (via property 1) that:

$$
D_x^3(D_{t_5} + D_x^5) f_2 \cdot f_2 = 40 k_1^2 k_2^2 (k_1 - k_2)^2 (k_1^2 - k_1 k_2 + k_2^2) \exp(\theta_1 + \theta_2) \qquad (34)
$$

It follows that $f_2(x,t,t_5)$ solves the 3-dimensional bilinear system:

$$\text{Bil}\, SK_7(f)\ : \quad \left[D_x(D_t + D_x^7) - \frac{7}{10}D_x^3(D_{t_5} + D_x^5)\right] f\cdot f = 0 \tag{35}$$

$$F_6(D_x, D_{t_5}) f\cdot f = 0 \tag{36}$$

and (via property 2) that $q_2(x,t,t_5) = 2\ln\, f_2(x,t,t_5)$ solves the primary system:

$$q_{x,t} + \frac{3}{10}P_{8x}(q) - \frac{7}{10}P_{3x,t_5}(q) = 0 \tag{37}$$

$$q_{x,t_5} + P_{6x}(q) = 0 \tag{38}$$

Elimination of the auxiliary variable t_5 from equ.(37) with the help of equ.(38), yields, after differentiation with respect to x, the following equation for the field $u(x,t) = \partial_x^2 q(x,t)$:

$$SK_7(u) \equiv u_t + u_{7x} + 21(uu_{4x} + u_x u_{3x} + u_{2x}^2 + 6u^2 u_{2x} + 3uu_x u_x^2 + 3u^4)_x = 0 \tag{39}$$

It is worth noticing that there is no other possible choice of a bilinear form of order 8, combining the required SK_7-dispersion with the auxiliary SK_5-dispersion than the above:

$$F_8(D_x, D_t, D_{t_5}) = D_x(D_t + D_x^7) - \frac{7}{10}D_x^3(D_{t_5} + D_x^5) \tag{40}$$

It implies that the equ.(39) is the only seventh-order candidate within the framework of a bilinearizable SK-hierarchy. It can be recognized as the actual seventh-order equation of the SK-hierarchy (it is identical with the seventh-order equation that can be obtained by applying the SK recursion operator[11)]).

Let us now look for a seventh-order equation - say $\widetilde{K}_7$ - belonging to the thranslated SK-hierarchy ($S\widetilde{K}$-family) with the resonance bearing two-soliton coupling factor (21). We presume that this equation should also correspond to a three-dimensional bilinear system which could be regarded as a translated version of Bil $SK_7(f)$. This bilinear system should contain the translated form $\widetilde{F}_6(D_x, D_{t_5})$ as well as a translated leading form $F_8(D_x, D_t, D_{t_5})$, consisting of a "main term" and a "counter-term". The "mian-term" must account for the correct $S\widetilde{K}_7$-dipersion as determined by the linear part of the translated version of equ.(39):

$$\widetilde{L}_7(u) = u_t + u_{7x} + 21cu_{5x} + 126c^2 u_{3x} + 252c^3 u_x, \tag{41}$$

the counter-term being determined by $\widetilde{F}_6(D_x, D_{t_5})$. This yields the following

$$\text{Bil}\, S\widetilde{K}_7(f) \equiv \begin{cases} \widetilde{F}_8(D_x, D_t, D_{t_5}) f \cdot f = 0 & (42) \\ \widetilde{F}_6(D_x, D_{t_5}) f \cdot f = 0 & (43) \end{cases}$$

with

$$\widetilde{F}_8(D_x, D_t, D_{t_5}) = D_x(D_t + D_x^7 + 21cD_x^5 + 126c^2D_x^3 + 252c^3D_x) - \frac{7}{10}D_x^3(D_{t_5} + D_x^5 + 15cD_x^3 + 45c^2D_x) \tag{44}$$

as the only possible candidate for $S\widetilde{K}_7$.

Again, it is a straightforward matter to check that the system (41,42) admits two-soliton generating solutions f_2 with $A_{12} = A_{12}^{S\widetilde{K}}$ and with:

$$\widetilde{\theta}_i = -k_i x + \widetilde{\omega}_i^{(7)} t + \widetilde{\omega}_i^{(5)} t_5 + \tau_i$$

$$\widetilde{\omega}_i^{(7)} = \widetilde{\omega}^{(7)}(k_i) = k_i^7 + 21ck_i^5 + 126c^2k_i^3 + 252c^3k_i \tag{45}$$

It is also easy to verify (by elimination of t_5) that the system (42,43) produces a seventh-order equation for $u(x,t)$ which corresponds to the SK_7 equation through the translation (10). It is clear that this equation admits resonant soliton triads on account of two-soliton resonances which are bound to occur on the above ellipse (22), where the resonance condition is satisfied:

$$\widetilde{\omega}^{(7)}(k_1 + k_2) - \widetilde{\omega}^{(7)}(k_1) - \widetilde{\omega}^{(7)}(k_2) =$$

$$7k_1k_2(k_1 + k_2)\left[k_1^2 + k_1k_2 + k_2^2 + 9c\right]\left[k_1^2 + k_1k_2 + k_2^2 + 6c\right] = 0 \tag{46}$$

It should be noticed that the resonance condition is also satisfied on the ellipse $k_1^2+k_1k_2+k_2^2-6|c| = 0$, but that this curve does not correspond to actual resonances.

The analysis of section 4 predicts the same range of soliton instability for $S\widetilde{K}_7$ as for $S\widetilde{K}_5$: solitons with amplitude between $\frac{9}{2}|c|$ and $6|c|$ are coupled to a decay channel.

Let us finally mention that the present analysis can be extended to higher orders. It is possible to show the existence of an infinite $S\widetilde{K}$-hierarchy of soliton equations of order $n = 2m + 1$, with $m \geq 2$ and $n \neq$ multiple of 3, which corresponds through translation to the SK-hierarchy. The $S\widetilde{K}$-equation of order n is given by an explicit (generic) bilinear system which can be derived from the generic bilinear form[13] associated with the SK-hierarchy. This generic bilinear representation, of order $n+1$, can be used to verify that similar soliton instabilities, generated by resonances occuring on the ellipse (22), are to be expected at each order in the hierarchy.

References

1. Ablowitz, M.J. and Clarkson, P.A. (1991) Solitons, nonlinear evolution equations and inverse scattering, Cambridge Univ. Press, Cambridge, Ch.1.
2. Newell, A.C. and Redekopp, L.G. (1977) "Breakdown of Zakharov-Shabat theory and soliton creation", Phys. Rev. Lett. 38, 377.
3. Miles, J.W. (1977) "Resonantly interacting solitary waves", J. Fluid. Mech. 79, 171.
4. Tajiri, M. and Nishitani, T. (1982) "Two-soliton resonant interactions in one spatial dimension: solutions of Boussinesq type equation", J. Phys. Soc. Japan 51, 3720.
5. Manoranjan, V.S., et al. (1984) "Numerical solutions of the good Boussinesq equation", SIAM J. Sci. Stat. Comput. 5, 946.
6. Fal'kovich, G.E., et al. (1983) "Destruction of stationary solutions and collapse in the nonlinear string equation", Phys. Lett. 99A, 271.
7. Flytzanis, N., et al. (1987) "Soliton resonances in atomic nonlinear systems", Physics 26D, 311.
8. Hirota, R. and Ito, M. (1983) "Resonance of solitons in one dimension", J. Phys. Soc. Japan 52, 744.
9. Musette, M., et al. (1987) "Soliton and antisoliton resonant interactions", J. Phys A20, 6223.
10. Lambert, F. and Musette, M. (1989) "On soliton instabilities for the nonlinear string equation", J. de Physique c3, 33.
11. Fuchssteiner, B. and Oevel, W. (1982) "The bi-hamiltonian structure of some nonlinear fifth- and seventh-order differential equations and recursion formulas for their symmetries and conserved covariants", J. Math. Phys. 23, 358.
12. Lambert, F. and Willox, R. (1991) "Searching for solitons with a direct binary operator method", in M. Remoissenet and M. Peyrard (eds.), Nonlinear coherent structures in physics and biology, Springer-Verlag, Heidelberg, 391.
13. Xing-Biao Hu and Yong Li (1991) "Some results on the Caudrey-Dodd-Gibbon-Kotera-Sawada equation", J. Phys. 24, 3205.

A CATASTROPHIC VIEW OF QUASI AND PSEUDOSPIN PHYSICS

E. S. Hernández

Departamento de Física, Facultad de Ciencias Exactas y Naturales, Universidad de Buenos Aires, 1428 Buenos Aires, Argentina

We show that for a class of systems with SU(2) or SU(1, 1) dynamical groups, the variational prescription with the coherent states as trial wave functions gives rise to a structurally unstable hamiltonian dynamics on the group manifold. The characteristics of the flow throughout parameter space can be established with simple geometrical aids and the structural phase transitions of the phase portrait can be related to Ginzburg - Landau - like phase transitions or to elementary catastrophes, in general. Illustrations are provided associated to heavy ion reaction theory and experiment, as well as to several problems in many - body physics.

1. Introduction

The purpose of this line of research is to investigate the universal dynamics of a class of systems with Spectrum Generating Algebra (SGA) or dynamical groups, when constrained to the group manifold where the associated coherent states lie.

Sources for this field and applications can be found in quantum optics, condensed matter physics and general many - body physics [1-4]. In most examples and applications, n-level models of the $SU(n)$ type are privileged [5] and among them, the $SU(2)$ - two - level systems exhibit a collection of geometrical characteristics which greatly simplify the investigation of the phase portrait, concerning [1-3]: location of fixed points, their nature, and characterization of the topologically invariant regions of phase space.

The technique to force a law of motion on the group manifold G/H, where G is a Lie group whose elements g can be locally parameterized as $g = exp(\vec{\hat{x}}.\vec{\hat{X}})$, (with $\vec{\hat{X}} = (\hat{X}_1, ...\hat{X}_N)$ a Lie algebra basis), and where H is the invariance subgroup of G with respect to some reference state $\mid ref >$, is given by Dirac's variational prescription [6],

E. Tirapegui and W. Zeller (eds.), Instabilities and Nonequilibrium Structures IV, 27–36.

$$\delta \int_0^t dt < \vec{x} \mid i\hbar \frac{\partial}{\partial t} - \hat{H} \mid \vec{x} >= 0, \tag{1}$$

where

$$\mid \vec{x} >= g(\vec{x}) \mid ref > . \tag{2}$$

A justification of (1) can be traced back to a path - integral analysis of the propagator [7], whose classical approximation corresponds to the extremum of the classical action,

$$S(\vec{x}\,\dot{\vec{x}}) = \int dt L(\vec{x}, \dot{\vec{x}}), \tag{3}$$

for the Lagrangian,

$$L(\vec{x}, \dot{\vec{x}}) =< \vec{x} \mid i\hbar \frac{\partial}{\partial t} - \hat{H} \mid \vec{x} >= i\hbar < \vec{x} \mid \dot{\vec{x}} > -H(\vec{x}). \tag{4}$$

The Euler - Lagrange equations are then [8,9],

$$\sigma^{\mu\nu} \dot{x}^{\nu} = \frac{\partial}{\partial x^{\mu}} H(\vec{x}), \tag{5}$$

with the symplectic form,

$$\sigma^{\mu\nu} = i < \frac{\partial \vec{x}}{\partial x^{\mu}} \mid \frac{\partial \vec{x}}{\partial x^{\nu}} > - < \frac{\partial \vec{x}}{\partial x^{\nu}} \mid \frac{\partial \vec{x}}{\partial x^{\mu}} >, \tag{6}$$

that endows the group manifold with a phase space structure. A canonical transformation $\vec{x} \longrightarrow (p_i, q_i)$ that carries $\sigma^{\mu\nu}$ into the symplectic identity $I_{ij} = (1 - \delta_{ij})(-)^i$, represents then a mapping of the group manifold onto a canonical phase space.

2. Special applications to many fermion physics

When $\mid ref >=\mid \Phi >$, the antisymmetrized hole vacuum or Slater determinant for N fermions, and when

$$\hat{X}_{\mu} = \alpha_p^{\dagger} \alpha_h, \tag{7}$$

where p and h subscripts respectively denote particle and hole states, Thouless's theorem [10] guarantees that $\mid \vec{x} >$ is the Slater determinant, the group manifold being thus the Grassmann manifold. The Hamilton's equations (5) are the $Time - Dependent - Hartree - Fock(TDHF)$ equations of motion [8].

The *TDHF* method became popular in the late 70's among nuclear physicists concerned with computations of the collective path in low-energy heavy ions collisions, where energy is transferred from the relative motion to the internal degrees of freedom. Much work has been devoted to investigate either model systems or realistic channels

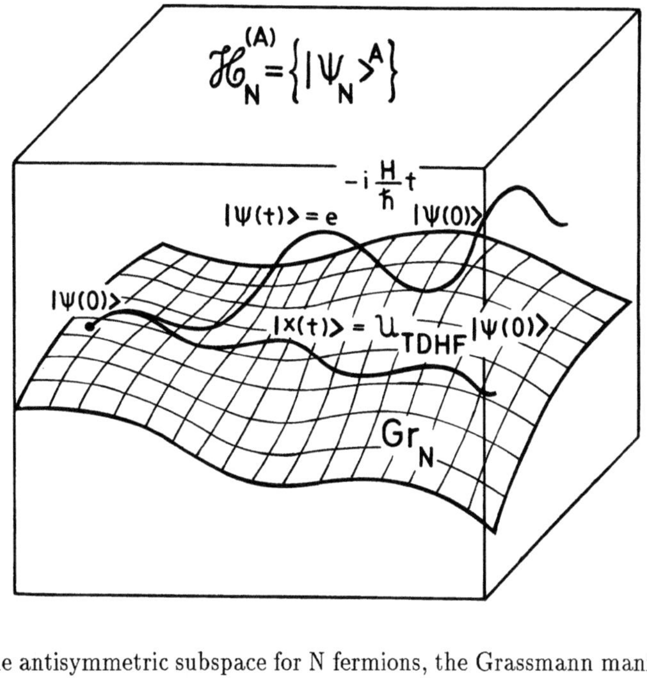

Fig. 1. The antisymmetric subspace for N fermions, the Grassmann manifold, the exact wave function evolving from a Slater determinant and the $TDHF$ wave function.

(see, for example, [11] and refs. therein), however some severe criticisms were raised, mainly addressing the underestimation of mass widths and the short time validity of the approach. The former point is related to the one - body character of the *TDHF* description and can be cured to some extent with the inclusion of two - body correlations. The second issue can be pictorially described as in Fig. 1, where H_N is the antisymmetric subspace of Hilbert's space for N fermions, $\mid \Psi(t) >$ the exact wave function propagated from an initial state $|\Psi(0)>\equiv|\vec{x}(0)>$ selected as a Slater determinant on Gr_N, and $|\vec{x}(t)>$ the *TDHF* evolving wave function. In this figure, it is indicated that the exact and approximate wave functions remain close to each other during a time interval of the order of the exact period.

In addition, it can be seen that in some cases, i. e., when the mean field approximation destroys some inherent symmetry, the loss of validity of the approach is not due to an improper choice of the wave function parametrization but to the adopted law of motion [9]. The short time restriction can be raised, just restoring broken symmetries within the frame of the variational description on Gr_N [9].

Among the many physical realizations of Lie algebras and systems with *SGA*, the $SU(2)$, $SU(1,1)$ and $SU(n)$ models are the stars. The first two are especially interesting in view of the fact that the dynamical hamiltonian system [5] is integrable [1–4,9,11]

3. SU(2) models

From the viewpoint of a many-body physicist, one is usually concerned with two - level models for N-fermions, for which the realization of the $SU(2)$ algebra with generators $\vec{\hat{J}} = (\hat{J}_x, \hat{J}_y, \hat{J}_z) = (\hat{J}_+, \hat{J}_-, \hat{J}_z)$ such that $\vec{\hat{J}} \times \vec{\hat{J}} = i\hbar\vec{\hat{J}}$, corresponds to bilinear functions of the fermion creation and annihilation operators $\alpha^\dagger_{p\sigma}$, where label p ranges between zero and N and label $\sigma = 1,2$ denotes the level. The traditional Lipkin - Meshkov - Glick (*LMG*) [12] hamiltonian is,

$$\hat{H} = \epsilon \hat{J}_z + \frac{V}{2}(\hat{J}_+^2 + \hat{J}_-^2) + \frac{W}{2}\{\hat{J}_x \hat{J}_-\}, \tag{8}$$

i. e., a monopole - monopole two - body interaction including both zero - and - two - exciton scattering processes. The model has been a theoretical laboratory for testing many - body techniques in nuclear physics, either pointing to structural or to dynamical problems (see, for example, [11] and refs. therein). It exhibits - as well as its $SU(n)$ generalization - a shape phase transition at $T \neq 0$ [13]. Its symmetry characteristic is parity conservation,

$$\hat{P} = exp[i\pi(\hat{J}_z + J)], \tag{9}$$

where $J = N/2$ and $[\hat{P}, \hat{H}] = 0$. This symmetry becomes broken in the *TDHF* picture where the variational state $\mid \vec{x} >$ is an $SU(2)$ coherent state, labelled by a complex parameter $\tau = -tg(\theta/2)exp(-i\phi)$, in other words,

$$\mid \vec{x} > \equiv \mid \tau > = \hat{R}(\tau) \mid -J, J >, \tag{10}$$

where $\hat{R}(\tau)$ is a rotation operator,

$$\hat{R}(\tau) = exp[\frac{tg^{-1} \mid \tau \mid}{\mid \tau \mid}(\tau \hat{J}_+ - \tau^* \hat{J}_-)], \tag{11}$$

and $\mid -J, J >$ is the ground state,

$$\hat{J}_z \mid -J, J > = -J \mid -J, J > . \tag{12}$$

The Grassmann manifold is in this case the Bloch sphere with radius J, $(S^2 \equiv SU(2)/U(1))$. Notice that the mean parity is [9],

$$\pi(t) = < \tau \mid \hat{P} \mid \tau > = cos^N \theta(t). \tag{13}$$

It has been shown [9] that when parity conservation is restored by a proper choice of the trial wave function, however parameterized on Gr_N, the variational dynamics follows the exact one for longer times.

The *LMG* model is not appropriate for modelling processes involving particle transfer through 1 - exciton events, as it may take place in heavy ion collisions. The most general $SU(2)$ hamiltonian is [2,3],

$$\hat{H} = \vec{\Omega}.\hat{\vec{J}} + \frac{1}{2}\hat{\vec{J}}.\alpha.\hat{\vec{J}}, \tag{14}$$

with symmetric α. For $\vec{J} = < \tau \mid \hat{\vec{J}} \mid \tau >$, Ehrenfest's theorem $i\hbar\dot{\vec{J}} = < \tau \mid [\hat{\vec{J}}, \hat{H}] \mid \tau >$ gives [2,3],

$$\dot{J}_i = \epsilon_{ijk}\Omega_j J_k + \frac{1}{2}\alpha_{jk}\epsilon_{ikl} < \tau \mid \{\hat{J}_j, \hat{J}_l\} \mid \tau > . \tag{15}$$

Now the essence of a mean - field (or variational, or coherent state (CS)) dynamics consists of the suppresion of two - body correlations other than quantal ones. In the CS picture it means [1],

$$< \tau \mid \frac{1}{2}\{\hat{J}_i, \hat{J}_j\} \mid \tau >= \frac{N-1}{N} J_i J_j + \frac{N}{4}\delta_{ij}, \tag{16}$$

then,

$$\dot{J}_i = \epsilon_{ijk}(\Omega_j + \frac{N-1}{N}\alpha_{lj} J_l) J_k, \tag{17}$$

i. e.,

$$\dot{\vec{J}} = \vec{\Omega}^{HF}(\vec{J}) \times \vec{J}, \tag{18}$$

where, in addition,

$$\vec{\Omega}^{HF} = \nabla_{\vec{J}} H(\vec{J}), \tag{19}$$

$$H(\vec{J}) \equiv H^{HF} =< \tau \mid \hat{H} \mid \tau > . \tag{20}$$

Equation (17) is consistent with the conservation of both H and $\vec{J}.\vec{J}$. Accordingly, the locus of $\vec{J}(t)$ is the intersection of the Bloch sphere with the energy quadrics [1–3],

$$H/\epsilon = J_z + \frac{1}{2J}\chi_{kl} J_k J_l. \tag{21}$$

There has been some amount of recent work centered on quasispin or pseudospin models with Eq. (17) as the dynamical law, however concerning hamiltonians belonging to the algebra (see, for example, [4] and refs. therein). It is interesting to notice that in the current frame, (17) is a nonlinear Bloch - like (or torque) equation whose origin lies in the mean - field approach with a quadratic coupling. In this spirit, it has been shown [1–4] that with simple geometrical aids one can easily visualize the complete features of the phase flow, including: a) location and character of the fixed points; b) the topologically invariant regions and c) the nonthermodynamic phase transitions undergone by the flow as one evolves in parameter space χ_{kl}.

4. Illustration

Consider the parity nonconserving hamiltonian,

$$\hat{H} = \epsilon \hat{J}_z + \frac{U}{2}\{\hat{J}_x \hat{J}_z\}, \tag{22}$$

with classical energy,

$$H(\vec{J})/\epsilon = J_z + \frac{\chi}{J} J_x J_z. \tag{23}$$

The appearance of the phase flow, for any parameter value, can be drawn by resort to a simple trick, namely, to superimpose a gradient dynamics on the nonlinear Bloch equation on the sphere [14]. Physical references can be found in models of magnetic relaxation in single domains [15]. One forces the drift of the flow towards the local minimum introducing a velocity component perpendicular to the equipotentials [14],

$$\dot{\vec{J}} = \vec{\Omega} \times \vec{J} + f(\vec{J}), \tag{24}$$

and choose $f(\vec{J})$ so that a) $\dot{\vec{J}}.\vec{J} = 0$; b) $\dot{H}(\vec{J}) \leq 0$; fixed points remain invariant and d) the dynamics is rotationally invariant. One gets [14]

$$\dot{\vec{J}} = (cos\delta)\vec{\Omega} x \vec{J} + (sin\delta)\vec{\Omega}_{perpen}, \tag{25}$$

with $0 \leq \delta \leq \pi/2$. This procedure is related to that proposed by Gilmore [13], namely a variational prescription in flat phase space out of a Lagrange - Lyapunov function,

$$LL(\delta) =< \tau \mid ie^{i\delta}\frac{\partial}{\partial t} - \hat{H} \mid \tau > . \tag{26}$$

Incidentally, one may observe that $\delta/2$ is the rotation angle of the eigenvectors at the saddle point with respect to the conservative flow.

The geometrical viewpoint above devised permits a fast recognition of the flow patterns. We have especially investigated [3,14] the $SU(2)$ hamiltonian,

$$\hat{H} = \epsilon \hat{J}_z + \frac{V}{2}(\hat{J}_+^2 + \hat{J}_-^2) + \frac{W}{2}\{\hat{J}_+ \hat{J}_-)\} + \frac{U}{2}\{\hat{J}_x \hat{J}_z\}. \tag{27}$$

In all these cases the flow transitions are nonthermodynamic phase transitions or catastrophes, that occur as the system evolves in parametr space and the fixed points bifurcate. Let us shortly recall that fixed points can be:

a) nondegenerate or Morse i -saddles. A gradient dynamics $\dot{\vec{x}} = -\partial V/\partial \vec{x}$ with $V(\vec{x};\chi)$ is stationary at an i - saddle $\vec{x}_0(\chi)$ in the neighbourhood of $\vec{x}_0$,

$$V(\xi,\xi) \approx -\xi_1^2 - .... - \xi_i^2 + \xi_{i+1}^2 + ... + \xi_N^2, \tag{28}$$

b) degenerate or non - Morse points, if $\partial^2 V/\partial x^\nu|_{\vec{x}_0}$ for given χ.

For fixed points of a vector field $\dot{\vec{J}}(\vec{J},\chi)$ an index theorem exists [16], establishing that the sum of indices of singular points on a sphere (Euler's characteristic) is equal to 2. In addition, one can show that the index of a singular point - actually the index of any circumference of arbitrarily small radius surrounding the point - is conserved under a bifurcation. The conservation of the characteristic poses then a limit on the number and the kind of bifurcations that can occur [3,4].

Morse fixed points of the flow generated by the hamiltonian (26) are those for which $\dot{\vec{J}} = 0$ and $| \partial\Omega^{HF}\partial J_i | \neq 0)$. On the bifurcation set in space χ_{kl} Morse points degenerate, and bifurcate as one proceeds with the variation of parameters. A detailed study of the kind of bifurcations that may appear can be found in refs. [2,3].

5. SU(1, 1) models

There is a large set of physical realizations of the $SU(1,1)$ algebra and of problems that can be modelled with linear hamiltonians [4]; harmonic motion, superfluid dynamics and excitations, Coulomb motion, two - photon processes, degenerate parametric amplification, s - states of the Morse oscillator, path integral representations of propagators and partition functions, anharmonic motion, squeezing of states in the electromagnetic field, and damped harmonic motion. Quadratic hamiltonians are demanded, for example, to treat anharmonic oscillatory motion with quartic perturbations in the mean field and superfluid dynamics in the presence of residual quasiparticle interactions [4]. The Lie algebra closure rules are $[\hat{K}_1, \hat{K}_2] = -i\hat{K}_3$ together with angular momentum commutation relations for the remaining pairs and Casimir operator $\hat{C}^2 = \hat{K}_3^2 - \hat{K}_1 - \hat{K}_2$. Choosing a convenient representation [17] with basis $| k, n >$ such that,

$$\hat{K}_3 \mid k,n >= (k+n) \mid k,n >, \tag{29}$$

$$\hat{C}^2 \mid k,n >= k(k-1) \mid k,n >, \tag{30}$$

one may build up coherent states,

$$| z >= exp(z\hat{K}_+ - z^*\hat{K}_-) \mid k,) >, \tag{31}$$

where the following factorization rule holds,

$$< z \mid \frac{1}{2}\{\hat{K}_i, \hat{K}_j\} \mid z >= \frac{2k+1}{2k} K_i K_j - \Delta_i \delta_{ij}, \tag{32}$$

with $\vec{\Delta} = (-k/2, -k/2, k/2)$.

Then a quadratic hamiltonian,

$$\hat{H} = \epsilon_i \hat{K}_i + \frac{1}{2}\beta_{ij}\hat{K}_i\hat{K}_j, \tag{33}$$

gives energy quadrics [4],

$$H(\vec{K}) = \epsilon_i K_i + \frac{1}{2}\beta^*_{ij} K_i K_j - \frac{1}{2}\Delta_i \delta_{ii}, \tag{34}$$

and Ehrenfest's theorem reads [4],

$$\dot{\vec{K}} = -\frac{1}{2}\nabla_{\vec{K}} H(\vec{K}) \times \nabla C^2, \tag{35}$$

being $C^2 \equiv K_3^2 - K_1^2 - K_2^2 = k^2$.

Equation (3) is coherence - preserving by construction; the phase flow will undergo catastrophes as before. However, the major feature is that the group manifold is noncompact, then there may not exist a conserved characteristic of the manifold, since for some singular points it will not be possible to find a closed finite curve surrounding it [4]. As a consequence of such nonconservation, a richer set of bifurcation diagrams appears, as compared to the $SU(2)$ systems.

6. Conclusions

For systems with SGA of the $SU(2)$, $SU(1,1)$ or $SU(n)$ type with quadratic hamiltonians, the variational prescription for dynamics on the group manifold gives rise to nonlinear Bloch equations for the quasispin or pseudospin vector. For integrable systems, the orbits are intesections of the group manifold with the energy quadrics.

The fixed points of the flow and the invariant regions can be located with simple geometrical considerations.

Bifurcations of the fixed points occur as the location or orientation of the energy quadrics qualitatively change. They can be classified according to elementary catastrophe theory. The type and kind of bifurcations that may take place are bounded for compact group manifolds due to the conservation of Euler's characteristic.

References

1. D.M. Jezek, E.S. Hernández and H.G. Solari, Phys. Rev. C34, 297 (1986).
2. D.M. Jezek and E.S. Hernández, Phys. Rev. C35, 1955 (1987).
3. C.E. Vignolo, D.M. Jezek and E.S. Hernández, Phys. Rev. C38, 506 (1988).
4. D.M. Jezek and E.S. Hernández, Phys. Rev. A, (1990).
5. Da Hsuan Feng and R. Gilmore, Phys. Lett. 90B, 327 (1980); R. Gilmore, J. Math. Phys. 20, 891 (1979).
6. P.A.M. Dirac, Proc. Cambridge Philos. Soc. 26, 376 (1830).
7. H. Reinhardt, J. Phys. G5, L91 (1879): H. Kuratsuji and T. Suzuki, Phys. Lett. B92, 19 (1980).
8. D.J. Rowe, A. Ryman and G. Rosenfeld, Phys. Rev. A22, 2362 (1879).
9. H.G. Solari and E.S. Hernández, Phys. Rev. C28, 2472 (1883).
10. D.J. Thouless, Nucl. Phys. 21, 225 (1960).
11. H.G. Solari and E.S. Hernández, Phys. Rev. C26, 2310 (1982).
12. H.J.Lipkin, N. Meshkov and H.J. Glick, Nucl. Phys. 62, 188 (1965).
13. R. Gilmore, Catastrophe Theory for Scientists and Engineers (Wiley, N.Y., 1978).
14. E.S. Hernández and D.m. Jezek, J. Phys. A21, 2899 (1988).
15. W.F. Brown, Phys. Rev. 130, 1677 (1863).
16. V.J. Arnold, Ordinary Differential Equations (MIT Press, Cambridge, 1973).
17. V. Bargmann, Ann. Math. 48, 588 (1947); Commun. Pure & Appl. Math. 14, 187 (1961) and 20, (1967).

AN APPROACH TO QUANTUM DISSIPATION

G. CRESPO, A. M. KOWALSKI, A. PLASTINO AND A. N. PROTO
Grupo de Sistemas Dinámicos. Centro Regional Norte.
Universidad de Buenos Aires
C.C. 2 (1638) Vicente López. Argentina

ABSTRACT: We show that the extension to dissipative situations of the usual quantization method, based on a formal analogy between the classical and quantal Hamiltonian dynamics, cannot be considered plausible since the Heisenberg principle is violated. Alternatively, we demonstrate that, by dealing with two oscillators, a quantum and classical one, coupled through a linear term, we can mimic the dissipating behaviour of a quantum system, without recourse to any quantization of classical equations and without violation of any quantum rule.

I. Introduction

During the last decades, much attention has been paid to the problem of the quantization of dissipative forces, due to abundant experimental evidence that has been accumulating. There have been numerous attempts in trying to quantify dissipation, but almost all of them have proved to be either misleading or non-consistent [1]. Two are the major approaches for the quantum mechanical handling, (as a one-body problem), of the damped motion of a particle [2-6]: a) the historical, Kanai [2] method, in which an explicitly time-dependent Hamiltonian is used, and b) introducing a "friction" potential which relies upon specially selected expectation values [3-4]. The main criticisms with respect to these Hamiltonians, allude to the fact that there exists an apparent violation of Heisenberg's uncertainty principle, and that, in general, the given Hamiltonians are not identical with the energy operator [6,8-10]. Recent literature devoted to the application of the Maximun Entropy Principle (MEP)[7] to special problems of Hamiltonian dynamics [8-10] shows that an Information Theory approach can indeed provide one with a new tool for the analysis of quantum dissipation. Other important contributions concerning general features of classical-quantum interactions are summarized in ref.1.

Our present work considers two different aspects of the problem: On the one hand, we reanalyze a well-known procedure for the quantization of dissipative effects, i.e, the quantization of a system starting from the generalized Hamilton's equations. This approach is based on the well-known analogy between the Hamiltonian description of classical dynamics and the Heisenberg description of quantum mechanics. Starting from this analogy, we explicitly demonstrate the unattainability of a consistent quantum description by recourse this procedure. On the other hand, we propose a time-dependent Hamiltonian which is able to mimic a dissipative quantum behaviour while, at the same time, preserving essential quantum characteristics like the non-violation of the uncertainty principle and the finiteness of the ground state energy.

E. Tirapegui and W. Zeller (eds.), Instabilities and Nonequilibrium Structures IV, 37–45.

In Section II the unattainability of a consistent quantum description of dissipative systems by recourse generalized Hamilton equations is discussed. In Section III our time-dependent alternative Hamiltonian is presented, and we briefly outline our conclusions in Section IV.

II. Classical and Quantal description of Dissipative Systems

A classical Hamiltonian description including dissipative forces can be achieved by starting from the generalized Lagrange equations where generalized nonconservative forces, non derivable from a generalized potential, are included. A dynamical consequence of applying this criterium is the equation of motion for the Poisson bracket of any two dynamical variables, A, B, for which (n degrees of freedom)

$$\frac{d}{dt}\{A,B\} = \{\frac{dA}{dt}, B\} + \{A, \frac{dB}{dt}\} + \sum_{i=1}^{n} \frac{\partial A}{\partial p_i}\{B, Q_i\} + \sum_{i=1}^{n} \frac{\partial B}{\partial p_i}\{Q_i, A\}, \quad (1)$$

where we have set $Q_i = Q_i(q,p,t) = F_i\{q,\dot{q}(q,p),t\}$. $F_i(q,\dot{q},t)$ stand for generalized nonconservative forces that cannot be derived from a generalized potential $U(q,\dot{q},t)$. If one computes the time derivative of any of the fundamental commutators, it is easy to prove that all of them vanish. In particular,

$$\frac{d}{dt}\{q_k, p_k\} = 0. \quad (2)$$

This last result, far from being unexpected, clearly exhibits the selfconsistency of the whole description.

Due to the close analogy between the classical description for conservative systems and the Heisenberg picture of quantum mechanics (obtained by replacing the dynamical variables by their corresponding quantum operators and the Poisson brackets by the corresponding commutators through

$$\{A,B\} \rightarrow \frac{1}{i\hbar}[\hat{A}, \hat{B}],$$

[11]),we can write down

$$\frac{d\hat{A}(t)}{dt} = \frac{1}{i\hbar}[\hat{A}(t), \hat{H}(t)] + \sum_{i=1}^{n} \frac{\partial \hat{A}(t)}{\partial \hat{p}_i(t)} \hat{Q}_i(t) + \frac{\partial \hat{A}(t)}{\partial t}, \quad (3)$$

and

$$\frac{d}{dt}[\hat{A}, \hat{B}] = [\frac{d\hat{A}}{dt}, \hat{B}] + [\hat{A}, \frac{d\hat{B}}{dt}] + \sum_{i=1}^{n} \frac{\partial \hat{A}}{\partial \hat{p}_i}[\hat{B}, \hat{Q}_i] + \sum_{i=1}^{n} \frac{\partial \hat{B}}{\partial \hat{p}_i}[\hat{Q}_i, \hat{A}]. \quad (4)$$

As a consequence, one can check that the time derivatives of the so-called fundamental commutators

$$[\hat{q}_k,\ \hat{p}_l] = i\hbar\,\delta_{kl};\ [\hat{p}_k, \hat{p}_l] = [\hat{q}_k, \hat{q}_l] = 0, \tag{5}$$

are all equal to zero, i.e.

$$\frac{d}{dt}[\hat{q}_k,\ \hat{p}_l] = \frac{d}{dt}[\hat{p}_k, \hat{p}_l] = \frac{d}{dt}[\hat{q}_k, \hat{q}_l] = 0, \tag{6}$$

if one consistently uses Eq.(4). It should be emphasized that no restrictions on $\hat{Q}_i$ are imposed.

So far, the main drawback of this quantization approach has not been discussed. If one considers, for the sake of clarity, the harmonic oscillator subject to a dissipative force proportional to its velocity, one should write (for the 1-dim case):

$$\hat{H} = \frac{\hat{p}^2}{2\,m} + \frac{1}{2} m\,\omega_o^2\,\hat{q}^2, \tag{7a}$$

$$\hat{Q} = -\gamma\,\hat{p}. \tag{7b}$$

Using the key equation of motion of this approach (Eq.(3)), it is easy to prove that the dynamical quantum evolutions that are obtained for the Heisenberg operators $\hat{q}(t)$ and $\hat{p}(t)$ are

$$\hat{q}(t) = \exp(-(\gamma/2)t)\ [A \exp(i\Omega t) + B \exp(-i\Omega t)], \tag{8a}$$

$$\hat{p}(t) = m \exp(-(\gamma/2)t)\ [(-\frac{\gamma}{2} + i\Omega)\ A \exp(i\Omega t) - (\frac{\gamma}{2} + i\Omega)\ B \exp(-i\Omega t)], \tag{8b}$$

where

$$\Omega^2 = \omega_o^2 - \gamma^2/4, \tag{9}$$

$$A = \frac{\hat{q}(0)}{2} + \frac{\hat{p}(0)}{2im\Omega} + \frac{\gamma\,\hat{q}(0)}{4i\Omega}, \tag{10}$$

$$B = \frac{\hat{q}(0)}{2} - \frac{\hat{p}(0)}{2im\Omega} - \frac{\gamma\,\hat{q}(0)}{4i\Omega}, \tag{11}$$

which leads one to conclude that

$$[\hat{q}(t),\ \hat{p}(t)] = e^{-\gamma t}\,[\hat{q}(0),\ \hat{p}(0)], \tag{12}$$

a gross violation of the uncertainty principle for $t \neq 0$. This undesired consequence can be traced to the following reason: The analogy between the Hamiltonian description of classical conservative mechanics and the Heisenberg picture of quantum mechanics must be considered a formal but very general one. This analogy is based upon the formal equivalence of the dynamical equation of motion for the dynamical variables for the first one and the dynamical operators for the second one, as well as the fact that either the Poisson brackets or the commutators verify similar fundamental equations, except for the $i\hbar$ factor. Nevertheless, the main difference between the two formalisms can be seen to lie on the very definition of, respectively, the Poisson brackets and the commutators. This fact becomes evident when one realizes that ***the fundamental Poisson brackets hold, by definition, for all times t independently of the dynamical behaviour of the classical variables q(t) and p(t).*** contrarywise, ***due to the definition of the commutator of two operators, the fundamental commutators do not necessarily retain their form for all times t in the case of arbitrary dynamical evolutions of the quantum operators*** $\hat{q}(t)$ ***and*** $\hat{p}(t)$, as shown in the example discussed above.

III. An alternative route

In order to give an alternative procedure to deal with the quantum dissipation problem we shall study the interaction between a quantum and a classical system described by a Hamiltonian of the form

$$\hat{H} = \hat{H}_q + H_{cl} + \hat{V}_{cl}^{q}, \tag{13}$$

where $\hat{H}_q$ and H_{cl} stand for, respectively, "quantal" and "classical" Hamiltonians and $\hat{V}_{cl}^{q}$ is the interaction potential between both systems (quantal and classsical variables belong of course to different spaces). This type of time-dependent Hamiltonian, as we shall show below, is able to mimic a dissipative quantum behaviour while preserving essential quantum characteristcs. The simplest Hamiltonian we can think off is that of two coupled harmonic oscillators

$$\hat{H} = \frac{1}{2m}\hat{p}^2 + \frac{1}{2M}p_s^2 + \frac{k}{2}\hat{x}^2 + \frac{k_s}{2}s^2 + \alpha\, s\, \hat{x}, \tag{14}$$

where $\hat{p}^2$, $\hat{x}$, and $\hat{x}^2$ are quantum operators and p_s, and s classical variables. Following a procedure discussed in well-known available literature concerned with the Maximum Entropy Principle [7-10],we look for a closure relationship of the form

$$[\hat{H}, \hat{O}_k] = i\,\hbar \sum_{j=0}^{q} \hat{O}_j G_{jk}. \tag{15}$$

The time evolution of the expectation values of the operators $\hat{O}_k$ can be found out new explicitly. An appropriate set of operators is $\{1, \hat{p}, \hat{x}, \hat{x}^2, \hat{p}^2, \hat{L}\}$ (where $\hat{L} = \hat{x}\hat{p} + \hat{p}\hat{x}$). For these operators,recourse to the generalized Ehrenfest's theorem [7], leads to the following system of coupled differential equations

$$\frac{d<\hat{x}>}{dt} = \frac{1}{m} < \hat{p} >, \tag{16a}$$

$$\frac{d<\hat{p}>}{dt} = - \left\{ k < \hat{x} > + \alpha\, s \right\}, \tag{16b}$$

$$\frac{d<\hat{x}^2>}{dt} = \frac{1}{m} < \hat{L} >, \tag{16c}$$

$$\frac{d<\hat{p}^2>}{dt} = - \left\{ k < \hat{L} > + 2\, \alpha\, s < \hat{p} > \right\}, \tag{16d}$$

$$\frac{d<\hat{L}>}{dt} = 2 \left\{ \frac{1}{m} < \hat{p}^2 > - k < \hat{x}^2 > - \alpha\, s < \hat{x} > \right\}, \tag{16e}$$

Eqs. 16 are the dynamical equations for the quantum relevant variables, which depend on the classical coordinate, s, which is of course time-dependent. The sets $\{1, \hat{p}, \hat{x}\}$ and $\{\hat{x}^2, \hat{p}^2\, \hat{L}\}$ correspond to the Heisemberg and S(1,1) groups respectively. Our main goal is that of obtaining a dissipative behaviour while avoiding the undesired consequences of the previously commented approaches. A crucial step is to prescribe for s (and of course p_s) the *canonical* evolution prescribed by classical mechanics for classical dissipative systems,considering the classical energy as the mean value of the Hamiltonian given by Eq.(14). This entails that the classical equations of motion acquire the form

$$\frac{ds}{dt} = \frac{1}{M} p_s. \tag{17a}$$

The essential trick, taken from the classical treatment of friction, is to introduce an ad-hoc, velocity-depent term in the equation of motion for p_s

$$\frac{dp_s}{dt} = - \left\{ \alpha < \hat{x} > + k_s\, s + \delta\, p_s \right\}, \tag{17b}$$

where δ is the so-called dissipation parameter. Eqs 16 and 17 constitute a set of coupled differential equations. The fact that the relationship given by the Eq. 15 is satisfied for all t, ensures the preservation of all quantum properties. The temporal evolution of the relevant operators can be now written down in an explicit fashion. For the sake of brevity we give below two of the corresponding relationships, namely

$$< \hat{x} >(t) = \sum_{k=1}^{4} A_k \exp(r_k t), \tag{18a}$$

$$< \hat{p} >(t) = m \sum_{k=1}^{4} r_k A_k \exp(r_k t)\,, \tag{18b}$$

The r_k's are the roots of the fourth degree equation

$$r^4 + \delta\, r^3 + (\omega^2 + \omega_s^2)\, r^2 + \omega^2\, \delta\, r + \omega^2\, \omega_s^2 - \frac{\alpha^2}{m\, M} = 0, \tag{19}$$

where $\omega = (\frac{k}{m})^{1/2}$ and $\omega_s = (\frac{k_s}{M})^{1/2}$ are, respectively, the frequencies of the quantal, and classical harmonic oscillators, respectively. In general, the roots of Eq.19 can be: i) two pair of complex conjugate roots, ii) two real roots and a pair of complex conjugate ones, iii) four.real roots. The real part of the complex conjugate roots does not vanish, except when $\alpha = 0$, which is not the case we are interested in. Negative real roots and negative real parts of the complex roots are obtained, for $\omega^2\, \omega_s^2 \geq \frac{\alpha^2}{m\ M}$. If instead, $\omega^2\, \omega_s^2 < \frac{\alpha^2}{m\ M}$ a non-dissipative dynamics is described. To solve analytically Eq. 19, for the most general situation, even if possible, is an awkward task. In order to obtain results amenable to an interesting interpretation, we have considered the particular case for which the relationship

$$\delta^2 = 4\,(\omega_s^2 - \omega^2), \tag{20a}$$

holds.For this case, we transform Eq. 19 to a cuadratic form by introducing

$$r = u - \frac{\delta}{4}. \tag{20b}$$

In Fig.1, the domains of each of the above mentioned cases are depicted, as respectively, regions I, II and III. Region IV is the domain of roots where at least one of them exhibits a positive real part (divergent behaviour), thus being of no interest. This type of solution can be avoided using an adequate set of initial conditions. Additional details are given in the figure caption. Both the temporal evolutions of the quantities, $(m\ \omega)^{1/2} < \hat{x} >$, $< \hat{p} >/(m\ \omega)^{1/2}$ relative to the relevants operators (given by Eqs. 18), and the quantal energy per unit ω, are plotted in Figs.2 for the physical situation corresponding to the point A in Fig. 1. The initial conditions are: $(m\ \omega)^{1/2} < \hat{x} > = 1$, $< \hat{p} >/(m\ \omega)^{1/2} = 3/2$, $(M\ \omega_s)^{1/2} s = 10$, $p_s/(M\ \omega_s)^{1/2} = 15$, $(m\ \omega) < \hat{x}^2 > = 7/2$, $< \hat{p}^2 >/(m\ \omega) = 9/2$, $< \hat{L} > = 1$ $(\hbar = 1)$. Other initial conditions such those corresponding to the energies and quantal correlations are deduced from the given ones. The $(\hat{p}, \hat{x})$-uncertainty can be written as (subindex zero means obviously, t = 0)

$$(\Delta x)^2(t)\, (\Delta p)^2(t) = \left\{ \frac{1}{2} m\, \omega\, [\frac{<\hat{K}_{22}>_o}{m^2\, \omega^2} + <\hat{K}_{11}>_o] \sin(2\omega t) + <\hat{K}_{12}>_o \cos(2\omega t) \right\}^2 + \\ + <\hat{K}_{11}>_o <\hat{K}_{22}>_o - <\hat{K}_{12}>_o^2, \tag{21}$$

where

$$\hat{K}_{11} = \hat{x}^2 - <\hat{x}>^2 = (\Delta x)^2, \tag{22a}$$

$$\hat{K}_{22} = \hat{p}^2 - <\hat{p}>^2 = (\Delta p)^2, \tag{22b}$$

$$\hat{K}_{12} = \frac{1}{2}\hat{L} - <\hat{x}><\hat{p}>, \tag{22c}$$

are the usual quantal correlations. As can be seen from Eq.21, $(\Delta x)^2(t)\,(\Delta p)^2(t)$ presents the oscillatory behaviour which corresponds to an harmonic oscillator. Thus, if the the Heisenberg's uncertainty principle is satisfied at $t = 0$, one has

$$(\Delta x)^2\,(\Delta p)^2 \geq (\hbar/2)^2, \tag{23}$$

for all t.

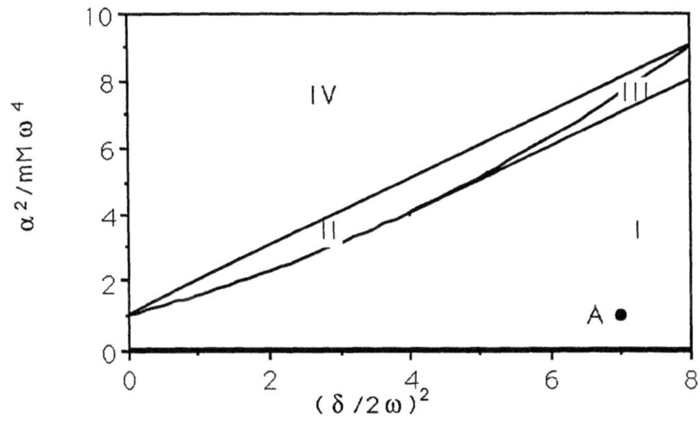

Fig.1: Domain of the different regimes for of the roots of Eq.19. Region I two pairs of complex conjugate roots. Region II: One pair of complex conjugate and two real roots. Region III: four real roots. Region IV: At least one of the roots is definite positive, corresponding to a divergent behaviour.

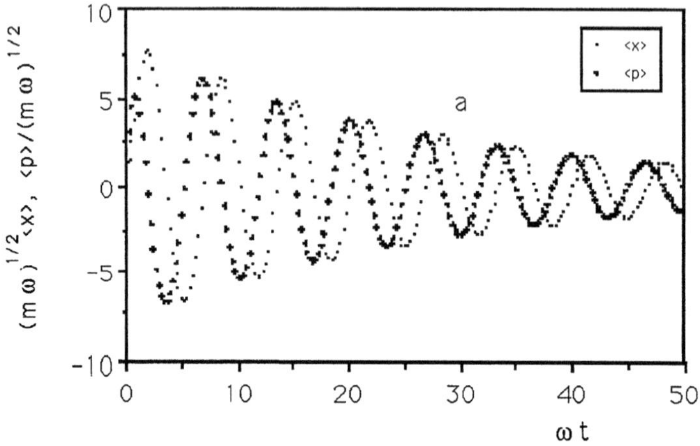

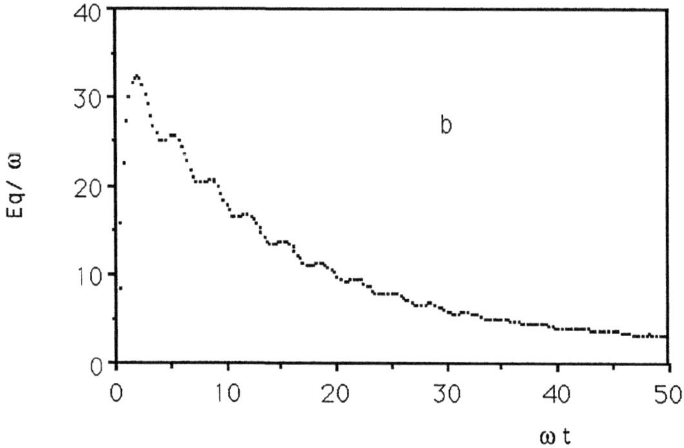

Fig.2 a) Temporal evolution of $(m\ \omega)^{1/2} < \hat{x} >$, and $< \hat{p} >/(m\ \omega)^{1/2}$ for point A of Fig.1. b) Quantal energy considered as the mean value of the quantal part of Eq.14, per unit ω, for the same point. Is easy to prove that this quantity is ever greater or equal than 1/2. As we take $\hbar = 1$, these quantities are dimensionless. The value of the parameters are: $(\delta/2\ \omega)^2 = 7$ and $(\alpha/(m\ M\ \omega^4)^{1/2}) = -1$.

IV. Conclusions

In an attempt to obtain a correct quantum description of dissipative phenomena,we demonstrate that, in spite of the fact that the usual quantization method is based on a formal analogy between the classical and quantal Hamiltonian dynamics, its extension to dissipative situations cannot be considered plausible since the Heisenberg principle is violated, *due the interpretation, for different times, between the definition of the Poisson bracket and the quantum commutator*. Alternatively, we can conclude that, by dealing with two oscillators, a quantum and classical one, coupled through a linear term, we can mimic the dissipating behaviour of a quantum system, without recourse to any quantization of classical equations. Due to the fact that the dissipating character of the system is introduced in the classical part of the hamiltonian, the quantum properties are not altered. The dynamical equations reflect dissipation through the s variable, whose temporal evolution is governed by Eqs.(17).

A.N.P., A.K. and G.C. acknowledge support from the Comisión de Investigaciones Científicas de la Provincia de Buenos Aires (CIC). A.P., who acknowledges suport from CONICET, Argentine, has as permanent address the Departament of Physics of the National University La Plata.

References

1) E.D. Harris, Phys. Rev **A42**,3685 (1990) and references therein
2) E.Kanai, Prog. Theor. Phys. **3**,440 (1948).
3) G. Ghosh and R. W. Hasse, Phys. Rev. **A** 24, 1621 (1981)
4) H. Dekker, Phys. Rep. **80**, 1 (1981)
5) D. M. Greenberger, J. Math. Phys. **20**, 762 (1979); **20**, 771 (1979)
6) M. D. Kostin,J. Stat. Phys. **12**, 145 (1974), R. W. Hasse, J. Phys. **A**, 1245 (1978).
7) A.N. Proto "Maximun Entropy Principle and Quantum Mechanics" Procc. XIII Int. Workshop on Condensed Matter Theories V. Aguilera-Navarro (Ed) Plenum Press 1989
8) J.Aliaga, G. Crespo, and A.N. Proto "Dissipative evolutions in Quantum Mechanics" Procc. XIII Int. Workshop on Condensed Matter Theories V. Aguilera-Navarro (Ed) Plenum Press 1989
9) J.Aliaga, G. Crespo, and A.N. Proto Phys. Rev. **A42**, 4325 (1990).
10) J.Aliaga, G. Crespo, and A.N. Proto Phys. Rev. **A43**, 595 (1991), and Phys. Rev. **A44**, (1991).
11) C. Cohen-Tannoudji, B. Diu, and F. Laloe, *Quantum Mechanics* (Wiley, New York, 1977)

ARNOLD TONGUES IN A PERIODICALLY PERTURBED LOGISTIC OSCILLATOR

JAIME RÖSSLER [a]

EDMUNDO LAZO [b]

MIGUEL KIWI [a,c]

[a] Departamento de Física, Facultad de Ciencias,
Universidad de Chile, Casilla 653, Santiago
[b] Departamento de Física, Universidad de Tarapacá
Casilla 7–D, Arica, Chile
[c] Facultad de Física, Universidad Católica de Chile
Casilla 6177, Santiago 22, Chile

ABSTRACT. A logistic oscillator, perturbed by a periodic force of amplitude D and frequency ω, is studied. We obtain the condition for phase locking between the non-linear oscillator, of frequency ω_0, and the external perturbation. The locked–unlocked boundary, $D = D_L(\omega)$, has a remarkable structure, with "V–shaped" minima appearing for every rational value of ω/ω_0.

1. Introduction

The phase locking (ph–l) phenomenon consists in a synchronization of the motion of two interacting non–linear oscillators, in such a way that the ratio of their frequencies becomes a "simple" rational number. Accordingly, they move almost in phase during a fraction of of their time evolution. There are several well known examples of ph–l: already Huygens, several centuries ago, observed that two back to back clocks synchronized their motion. Ph–l is also manifest in Josephson junctions, forced thermoacustic [1] and chemical oscillators, [2] Celestial Mechanics (like the commensurable relation $1/5$ between the rotation and synodic periods of Venus), etc... Moreover, the use of a periodic external "driving" signal in order to "entrain" otherwise chaotic dynamics [3] is closely related to the concept of ph–l.

The circle map provides a simple mathematical model to understand ph–l. It has has been intensively investigated during the last three decades. [4,5] Its properties fit within the general behavior of several physical models, [4,6] in all of which the characteristic structure of "Arnold tongues" demarking the ph–l regions of parameter space, do appear.

In this paper we tackle a closely related problem: a logistic map perturbed by an additive periodic signal. To be more specific, our system is governed by the discrete–

E. Tirapegui and W. Zeller (eds.), Instabilities and Nonequilibrium Structures IV, 47–52.

time evolution equation

$$\begin{aligned} X_{t+1} &= AX_t(1 - X_t) + DU(\Phi_t) \equiv g(X_t, \Phi_t) \\ \text{with} \quad U(\Phi) &= U(1 + \Phi) \quad \text{and} \quad \Phi_t = \Phi_0 + t\omega, \end{aligned} \tag{1}$$

where X_t is the logistic map coordinate at times $t = 0, 1, 2, 3...$. The perturbation plays the role of an "external force" $F_t = DU(\Phi_t)$ acting on the "logistic oscillator". We assume that the extrema of the perturbation are $U(\Phi = 0) = -\frac{1}{2}$ and $U(\frac{1}{2}) = \frac{1}{2}$; hence, the amplitude and frequency of the "force" are D and $\omega \in [0, 1]$, respectively. For specific calculations we will choose $U(\Phi)$ as a triangular wave.

This model is related to several contributions in which stocastic [7] or periodic [7,8] perturbations are applied on a logistic map. Both the additive perturbation of Eq. (1) and the modulation of the "bifurcation parameter" A , studied in reference [8], lead to qualitatively similar behavior. However, the present model can be related more directly to an actual nonlinear physical system, driven by an external force. The present study allows to observe new phenomena, not previouly reported, although contained in the model of reference [8].

2. Attractor Shape and the Phase Locking Phenomenon

The non–autonomous character of the one–dimensional map (1) can be mended by introducing the phase Φ_t as a dynamical variable. Then, the evolution of our system is described by the autonomous two–dimensional map $\{X_t, \Phi_t\} \longrightarrow \{X_{t+1}, \Phi_{t+1}\}$. We choose $-\frac{1}{2} \leq \Phi \leq \frac{1}{2}$ as the phase range, while $0 \leq X_t \leq 1$. In order to fill densely the Φ axis, we use non–rational values of ω .

When $D \to 0$ the unperturbed logistic map is recovered. In that limit, and for the case when the control parameter A lies in a (stable) c–cycle of the logistic map (say, its frequency is $\omega_0 = 1/c$), the attractor $\{X_t, \Phi_t\}$ simply reduces to c horizontal lines $\{X_t = K_\nu , \mid \nu = 1, 2, \ldots c\}$; here K_ν is a point of the c–cycle. For a small, but non–vanishing perturbation, $D \neq 0$, the c straight lines distort into c bands, which we denote by $\{\xi_\nu(\Phi)\}_{\nu=1}^{c}$. The whole set of points $\{X_t, \Phi_t\}$ of the attractor lie on these bands, say, $X_t = \xi_\nu(\Phi_t)$ for some ν.

Obviously, the attractor is invariant under the mapping $g(X_t, \Phi_t)$ of Eq. (1). It holds that $g(\xi_\nu(\Phi), \Phi) = \xi_{\nu+1}(\Phi + \omega)$, in analogy with the relation $K_{\nu+1} = f(K_\nu)$; here $f(X) = AX(1 - X) = g(X, \Phi \mid D = 0)$ is the unperturbed logistic map. Moreover, if $X_t \in \xi_\nu(\Phi)$ then $X_{t+c} \in \xi_\nu(\Phi)$ for the same ν, as c is the period of the unperturbed map.

A breakdown of the above scheme, which may occur on increasing the perturbation amplitude D, constitutes a clue indicating that a very essential change in the dynamics of the system has occurred. For example, the bands $\xi_\nu(\Phi)$ may fuse among them, or break into a rather diffuse set of points which densely cover the plane $[X, \Phi]$. The last case corresponds to chaotic dynamics, while the former one is a ph–l process, as we shall show shortly.

Nearly coincident external and natural frequencies In order to be precise, let us now concentrate in the special case when the unperturbed map has period 2, and its natural frequency ($\omega_0 = \frac{1}{2}$) is nearly equal with the external one, ω . Say, $\omega = \frac{1}{2} + \epsilon$, with $|\epsilon| \ll 1$. In this limit, the phase $\Phi_{2t} = 2t\epsilon$ changes slowly in time. Therefore, it is plausible to assume that $X_{2t} \approx \eta_1(\Phi_{2t})$ and $X_{2t-1} \approx X_{2t+1} \approx \eta_2(\frac{1}{2} + \Phi_{2t})$; here the condition $X_{2t+1} \approx X_{2t-1}$ rests on the underlying period 2 cycle. Using Eq. (1) it follows that $\eta_2 = g[\eta_1(\Phi), \Phi]$, and

$$\eta_1 = g(\eta_2, \frac{1}{2} + \Phi) \equiv G(\eta_1, \Phi) , \tag{2}$$

where $\eta_1 \equiv \eta_1(\Phi)$, $\eta_2 \equiv \eta_2(\Phi + \frac{1}{2})$ and $G(\eta, \Phi) \equiv g[g(\eta, \Phi), \frac{1}{2} + \Phi]$.
It is obvious that the static band $\eta_2(\Phi)$ also satisfies Eq. (2). Although these "static" solutions $\eta_1(\Phi), \eta_2(\Phi)$ have been derived by taking the extreme adiabatic limit in the dynamical map (1), the basic hypothesis $X_t \approx X_{t+2} \approx \eta_\nu(\Phi_t)$ can break down in some special cases, as we shall see below.

The fixed points $\eta(\Phi)$ of the map $G(\eta, \Phi)$ are shown in Figs. 1, for two values of D. The stable (unstable) fixed points correspond to the continuous (dotted) curves. When D is lower than a critical value, say D_{cr}, two stable fixed points exist for each value of Φ . They constitute the adiabatic bands $\eta_\nu(\phi)$; $\nu = 1, 2$, as illustrated in Fig. 1a. For $D > D_{cr}$ (see Fig. 1b) each stable band exists only in a range of values of Φ, with gaps appearing between them. Let $-\alpha < \Phi < \alpha$ be the gap of η_1 , while the gap of η_2 is obtained by shifting the former gap by $\frac{1}{2}$, say $\frac{1}{2} - \alpha < \Phi < \alpha + \frac{1}{2} \equiv \alpha - \frac{1}{2}$.

The phase locking phenomenon The above mentioned gaps play an essential role in the understanding of the dynamics of the system in the case $|\epsilon| \ll 1$ and $D > D_{cr}$. To be specific let us assume $\omega - \frac{1}{2} = \epsilon < 0$, and consider the point X_{2t} moving slowly, from right to left over the band η_1 , as t increases. Meanwhile, the point X_{2t+1} moves over η_2, with a phase delay of $\frac{1}{2}$. When X_{2t} reaches the gap, it "falls" into the unstable zone, reaching the band η_2 some time thereafter. Henceforth X_{2t} paces the reference band $\eta_2(\Phi)$, while X_{2t+1} treads over the band η_1, until this attraction band also comes to an end. After that the X_{2t} orbit becomes unstable again, until it returns to $\eta_1(\Phi)$, thus closing the cycle (we recall that points $\Phi = \frac{1}{2}$ and $\Phi = -\frac{1}{2}$ coincide). But, the the latter cycle, which treads over the *two "adiabatic bands"* of Fig. 1b, just corresponds to the dynamical attractor depicted in Fig. 2, where a *single band* appears. This way, the existence of a gap in the adiabatic bands η_ν implies that the attractor $\{X_t, \Phi_t\}$ reduces to a single band $\xi(\Phi)$.

The case of a single band attractor corresponds to the regime of ph–l between the logistic oscillator and the external "force". [8] In fact, let us associate the bands η_1 , η_2 with the two extreme amplitudes of our "logistic oscillator". According to the previous discussion, for a negative detuning $\epsilon < 0$, the point X_{2t} remains in the lower band η_2 when the phase $\Phi_{2t} \in [-\frac{1}{2}+\alpha\,,\alpha]$. Otherwise it stays in the upper band. Therefore, the band alternancy is ruled by the phase of the *external* perturbation $\Phi_t = t\omega$; consequently, the logistic oscillator alternates between the reference bands η_1 and η_2 (its "turning points") with the frequency of the "external driver" ω, instead of the "internal" one, $\omega_0 = \frac{1}{2}$.

Conversely, if the attractor is formed by two separate bands, the logistic oscillator retains its "natural" period $\omega_0 = \frac{1}{2}$, as X_t alternates in time between the two bands, say $X_{2t} \in \xi_1$ and $X_{2t+1} \in \xi_2$. Therefore, in this case the oscillator is unlocked.

The Locked–Unlocked Boundary Let us consider the case $D > D_c$, in which a gap appears in the adiabatic bands η_ν . When the detuning $|\frac{1}{2}-\omega| = |\epsilon|$ increases, the attractor $\xi(\Phi)$ leaves the reference bands η_ν . If ϵ exceeds a critical value, say $\epsilon_L(D)$, then the point X_{2t} carries a "speed" large enough to jump across the gap of the reference band η_1 , and thus remains in that band indefinitely. At the same time X_{2t+1} remains in η_2 , and the *unlocked regime is reached,* since the attractor is made up by two bands. In Fig. 3 we show the attractor in the neighborhood of the locked–unlocked (l–u) transition, for $D = 0.1474$. Fig. 3a, where $\omega = 0.4166603 \approx 5/12$, corresponds precisely to the l–u boundary. The bands separate if ω slightly decreases (see Fig. 3b, with $\omega = 0.4164032$).

Let us now consider a fixed value of the parameter A . It is convenient to describe the l–u boundary by the function $D_L(\omega)$. The ph–l region of the $[\omega\,,D]$ plane corresponds to $D > D_L(\omega)$. Since the gap in η_ν disappears for $D < D_c$, $D_L(\omega=\frac{1}{2}\,,A) = D_c(A)$ is the absolute minimum of $D_L(\omega)$.

The boundary $D_L(\omega)$ was evaluated numerically; the results are shown in Fig. 4. In agreement with the discussion above the plot shows a deep minimun for $\omega = \frac{1}{2}$. However, Fig. 4 also contains a whole wealth of secondary "V–shaped" minima, located at *every rational value of* ω/ω_0 , as can be seen from the enlargement of the neighborhood around $\omega = 3/8$. We identify these minima as the *Arnold tongues* [4,5] of our dynamical system. They can be ordered in a hierarchical structure, according to the level of the associated rational in the Farey tree. [5] Tongues associated with "simple" rational numbers are the most relevant. However, additional rules do exist; for example, our "Arnold tongues" are specially conspicuous if $\omega/\omega_0 = M/N$, with M, N odd integers. The latter properties can be explained in terms of the peculiar features of the attractor in the neighborhood of a "simple" rational frequency (see Fig. 3a), details of which will be provided in a forthcoming paper.

The present model does not fit within the universality class of the circle map, [4,6] since it requires of a finite external perturbation to yield ph–l, while the circle map can achieve ph–l for a vanishingly small perturbation. In this respect the present system is similar to the damped Mathieu equation. [9] In addition, in our case the frequency of the unperturbed oscillator is fixed at $\omega_0 = 1/2$, while the external driving frequency is continuously changed in the diagram of Fig. 4. Accordingly, the frequency of the dynamical process is not rigurously fixed in each tongue, and therefore the tongues fuse among themselves without conflict, as can be verified in Fig. 4. In contrast, in the circle map each tongue is characterized by a fixed rational, and hence the fusion

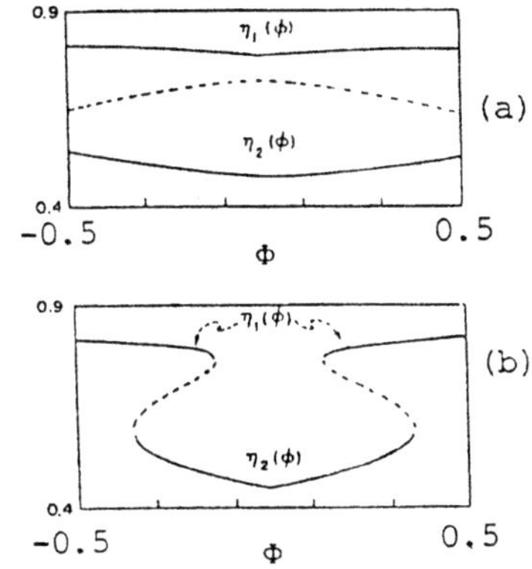

Fig. 1

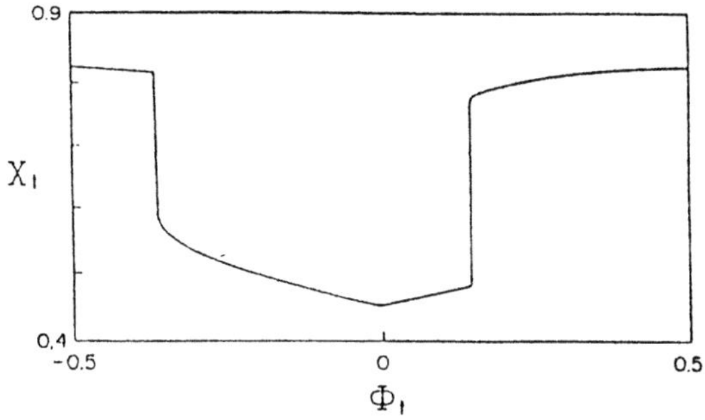

Fig. 2

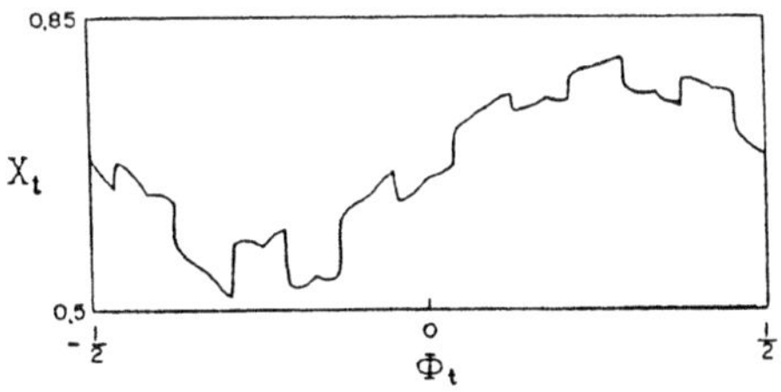

Fig. 3a

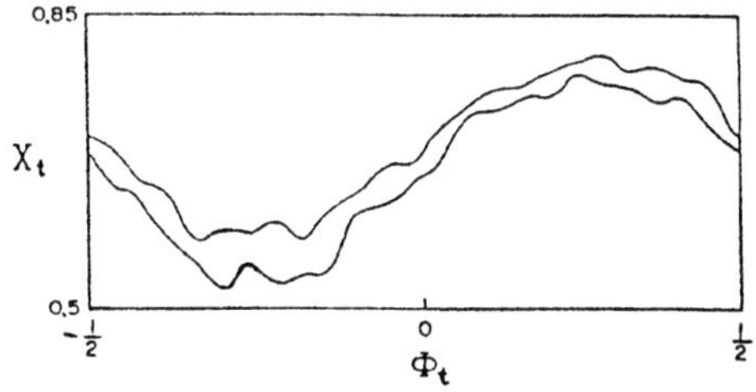

Fig. 3b

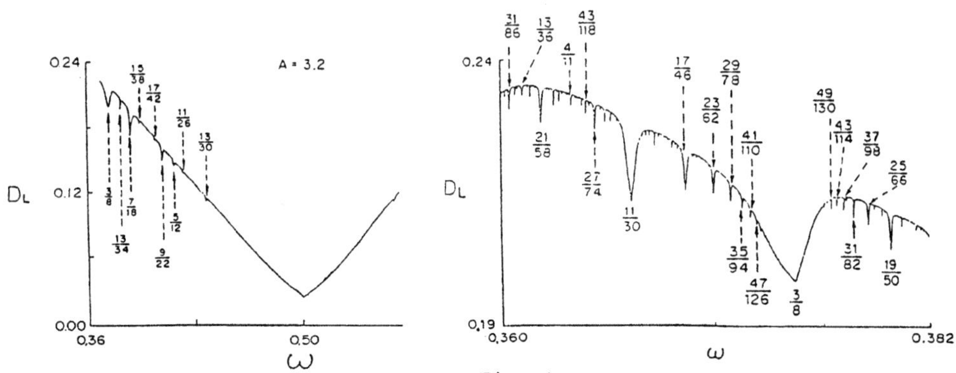

Fig. 4

of tongues becomes meaningless.

Acknowledgements

This work was supported by the DTI of the Universidad de Chile, DIUC of the Universidad Católica de Chile, and FONDECYT (Proj. 91–0832 & 92–0753)

References

[1] T. Yazaki, S. Sugioka, F Mizutani and H. Mamada *Phys. Rev. Lett.***64**, 2515 (1990)

[2] M. Eiswirth and G. Ertl *Phys. Rev. Lett.***60**, 1526 (1988)

[3] E. A. Jackson and A. Hübler *Physica* **D 44**, 407 (1990)

[4] P. Bak, T. Bohr, and M. Jensen, *Circle maps, Mode–locking and Chaos* in: Directions in Chaos, Vol. **2**, ed. Hao Bai–Lin (World Scientific, Singapore, 1988). See also *Physics Today*, 39 (Dec. 1986)

[5] A. S. Pikovsky and M.A. Zaks *Phys. Lett.***A**, 373 (1991)

[6] D. Gonzalez, and O. Piro; *Phys. Rev. Lett.* **50**, 870 (1983).

[7] J. Rössler, M. Kiwi, B. Hess, and M. Markus, *Phys. Rev.* **A39**, 5954 (1989).

[8] J. Rössler, E. Lazo and M. Kiwi, *Nonlinear Phenomena in Fluids, Solids and Other Complex Systems*, ed. P. Cordero & B. Nachtergale (Elsevier, Amsterdam), 419 (1991).

[9] P.Bergé, Y. Pomeau, and C. Vidal, *L'Ordre dans le Chaos* (Hermann, Paris, 1984).

AXISYMMETRIC COHERENT VORTEX STATES IN AC DRIVEN JOSEPHSON JUNCTIONS ARRAYS

DANIEL DOMÍNGUEZ* and JORGE V. JOSÉ
Department of Physics
Northeastern University
Boston, MA 02115, USA

Abstract

We study the dynamic response of disordered Josephson junction arrays driven by dc and ac currents. The arrays are modelled by a set of coupled non-linear overdamped oscillators driven by the external current. The effect of Johnson thermal noise is modelled by a Langevin term in the equations of motion. The disorder is introduced by randomly displacing the bonds of a square lattice. We find a nucleation of vortex pairs at the defects produced by the external currents. After a transient time these vortices unbind and *radiate* other vortices. These radiated vortices settle into a parity-broken time periodic *axisymmetric coherent vortex state* characterized by rows of vortices lying along a tilted axis. This state has properties that are suggestive of similar far from equilibrium states found in hydrodynamic systems. The $ACVS$ produces steps in the current voltage characteristics typical of giant Shapiro steps. Connections of these results to experiments done in Nb-Au-Nb arrays are discussed.

1. Introduction

The Josephson effect is one of the most important manifestations of superconductivity [1]. In recent years there have been a considerable interest in the study of two dimensional Josephson junctions arrays (JJA) [2]. Recent advances in photolitographic techniques have allowed the fabrication of these arrays with tailor made properties. The initial studies in these systems were devoted to testing the Berezinskii-Kosterlitz-Thouless theory of topological long-range order [3,4]. More recently a growing interest has centered about studying their dynamical properties [5]. In this case the JJA are described in terms of a large set of coupled non-linear oscillators. Novel non-equilibrium coherent vortex states appear in the arrays when they are driven by time periodic currents. Experimentally, giant Shapiro steps (GSS) have been measured in the I-V characteristics of proximity-effect JJA in zero [6] and rational frustration magnetic fields [7]. These steps have been interpreted as coherent oscillations of ground state field-induced vortex configurations. This interpretation was successfully verified in numerical simulations when modelling the arrays by a resistively shunted junction (RSJ) model [8,9], as well as from analytic studies [10]. Moreover, the experiments in GSS have encouraged investigations of JJA as coherent radiation sources [11].

*Address after October 1992: International Centre for Theoretical Physics, Trieste, Italy.

E. Tirapegui and W. Zeller (eds.), Instabilities and Nonequilibrium Structures IV, 53–67.

Recent experiments on Nb-Au-Nb JJA[12,13] have found, however, half-integer GSS (and also some evidence of higher subharmonic steps) in zero magnetic field. These steps can not be explained within the RSJ model for *periodic* arrays. Recently, we found that the addition of disorder can in fact induce extra half-integer steps in zero magnetic field [14,15] within the RSJ model. These extra steps were found to be related to a coherent oscillation of current-nucleated vortices. Previously, the nucleation of vortex-antivortex pairs at the defects, induced by an applied dc current, was discussed by Xia and Leath[16]. In our studies to the dc current we added an ac component. The addition of the ac current changes the vortex dynamics in a fundamental way: It leads to the formation of a novel, far from equilibrium, *axisymmetric coherent vortex state* (ACVS). The ACVS corresponds to an oscillating pattern of tilted vortex rows that produce the extra half-integer GSS. The relevance of the disorder is to act as nucleating centers of vortex pairs thus providing the relevant initial conditions in the dynamics to generate the ACVS.

We present here a brief recap of our main results pertaining to the ACVS, stressing the fact that this vortex coherent state is essentailly a stationary state far from equilibrium. We also present new evidence for the existence of higher order subharmonic steps which appear when the size of the system is increased further.

2. Dynamics of Josephson Junctions Arrays

2.1. THE JOSEPHSON EFFECT

Let us briefly review the Josephson effect. In 1962 B. D. Josephson [1] predicted that a supercurrent can flow between two superconductors separated by a thin non-superconducting material. The supercurrent is

$$I_J = I_0 \sin(\theta_2 - \theta_1) \ , \tag{1}$$

where θ_1, θ_2 are the phases of the order parameter in superconductors 1 and 2 respectively. The critical current I_0 is the maximum current that can flow through the junction. The voltage drop between the superconductors is

$$V = \frac{\Phi_0}{2\pi}\frac{d}{dt}(\theta_2 - \theta_1) \ , \tag{2}$$

where $\Phi_0 = h/2e$ is the quantum of flux. Eqs.(1) and (2) are known as the Josephson equations. In real junctions there is always a source of dissipation. This dissipative effect has been succesfully modelled by the McCumber resistively shunted junction (RSJ) model [17]. In a current biased junction, the bias current I flows in parallel with an ideal Josephson junction and a resistor R so that the total current is given by,

$$I = I_J + V/R = I_0 \sin(\Delta\theta) + \frac{\Phi_0}{2\pi R}\frac{d\Delta\theta}{dt} \ , \tag{3}$$

where we have used Eqs.(1) and (2). When the current I is time-independent, Eq.(3) can be solved analytically. In this case, the time averaged voltage $< V >$ as a function of I,

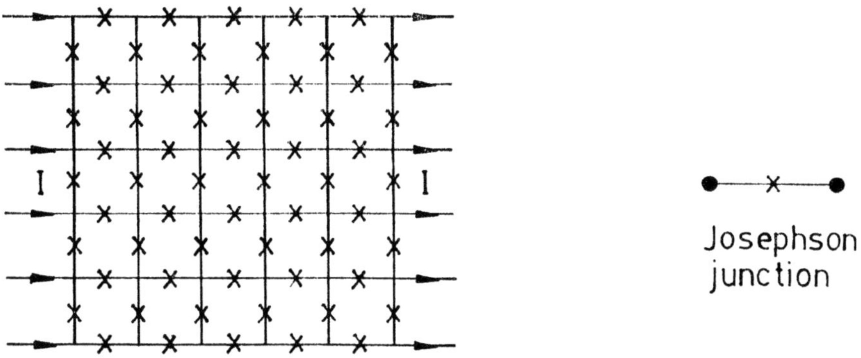

Figure 1: Schematic representation of a Josephson junction array driven by an external current I.

which defines the I-V curve of the junction, is given by,

$$\begin{aligned} < V > &= 0 \qquad\qquad \text{for } I < I_0 \\ < V > &= R\sqrt{I^2 - I_0^2} \quad \text{for } I > I_0 \end{aligned} \tag{4}$$

When the junction is driven by a time-periodic current $I = I_{dc} + I_{ac}\sin(2\pi\nu t)$, the $< V >$ vs I_{dc} curve shows plateaus at quantized voltages

$$V_n = n\frac{h\nu}{2e}, \quad n = 1, 2, 3, \ldots \ . \tag{5}$$

These are the Shapiro steps that have allowed the most precise measurements of the voltage unit[18]. In this paper we discuss the existence and properties of giant Shapiro steps in JJA.

2.2 JOSEPHSON JUNCTIONS ARRAYS

A Josephson junction array (JJA) is made of an $N \times N$ network of superconducting islands connected by Josephson currents [2,4]. For example, the square arrays of Benz *et al.* [7] are made of 1000×1000 Nb-Cu-Nb proximity effect junctions with a lattice constant of $a = 10\,\mu m$. A schematic representation of a JJA driven by an external current is shown in Fig. 1.

To model the dynamical behavior of this system we extend the RSJ model to a square JJA network [5]. The current $I_\mu(r)$ along the μ direction (with $\mu = \hat{e}_x, \hat{e}_y$) between the islands at site r and the island at site $r + \mu$ is given by:

$$I_\mu(r) = I_0 \sin(\Delta_\mu\theta(r) - A_\mu(r)) + \frac{\phi_0}{2\pi R}\frac{d}{dt}(\Delta_\mu\theta(r) - A_\mu(r)) + \eta_\mu(r,t)\,, \tag{6}$$

where $\Delta_\mu\theta(r) = \theta(r+\mu) - \theta(r)$. The external magnetic field produces frustration with the link variable $A_\mu(r) = \frac{2\pi}{\phi_0}\int_r^{r+\mu} \vec{A}\cdot d\vec{l}$, and:

$$2\pi f = A_x(r) + A_y(r+\hat{e}_x) - A_x(r+\hat{e}_y) - A_y(r)\,; \tag{7}$$

where the frustration parameter $f = \frac{Ba^2}{\Phi_0}$ is the average number of quantum of flux per unit cell. Thermal noise is included by adding the gaussian random variable $\eta_\mu(r,t)$ to the equations of motion with covariance, $< \eta_\mu(r)\, \eta_{\mu'}(r') >= \frac{2kT}{R}\delta_{r,r'}\delta_{\mu,\mu'}\delta(t-t')$.

Eq. 6 together with Kirchoff's current conservation law

$$\Delta_\mu \cdot I_\mu(r) = I_x(r) - I_x(r-x) + I_y(r) - I_y(r-y) = I^{ext}(r), \tag{8}$$

fully defines the dynamical evolution of the phase $\theta(r)$ as a function of time. The resulting explicit expression for $\frac{d\theta}{dt}$ derived from these equations [5] is,

$$\frac{d\theta(r)}{dt} = -\frac{2\pi R}{\Phi_0}\sum_{r'} G(r,r')\{I^{ext}(r) - \Delta_\mu \cdot [I_0 \sin(\Delta_\mu\theta(r') - A_\mu(r')) + \eta_\mu(r',t)]\}, \tag{9}$$

with $G(r,r')$ the two dimensional Green function. Eq.(9) defines the set of dynamical equations studied in this paper. In Benz's experiment this would be a set of 10^6 non-linear coupled oscillators driven by the external current.

Of course, solving this large set of coupled non-linear equations analytically is out of the question. We resort then to an efficient numerical algorithm to solve (9) numerically. In our simulations the external currents are injected along the y-direction with $I^{ext}(y=0) = I$ and zero elsewhere. The potentials for $y = N_Y$ are fixed to zero by setting the phase $\theta(y = N_Y + 1) = 0$ (this boundary condition removes a singularity in the Green function matrix). Along the x-direction we choose periodic boundary conditions (PBC) in most calculations, but use also free boundary conditions (FBC) for comparison purposes.

A direct numerical evaluation of Eq. (9) grows as N^4 [5], which is too slow and limits the sizes of the lattice sizes that can be simulated. Instead, we use a very efficient method [14,15] based in the use of the Fast Fourier Transform (FFT), which involves doing the FFT of Eq. (9) along the x-direction and then solving the resulting tridiagonal equation for the y-direction. This alorithm grows as $N^2\log(N)$ which is appreciably faster than the direct method. Eikmans and van Himbergen [19] have also used a similar method doing FFT along both the x and y directions. Our method shows an improvement over theirs of about 30%. The time integration is carried out using a fixed step fourth order Runge-Kutta (RK) method. Typical integration steps are $\Delta t\nu_0 = 0.01 - 0.1$, with characteristic Josephson frequency $\nu_0 = \frac{2\pi R I_0}{\Phi_0}$. At finite temperature we use the extended second order RK method for stochastic differential equations developed by Helfand and Greenside [20].

2.3. ORDERED ARRAYS AND GIANT SHAPIRO STEPS

For ordered arrays, Eq.(9) is invariant against the transformations $f \to f + n$, with n an integer, and under $f \to -f$. Then, it is enough to analyze the the properties of the system in the interval $f = [0, 0.5]$. The JJA can have topological current vortex excitations defined from

$$\circlearrowleft\!\!\!\!\!\!\sum \; \Delta_\mu\theta(r) = 2\pi n(R), \tag{10}$$

where $n(R)$ is an integer that gives the number of vortices at plaquette R. In zero field, vortices play a crucial role in triggering the Kosterlitz-Thouless topological phase transition

[3]. For fractional frustrations $f = p/q$, the ground states of the arrays consist of superlattices of field-induced vortices with unit cell size of $qa \times qa$ [21]. These states lead to non-monotonic behavior of the magnetoresistance and $T_c(f)$ studied both experimentally and theoretically [2,4,21].

What is measured experimentally are IV characteristics that we can calculate from solving Eq. (9). The calculation gives the time averaged total voltage drop $< V >$ in the array as a function of the dc current per row I_{dc}. For periodic arrays, the $f = 0$ dynamics of the system is equivalent to the superposition of N single Josephson junctions along the direction of the current[5]. This is so because the total current flows along the y-direction, and the system behaves as N_x rows with N_y junctions in series. When $I_{ac} = 0$, the IV characteristics are given by Eq. 4 multiplied by N_y. The same is true when $I_{ac} \neq 0$, with $I(t) = I_{dc} + I_{ac}\sin(2\pi\nu_a t)$, leading to *giant* Shapiro steps at voltages [6]:

$$V_n = N\frac{nh\nu}{2e},\ n = 1,2,3,\dots, \tag{11}$$

When a magnetic field is applied, the symmetry that allowed to separate the system into N_x independent arrays in series for $f = 0$ is broken. No exact solution of the IV characteristics is known in this case. It was first found experimentally that there are fractional GSS at the voltages [7,8,9,10] :

$$V_{n,q} = \frac{N}{q}\frac{nh\nu}{2e},\ n = 1,2,3,\dots,\ \text{and } q = 1,2,3,\dots, \tag{12}$$

when $f = p/q$. These fractional steps were explained as due to the collective motion between different $f = p/q$ ground state vortex configurations [8,9,10]. The vortex-superlattice oscillates in syncrhony with the ac current with frequency ν/q.

To get extra subharmonic GSS in zero magnetic field we need a mechanism that breaks the symmetry that reduces the JJA to effectively one-dimensional systems. One such mechanisms can be produced by adding disorder which we now discuss.

2.4. DEFECT NUCLEATED VORTICES

We introduce defects in an square array by moving its lattice sites radially away from their periodic position. We assume that the I^c's and the R's are all equal and thus the disorder is evident only in the $A_\mu(r)$ integrals. One defect in the array produces a change only in the the four A_μ's connecting the bonds linked to the displaced lattice site. Since the A_μ's are proportional to the external field, varying f increases the deformation of the lattice, i.e. increases the disorder. Because of the periodicity of the equations of motion with respect to f, if we take $f = n$ an integer, only the plaquettes around the defect will feel the presence of the field. The n value will then represent the "strength" of the defect. As we shall discuss later the specific nature of the defect is not important in producing the $ACVS$. For example the $ACVS$ appears as well when we cut bonds with $f = 0$.

The problem of a single defect in a JJA under a constant dc current was first studied by Leath and Xia [16]. They found that there is a tendency to create a vortex-antivortex pair pinned to the defect for $I_{dc} \neq 0$. When I_{dc} is large enough, the Lorentz force acting on the

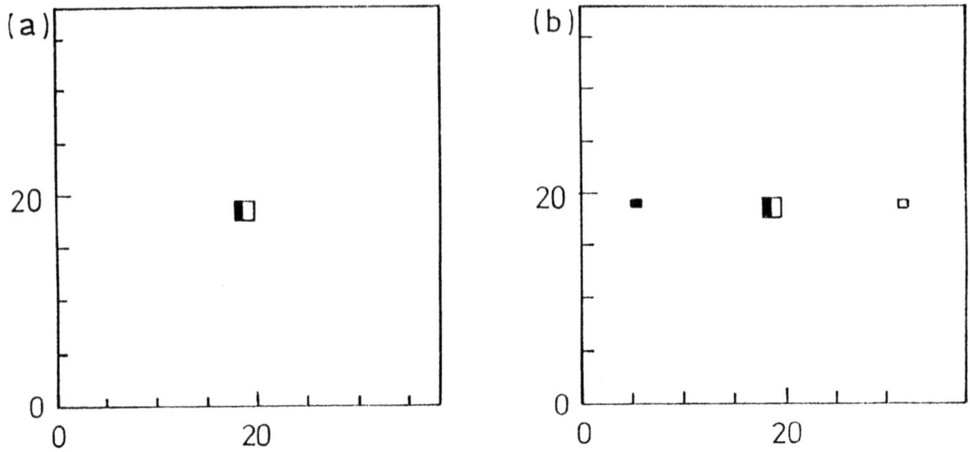

Figure 2: Vortex configurations for an array with one defect defined by $f = 10$, $\delta_x = 0.1$, $\delta_y = 0$. White and black squares indicate 1 and -1 vorticity per plaquette respectively. (a) For a current $I_{dc} = 0.8 I_0 < I_c$; (b) for a current $I_{dc} = 0.87 I_0 > I_c$ ($I_c = 0.84 I_0$).

vortices can break them away from the defect. As they move along a direction perpendicular to I_{dc}, because of Fraday's law, they produce a dissipative voltage that leads to a resistance in the array. There current above which vortices start moving, is the critical current I_c of the array, with $I_c < I_0$. For $I \gg I_c$, a larger number of vortices and antivortices are generated, leading to a more complex vortex motion in the array.

In Fig. 2 we show an example of this situation for the kind of topological defect defined above. We show the vortex distribution $n(R)$ for a 40×40 array, which was the typical size studied, although we will discuss results for larger lattices as well. For $I < I_c$ we see a vortex-antivortex pair pinned at the defect. When $I > I_c$ vortices and antivortices move away from the defect as found in Ref. [16].

3. Axisymmetric Coherent Vortex States

3.1. I-V CHARACTERISTICS

In this section we discuss the IV characteristics of JJA with defects and driven by a current $I = I_{dc} + I_{ac} \sin(2\pi\nu t)$. Let us start considering the one-defect case. Later we will analyze the many defect case. In Fig. 3(a) we show the IV curve for the periodic case. The IV curves are obtained by taking as initial conditions for the phases, for a given current, the final configuration of the preceding current value. In Fig. 3(a) we see the expected integer GSS (Eq. 11), with the IV curve scalable to the corresponding one Josephson junction IV. In Fig. 3(b) we show the IV results for a 40×40 lattice whith one defect located at

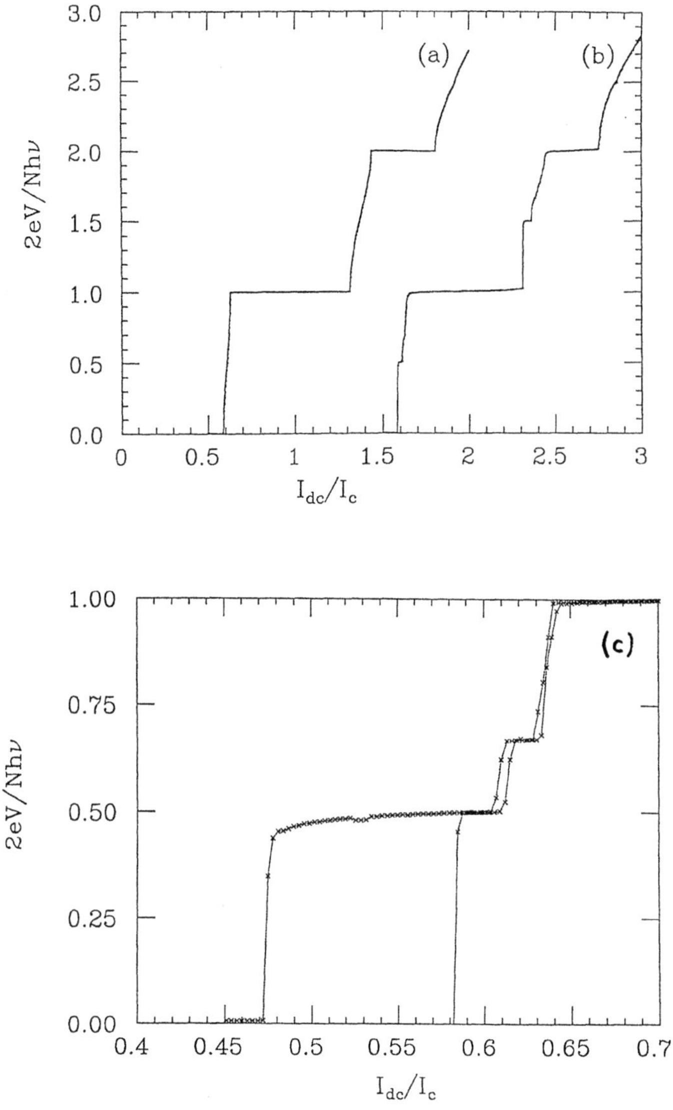

Figure 3: IV characteristics for a 40×40 square array with $I_{ac} = I_0$, $\nu = 0.1\nu_0$ and PBC. (a) Ordered array; (b) one defect in the center of the lattcie with $\delta_x = 0.1$, $\delta_y = 0.$ and $f = 2$ (the curve is shifted by one unit); (c) blowup of (b) for a 80×80 array showing hysteretic behavior and more subharmonic steps.

the center of the array. It turns out that the location of the defect does not affect the final stationary results. We clearly see new half integer steps in the IV curve at voltages $V_{n/2} = \frac{n}{2}\frac{N_y h\nu}{2e}$, in addition to the expected rounding of the integer steps due to disorder. In Fig. 3 (c) we show a blow up of the 1/2-step region, but for a 80×80 lattice. The 1/2 step is found to be hysteretic. The same hysteretic behavior is observed in the 3/2-step. We can define two critical currents, I_c^+ for increasing currents, and I_c^- for decreasing ones. For $I > I_c^+$ the 1/2-step is reversible, and it can be reached independent of the initial conditions. For $I_c^- < I < I_c^+$ the response depends on the initial conditions. There is also another subharmonic step at $V_{2/3} = \frac{2}{3}\frac{N_y h\nu}{2e}$, which was not present in the 40×40 lattice simulations. We will discuss this step in more detail in Sec. 3.3.

We carried out a finite size analysis of the 1/2-step width[14,15] and found that there is a minimum critical size $N_c \sim 16$ to be able to see the step. For the 1/2-step, the IV step width reaches an asymtotic plateau for lattices larger than 32×32. We believe that the need to have a minimum lattice size to see the 1/2-step is related to the nature of the vortex structures stabilized in the non-equilibrium stationary state. We discuss this point further in the next section.

We also have calculated the IV's for JJA with free boundary conditions. We find that the step is still present but rounded for small lattices but becomes sharper for larger lattices, approaching the same form as in the PBC calculations[15]. Therefore, the boundary conditions are not essential for the existence of this the novel fractional steps.

One way to characterize the Shapiro steps is studying their width dependence on the frequency ν and I_{ac}. In the single junction case the step width goes like $\Delta I_n = 2I_0 J_n(\frac{2eRI_{ac}}{h\nu})$, with J_n the Bessel function of integer order[22]. In Fig. 4(b) we show results for the 1/2-step width for one defect, with $\nu = 0.1\,\nu_0$, as a function of I_{ac}. We found an oscillatory behavior for $\Delta I_{1/2}$ which, in principle, could be associated to some Bessel function, but the data is not good enough to make a precise determination. What is more important, is the fact that wefind a critical value for I_{ac} above which the step appears. In Fig. 4(a) we show the dependence of the 1/2-step width as a function of ν when $I_{ac} = I_0$. Here we note that there is also a critical frequency above which the step width increases with ν, until it reaches a maximum around $\nu \approx 0.22\,\nu_0$, and then goes to zero rapidly for $\nu > 0.35\nu_0$. Experimentally, we then expect a "window" of frequencies and ac currents for which the 1/2-step can be seen.

To compare with experiment, we need to consider the effect of having finite temperatures. We have studied the stability of the 1/2-step as a function of temperature[14,15]. We found that the 1/2-step get rounded with increasing T and the hysteresis width is reduced. The step tends to disappear for $T > 0.04\frac{\Phi_0 I_0}{2\pi k_B}$, which is a small temperature as compared to the Berezinskii-Kosterlitz-Thouless critical temperature [3]. This is consistent with the fact that the width of the 1/2-step is small when compared to the critical current of the array.

We also found that when we increase the disorder, by adding a few more defects to the array, the results are qualitatively similar to those for one defect[15]. When all the lattice sites are deformed the 1/2-integer step is still there, but more rounded and with the hysteresis loop reduced in size. Therefore, a fully disordered configuration is not enough to destroy the extra step. However, a study as a function of the strength of the disorder [15] shows that there is a critical value above which the 1/2-steps disappears.

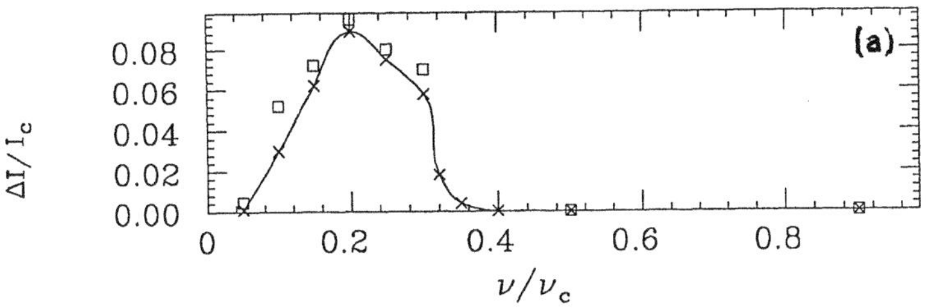

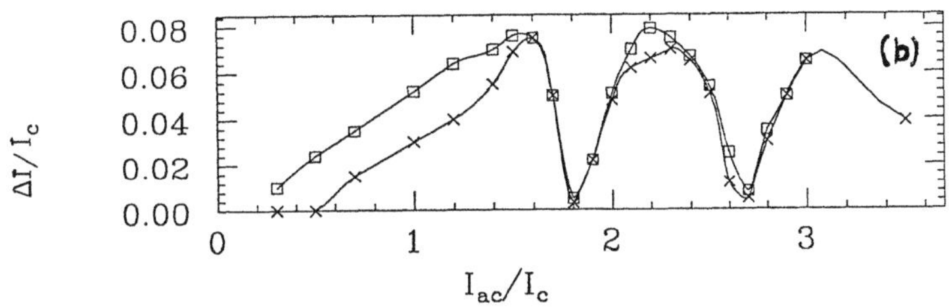

Figure 4: Step width for of the 1/2-step for arrays with one defect given by the same parameters as in Fig. 3(b). (a) As a function of the frequency, ($I_{ac} = I_0$); (b) as a function of the ac current, ($\nu = 0.1\,\nu_0$).

3.2. NON-EQUILIBRIUM COHERENT STATE

In the previous section we have studied the different conditions under which the extra half-integer steps appear in the IV's characteristics. Here we want to discuss the physical origin of these steps based in an analysis of the vortex dynamics.

When analyzing the spectral function $S(\nu) = \lim_{T\to\infty} |\frac{1}{T}\int_0^T V(t)e^{i2\pi\nu t}dt|^2$ it is found that it shows resonances at frequencies $n\nu/2$ for 1/2-steps [15]. This indicates that there is a state that oscillates coherenttly with twice the period of the ac current. These resonances only exist for currents in the 1/2-steps, while for other values of I_{dc} there is a broad band spectrum, except at integer steps where there are resonances only for integer multiples of ν's.

The presence of giant *fractional* Shapiro steps (GFSS), found when $f = p/q$, was explained in terms of coherent oscillations between vortex lattice ground states. At zero field there are no vortices in the array. However, as it was shown in Sec. 2.4 for the dc case, the presence of defects can produce current-induced vortices. Therefore, we computed the total absolute number of vortices in the array $N_a(t) = \sum_R |n(R,t)|$, as well as the total vorticity $N_T(t) = \sum_R n(R,t)$, where $n(R,t)$ was defined in Eq. (10). We started by choosing I_{dc} within the reversible part of the 1/2-step. Typical results for $N_a(t)$ and $N_T(t)$ in the single-defect case are shown in Fig. 5. We can identify two characteristic times t_d and t_{ACVS}. For $t < t_d$ there is a vortex pair (as in the dc case, Fig. 2) but with its $\pm$ polarity oscillating in phase with the current. The size of the vortex-dipole increases with time. At

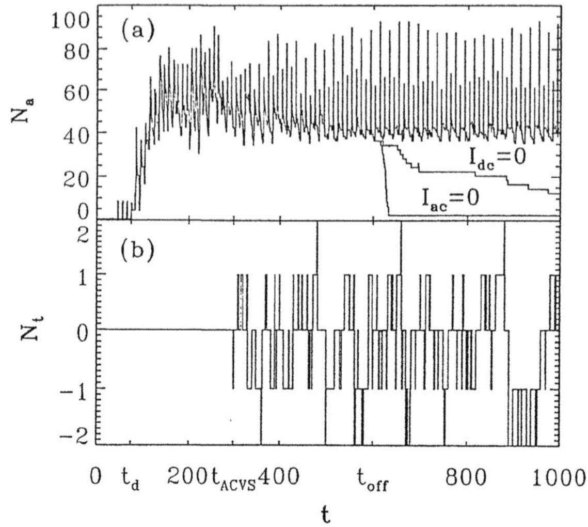

Figure 5: Number of vortices as a function of time for an array with one defect with $I_{dc} = 0.586I_0$ (corresponding to a 1/2-integer step). The size, frequency and ac current are the same as in Fig. 3 (b). (a) Absolute number of vortices $N_a(t)$; (b)total number of vortices $N_T(t)$. See text for definitions of t_d, t_{ACVS} and t_{off}. Time is measured in units of $1/\nu_0$.

t_d the vortex-dipole breaks away from the defect. After that time the number of vortices increase steeply. There is a mechanism that generates vortex pairs from the moving ones. This kind of vortex "radiation" phenomenon appears to be induced by the presence of an oscillating field due to the dc+ac current. The vortex motions in this intermediate phase do not follow a regular pattern. In the one-defect case this intermnidiate state has a left-right antisymmetry about an axis centered about the defect and along the current direction. This antisymmetry is reflected in N_T, which is exactly zero in this time interval. After a characteristic time t_{ACVS} the antisymmetry is broken, and an *axisymmetric coherent vortex state* (ACVS) starts to form. This is seen in the $N_a(t)$ curve as periodic variation of the number of vortices, and in the $N_T(t)$ as fluctuations about zero. In Fig. 6(a) we show the distribution of vortices and antivortices in the $ACVS$ state. The vortex configurations are formed by rows of vortices of alternating sign lying along a tilted axis with angle of about 27° measured clockwise with respect to the x-axis. The rows are separated along the x-axis by a distance of about $N_x/2$. In total there are $N_y/2$ vortices and $N_y/2$ antivortices. This state oscillates in sign in phase with the ac current. After a time $1/\nu$ the vortex rows interchange their signs, and after $2/\nu$ they are back to the original vortex state. Therefore, the vortex configuration changes periodically in time with a period twice that of the ac current. During most of the time the vortex patterns correspond to one or the other possible ACVS configuration. The change of sign of the vortex rows happens in a very short time, corresponding to the peaks in the $N_a(t)$ curve of Fig. 5(a). At that instant of time the vortices and antivortices cross from one row to the other. The collisions between them

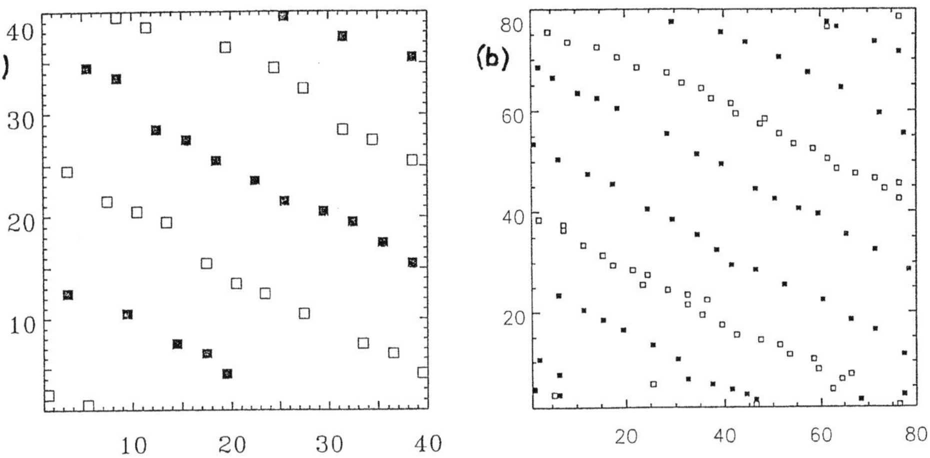

Figure 6: Axisymmetric coherent vortex states. (a) Stationary oscillating vortex configuration for an ACVS in the 1/2-step, for the same parameter values as in Fig. 3(b) with $I_{dc} = 0.586 I_0$. (b) ACVS configuration in the 2/3-step with $I_{dc} = 0.622 I_0$ in a 80×80 lattice. We use the same conventions as in Fig. 2.

produce additional vortices and antivortices, which annihilate each other rapidly before a new $ACVS$ of opposite sign is formed.

The $ACVS$ remains stable for the largest times considerd $(10^4/\nu)$. To study the stability of the ACVS, we turned off the ac and the dc currents at times t_{off} as shown in Fig. 5. In both cases, after turning off one of the currents with the other one still on, the ACVS decays into the initial two-vortex pinned state. The vortex number decreases by direct vortex-antivortex annihilation. However, the process of annihilation is much slower when the ac current is still on while the dc current is turned off. This means that the ac current strongly affects the collisions between vortices by actually delaying the direct annihilation of vortices by antivortices. The ACVS is also stable against small changes in the frequency and magnitude of the ac current., while making sure to be inside the 1/2-step.

A remarkable aspect of the ACVS state is that its structure, characterized by the angle, the distance between vortex rows and the number of vortices, appears to be a very robust non-equilibrium fixed point attractor for the dynamics. The ACVS shows the same vortex pattern when changing the number of defects, value of the ac current, frequency, defect "strength" f and temperature. The periodically oscillating vortex pattern is also the same for higher order $n/2$-steps with period $(n+1)/\nu$. When the boundary conditions are changed to FBC, the difference is that we need larger lattices to "fit" the ACVS in teh lattice. We need at least a 64×64 lattice to clearly see the ACVS state. The ACVS forms even when it appears to be more "disordered" than the one with PBC. Therefore, boundaries affect the coherence of the state, but for samples large enough the $ACVS$ forms and it is stable.

The one difference we found in the vortex patterns is that the angle can be either 27°

or its complimentary $\pi - 27°$, which are the two possible broken-symmetry states. Also in large rectangular lattices (when $N_y >> N_x$) it is possible to have the two possible states coexisting with each other[15].

We have also considered the cases when two, three, or all the lattice sites are disordered. We find that in all cases, although the length of the transient and the number of disorder-induced vortices is different, the stationary oscillatory state is essentially the same ACVS.

At this point one would like to know what are the essential ingredients to form the *ACVS*. A naive approach would suggest that the steady state is reached when the rate of vortex generation by the defect is equal to the rate of vortex annihilation. However, we saw that most of the vortices were generated away from the defect during the transient time ($t_d < t < t_{ACVS}$). If the disorder were essential for the formation of the ACVS, then when we switch off the disorder the state should collapse. However, the ACVS remained stable when the defects were removed. We also tried switching off the defects for times $T_d < t < t_{ACVS}$ an the state remained stable. On the other hand for times $t < t_d$ the state collapses. That means that once we have a vortex-antivortex pair large enough so that they can not annihilate each other, they are then capable of generating the whole ACVS by themselves. This proves that what is actually essential is the "radiation" mechanism produced in the collisions of the current-induced vortices under an ac current. We have also checked that starting initially with a pair of a vortex and antivortex in an ordered lattice, the ACVS is generated for the same I_dc, I_ac and ν as when the defect is present.

This means that *any physical mechanism (not only the disorder) that is capable to induce at least one pair of vortex-antivortex sufficiently apart can generate the ACVS.*

3.3. HIGHER ORDER SUBHARMONIC STEPS

When increasing the lattice size further we found extra subharmonic steps. As was shown in Fig. 3 (c), an additional $V_{2/3} = \frac{2}{3}\frac{N_y h\nu}{2e}$ step appears for a lattice size of 80×80. There is also a critical size for the appearance of this subharmonic step, which is about $N_c \sim 70$.

When we analyzing the time evolution of the vortex distributions in the 2/3-step, as we did in Sec. 3.2, we found that it is also generated by an ACVS but now oscillating with period $3/\nu$. This state becomes stable only after a very complex transient with a longer t_{ACVS} than in the 1/2-step case. In Fig. 6(b) we show the stationary vortex distributions for a current within the 2/3-step. Again, it consists of rows of vortices (and antivortices) lying along a tilted axis with the same angle of about 27° with respect to the x axis and separated by a distance of $N_x/3$ along the x-direction. The total number of vortices is now $\frac{3}{2}N_y$. To have period 3 there is one row of vortices and two of antivortices. They interchange positions in synchrony with the ac-current. The total vorticity is still zero on the average, because the rows with vortices have double density than the ones with antivortices. This implies that now there are more possibilities for the broken symmetry states: either with angle 27° or $\pi - 27°$, either with one vortex row and two antivortex rows, or with two vortex rows and one antivortex row.

4. Discussion

Let us first discuss the relation of our results to recent experiments. We shall mainly refer to the experiment by Ref.[12] and also to the experiment of Ref.[13] both done in periodic Nb-Au-Nb proximity arrays. They observed subharmonic steps at fractions $n/2$ and $n/3$ with zero field. These arrays have large critical currents, which means that screening effects are important. In fact, it was shown in the experiment of Ref.[12] that currents flowing in the array ($f = 0$) generate induced antisymmetric magnetic field distribution in the arrays. The magnetic field is very strong at the edges of the array, with magnitude $B_{edge} \sim \pm M I^{ext}/a^2$, where M is the mutual inductance between plaquettes in the array. We have recently included these self-induced edge field in our numerical studies by adding a field with flux $\phi = \mu I^{ext}\phi_0$ to the plaquettes in one of the edges parallel to the current direction, and $\phi = -\mu I^{ext}\phi_0$ to the plaquettes in the opposite edge [15,23]. We found extra 1/2-steps in the IV characteristics as we found before in arrays with defects. Again, these steps are due to the presence of an ACVS oscillating with period $2/\nu$. In this case the state was originated from vortices and antivortices nucleated at the edge of the array by the imposed edge magnetic fields.

This result is in agreement with the conclusion we reached at the end of Sec. 3.2 in that what is needed to produce the *ACVS* is a mechanism to nucleate vortices and antivortices in the array.

In this paper we have presented evidence for the formation of a non-equilibrium stationary vortex pattern formed in proximity effect JJA. We characterized some the properties of this state, and have shown the experimental manifestations for the existence of this coherent vortex state in terms of IV characteristics. However, many basic questions remain unanswered, to wit:

How is the ACVS originated dynamically?. It appears that the vortex "radiation" mechanism during the transient time is essential for its formation. This vortex generation phenomenon under ac fields does not appear to have been studied before, though.

We also need an explanation of why the particular pattern of the ACVS. Due to its robustness under changes in frequency, ac current, disorder and boundary conditions, it is evident that this state is a global attractor of the dynamics in the system. Moreover, the ACVS is a truly non-equilibrium state since it disappears when the driving currents are turned off. This implies that the ACVS can not be obtained from a minimum energy principle. This is a basic difference with respect to the fractional GSS observed with $f = p/q$ [8,9,10]. What appears to be important in the structure of the ACVS is the effect of the Lorentz force and the logarithmic interactions between vortices.

The model of Eq. (9) neglects the dynamics of the magnetic field. When this is taken into account, the interaction energy is logarithmic for distances smaller than a certain distance λ (the London penetration depth of the array), and then it falls exponentially for distances larger than λ. We found[23] that when $\lambda < Na$ the angle of the vortex rows changes, becoming parallel to the current direction, and the number of vortices per row increases. This is the only case in which we found a change in the structure of the vortex configuration for the 1/2-steps.

Finally, let us mention that the observation of more subharmonic steps as the lattice size increases would suggest that in the thermodynamic limit the IV curve can have a rather complicated structure. However, it appears that the width for larger lattices and higher harmonics decrease in size as the lattice grows. So in the infinite size limit the steps may become inmeasurably small.

We leave these and other questions for future studies.

Acknowledgements

Part of the research work presented here is the result of a collaboration with A. Karma and C. Wiecko. Fruitful discussions with R. Markiewicz, D. Stroud and M. Rzchowski are greatly acknowledged. Support from: CONICET, Argentina; NSF grant DMR-9211339 the NSF Pittsburgh Supercomputing Center, the Donors of the Petroleum Research Fund, administered by the American Chemical Society, under grant ACS-PRF#22036-AC6; ICTP, Trieste, are also gratefully acknowledged.

References

[1] B. D. Josephson, Phys. Lett. **1**, 251 (1962).

[2] For several papers on the subject see Physica (Amsterdam) **152B**, 1-302 (1988).

[3] J. M. Kosterlitz and D. J. Thouless, J. Phys. C **6**, 1181 (1973). V. L. Berezinskii, Zh. Eksp. Teor. Fiz. **61**, 1144 (1971), *ibid.* **59**, 907 (1970)

[4] For a review see C. D. Lobb, Physica (Amsterdam) **126B& C**, 319 (1989).

[5] J. S. Chung, K. H. Lee, and D. Stroud, Phys. Rev. B **40**, 6570 (1989); K. K. Mon and S. Teitel, Phys. Rev. Lett. **62**, 673 (1989).

[6] Ch. Leeman, Ph. Lerch, and P. Martinoli, Physica (Amsterdam) **126B**, 475 (1984); T. D. Clark, Phys. Rev. B **8**, 137 (1973).

[7] S. P. Benz, M. S. Rzchowski, M. Tinkham, and C. J. Lobb, Phys. Rev. Lett. **64**, 693 (1990).

[8] K. H. Lee, D. Stroud, and J. S. Chung, Phys. Rev. Lett. **64**, 692 (1990); K. H. Lee and D. Stroud, Phys. Rev. B **43**,5280 (1991).

[9] J. U. Free, S. P. Benz, M. S. Rzchowski, M. Tinkham, C. J. Lobb, and M. Octavio, Phys. Rev. B **41**,7267 (1990); M. Octavio, J. U. Free, S. P. Benz, R. S. Newrock, D. B. Mast, and C. J. Lobb, *ibid* **44**,4601 (1991).

[10] M. Kvale and S. E. Hebboul, Phys. Rev. B **43**, 3720 (1991); M. S. Rzchowski, L. L. Sohn, and M. Tinkham, *ibid* **43**, 8682 (1991); T. Halsey *ibid* **41**, 11634 (1990).

[11] S. P. Benz and C. J. Burroughs, Appl. Phys. Lett. **58**, 2162 (1991).

[12] H. C. Lee, D. B. Mast, R. S. Newrock, L. Bortner, K. Brown, F. P. Esposito, D. C. Harris, and J. C. Garland, Physica B **165& 166**, 1571 (1990); H. C. Lee, R. S. Newrock, D. B. Mast, S. E. Hebboul, J. C. Garland, and C. J. Lobb, Phys. Rev. B **44**, 921 (1991).

[13] S. E. Hebboul and J. C. Garland, Phys. Rev. B **43**, 13703 (1991).

[14] D. Domínguez, J. V. José, A. Karma, and C. Wiecko, Phys. Rev. Lett. **67**, 2367 (1991).

[15] D. Domínguez and J. V. José to be published in Phys. Rev. B.

[16] W. Xia and P. L. Leath, Phys. Rev. Lett. **63**, 1428 (1989); P. L. Leath and W. Xia, Phys. Rev. B **44**, 9619 (1991).

[17] D. E. McCumber, J. Appl. Phys. **39**, 3113 (1968); W. C. Stewart, Appl. Phys. Lett. **10**, 277 (1968).

[18] S. Shapiro, Phys. Rev. Lett. **11**, 80 (1963)

[19] H. Eikmans and J. E. van Himbergen, Phys. Rev. B **41**, 8927 (1990), and *ibid* **44**, 6937 (1991).

[20] H. S. Greenside and E. Helfand, Bell. Syst. Tech. J. **60**, 1927 (1981).

[21] S. Teitel and C. Jayaprakash, Phys. Rev. Lett. **51**, 1999 (1983); T. C. Halsey, Phys. Rev. B **31**, 5728 (1985).

[22] T. Van Duzer and C. Turner, *Principles of Superconductive Devices and Circuits* (Elsevier North Holland, New York, 1981)

[23] D. Domínguez and J. V. José, Phys. Rev. Lett. (1992).

Comments on the topological organization of 3d-flows and 2d-maps

H. G. Solari[1], M. A. Natiello[2], B. G. Mindlin[3] and R. Gilmore[4]
[1] *Department of Physics, FCEN - University of Buenos Aires. Pabellón I, Ciudad Universitaria, 1428 Buenos Aires, Argentina.*
[2] *Department of Quantum Chemistry, Uppsala University Box 518, S751 20 Uppsala, Sweden*
[3] *Department of Physics, School of Science, Navarra University, Pamplona, Spain*
[4] *Department of Physics and Atmospheric Science, Drexel University Philadelphia 19104, USA*

Abstract

The main concern of this paper is the problem of model validation. In this context we discuss recent advances in the practical implementation of topological methods for the characterization of three dimensional flows and show how this analysis can be economically performed for bidimensional maps. From this new perspective we explore the relation between braids and heteroclinic tangles. Several remarks and directions for further research follow naturally.

1 Introduction

The problem of model validation is essential to the natural sciences, for models and theories have to be compared with experimental data if they are going to have any relevance to our understanding of the natural world.

The chaotic paradigm poses in this respect a challenge since direct comparisons between modelled and measured variables cannot be done directly by comparison of the time traces. Sensitive dependence of the trajectories to initial conditions is expected to render such comparisons meaningless, furthermore, even the reproducibility of experiments have to be thought in a more general sense since, in general, the initial condition for the system is not fully controlled.

A second constraint on the elements considered for a meaningful comparison between models and experiments is the realization of continuity of the observables with respect to parameter space or some sort of smooth parameter dependence. As long as models are *idealizations* of experiments, i.e. descriptions leaving (large) parts of the experiment unaccounted for, and chaos is in most (all?) physical cases associated with structural instability of the attractor, the most we can expect for our models is to reproduce the experiments up to some scale and disregarding details; the use for this purpose of quantities based on infinite precision measurement, like the various dimensions[1], have to be viewed

E. Tirapegui and W. Zeller (eds.), Instabilities and Nonequilibrium Structures IV, 69–76.

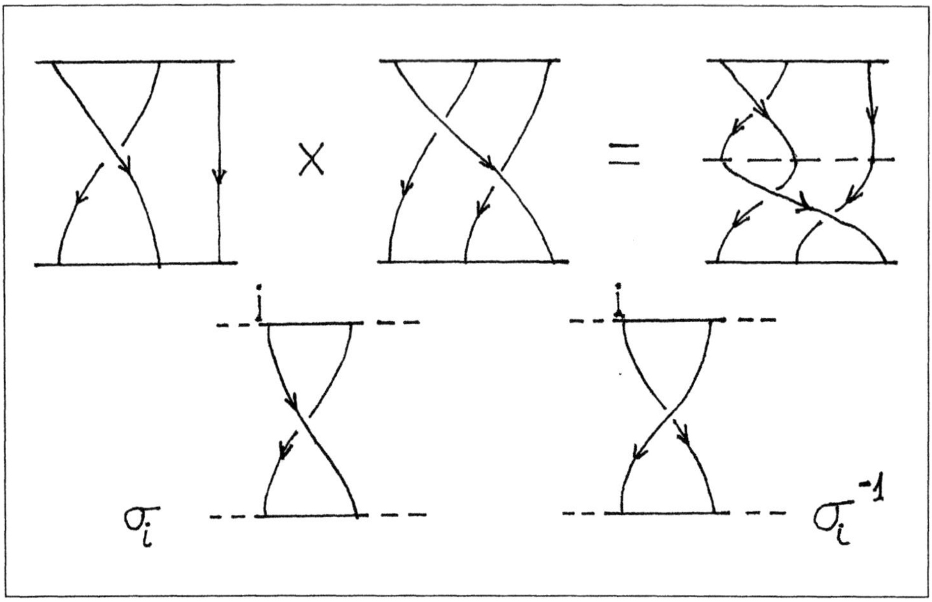

Figure 1: Braid composition law and elementary braids

with skepticism unless they can be shown to change smoothly with perturbations of the model.

The characterization and classification of chaotic systems has yet to be developed in the general case. For three dimensional flows such classification is tied to the topological organization of periodic orbits and their characterizations as links and knots. For two dimensional maps, and 3d-flows admitting a global Poincaré section, the fundamental element is the braid structure of periodic orbits[2, 3, 4].

There has been only a few attempts to relate experimental data to models based in the topological organization of periodic orbits[5, 6]. The method used consist in constructing a faithful 3d-representation of the flow, extracting segments of the chaotic attractor that closely resemble periodic orbits, studying the structure of these orbits and finally relating this organization with known hyperbolic maps comparing different characteristics of the links in the model and data such as linking numbers, relative rotation rates[7] and knots polynomials[5, 8, 6].

This procedure encounters at least two problems, first, the reconstruction of the flow for all times in a three dimensional embedding is a craft that has proven not to be easy to perform given the fact that all too often the experiment evolves for some time in the direction perpendicular to the variable measured and accordingly the time trace gives no information about the evolution at these times; second the restriction to a few hyperbolic models or templates is certainly a first step but sharper tools will have to be developed in order to make this program fully successful [6].

In this note we show how to lift the practical restriction to 3d-flows and work on the Poincaré section with the Poincaré map (Section 3). Within this representation of the Braid group it is immediate to realize that the braid type of the set of orbits considered is written in the manifold organization of the attractor and partially determines its organization (Section 4).

2 Braids and periodic orbits

The Braid group was introduced by Artin[10] and its internal operation consists in piling one braid on top of another erasing the middle surface (for this procedure to work it is clear that the number of strands in each braid must be the same) Fig.1. The Braid group has an algebraic representation in term of the generators, σ_i , $i = 1 \cdots n$ of a free group. The relations that this generators have to satisfy are

$$\sigma_i \sigma_{i+1} \sigma_i = \sigma_{i+1} \sigma_i \sigma_{i+1} \tag{1}$$

$$\sigma_i \sigma_j = \sigma_j \sigma_i \text{ if } |j - i| > 1 \tag{2}$$

as is illustrated in the Fig.1.

Consider a periodic orbit in a three dimensional flow admitting a Poincaré section. The periodic orbit intersects the Poincaré surface at p points defining the periodicity of the orbit with respect to the first return map. We can imaginarily split the Poincaré surface in two identical sheets and open up the flow cutting the phase space that now proceeds from the "top" surface to the "bottom" surface. With this surgical procedure we have transformed the periodic orbit into a braid, the inverse process consist in identifying both halves of the Poincaré surface going back to the periodic orbit Fig.2. It is then natural to represent periodic orbits as elements of the Braid group.

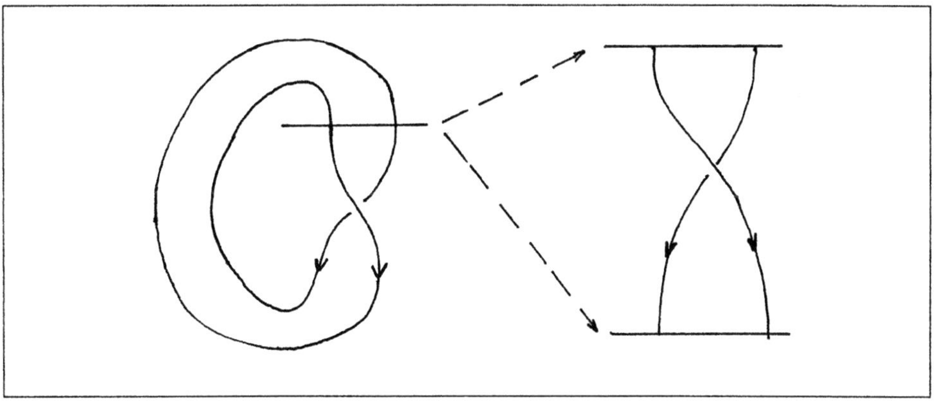

Figure 2: Periodic orbit, Poincaré section and braids

It should be noticed that the representation of the orbit by an element of the Braid group depends on the election of the Poincaré control surface, and there is a certain degree of arbitrariness. The braid representative of a set of orbits can only be determined up to conjugations, or what is the same, the orbit representatives are the classes of braids (having the conjugation as the class defining relation) Fig. 3.

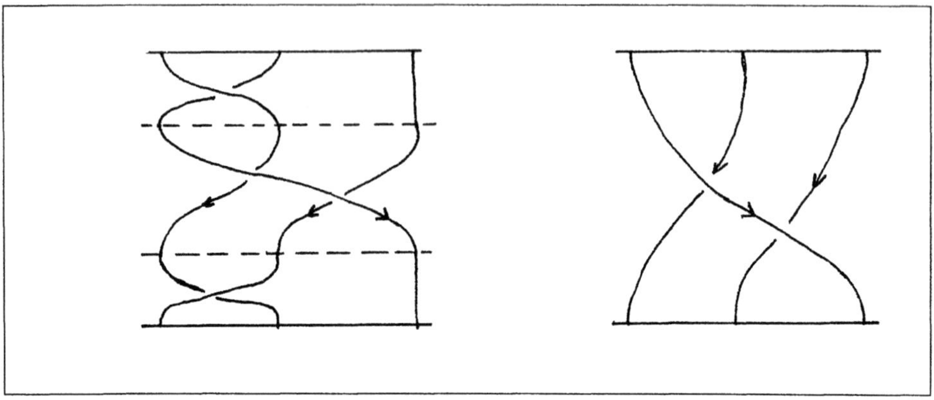

Figure 3: The same orbit in two different (conjugated) presentations.

Braids are directly associated with the classification of diffeomorphisms of a surface[2] and are able to tell us whether a map has associated positive topological entropy or not. Conversely, maps with zero topological entropy are required to have a special organization of the periodic orbits in terms of families of rotation compatible orbits[4]. It is then relatively straightforward to establish whether a periodic orbit implies positive topological entropy or not[3], for example, the detection of a period seven orbit knotted as a pretzel knot has been enough to establish that the Belousov-Zhabotinskii data set in Ref.[5] is chaotic.

3 A representation of braids on the Poincaré surface

Despite being the Braid group the fundamental group of a punctured disk[9], the direct computation, following this definition, of the braid type of a set of periodic orbits of a 2d-map is a complicated task from a practical point of view implying the computation of images of loops corresponding to each generator.

A simpler procedure consists in connecting the periodic points with a closed curve without self intersections. For every image that keep the periodic points fixed (considered up to deformations) there is one and only one element of the Braid group singled out[11]. Clearly, this procedure requires taking the image of just one loop becoming a more efficient tool. In Fig.4 the images is depicted.

The intuition for this representation of the Braid group can be constructed as follows: In the standard pictorial presentation of a braid we draw the original circle going through all the intersections of the periodic orbit with the Poincaré section. As "time" proceeds this circle moves down towards the bottom image of the Poincaré section with the circle always intersecting transversely the periodic orbit. Each strand of the periodic orbit can be thought as a wire and each arc of the topological circle as a rope. Moving the rope down the wires forces a deformation of rope (the topological circle) which depends only on the braid type of the orbit. This idea can be formalized by means of a fibering projection allowing to show that the association is one to one[11].

In the case of experimental data this procedure requires the reconstruction of the

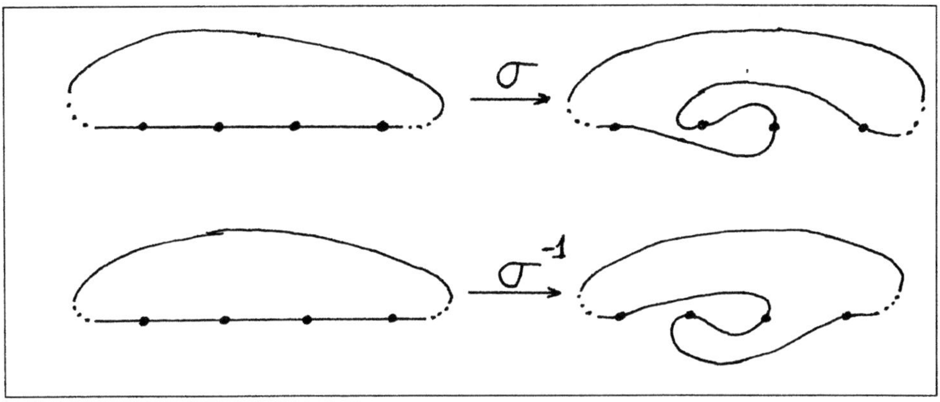

Figure 4: Braids in the plane: presentation of the fundamental braids.

Poincaré map at a single Poincaré surface. A method that leaves enough room to avoid the regions of phase space poorly reconstructed. More over, the reconstruction of the Poincaré map needs not to be extremely precise for it is going to be used in a single iteration, hence with no cumulative error.

4 Braids and manifolds

The association of braids with periodic orbits in 3d-flows is naturally made by the orbit directly representing the braid. On the other had, the association in the case of the Poincaré section is made with an auxiliar construction. The question becomes whether a closer association with invariant sets, and then with the dynamics, can be achieved or not.

In the context of chaotic attractors it is expected (given the topological transitivity of the attractors) that the stable manifold of every saddle "in" the attractor intersect transversely the unstable manifold of the every saddle. In such a case, the topological circles connecting the periodic orbits will be realized by the heteroclinic connections.

A second observation is that, cyclic homoclinic tangencies[12] for periodic orbits with a prime period, p, can only happen if the braid type is of the form $(\sigma_1\sigma_2\cdots\sigma_p)^k$ since the homoclinic connection carries the braid type written in. The comparison of two period five orbits present in the horseshoe illustrates this point Fig.5and Fig.6.

Third, we notice that it is not possible to construct a homoclinic tangle, with the proper braid type, connecting all points of the period five orbit xxyxy in Fig.6 without generating more intersections. Observation that prompt the following questions: In the process of complexification of a map, for example by increasing the non-linearity in the Henon map, how are cyclic homoclinic connections created in the general case of a non cyclic saddle? Do they exist whenever the orbit exists? (Note that if the answer is yes, this will contribute to a direct understanding of Thurston Theorem[2] in terms of dynamical elements.) Actually, all our actual numerical observations indicate that this is indeed the case.

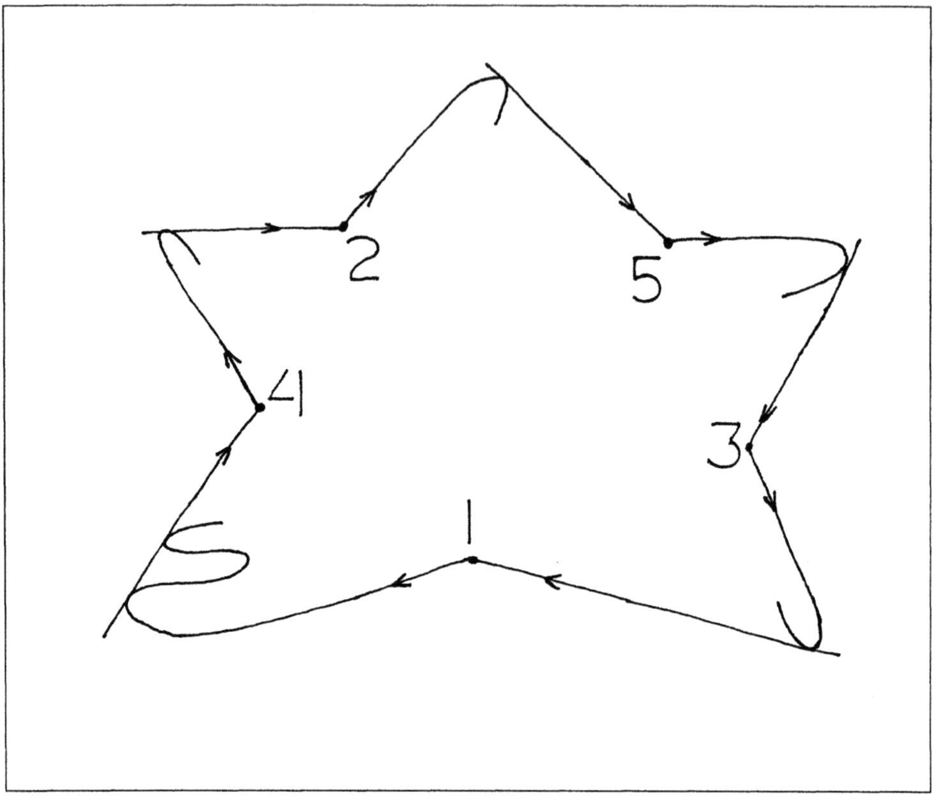

Figure 5: Manifold organization in the cyclic periodic orbit xyyyy

5 Summary

The topological organization of periodic orbits in 3d-flows and 2d-maps serves as the basis of comparison for models and data, and can be used in model validation.

It is possible to find the braid type of a set of periodic orbits in 2d-maps within a representation that does not require the explicit suspension of the map. This representation is specially appropriated for flows in which some part of the dynamics is not directly accessible.

The relation between the representation of the Braid Group by mean of images of topological circles passing through the points in the periodic orbits and the manifold organization presents promising aspects and deserves further inquiry, specially, considering the possibility of realizing the topological circle with pieces of invariant sets.

Acknowledgement

We would like to thanks Francesco Papoff for fruitful and stimulating discussions.

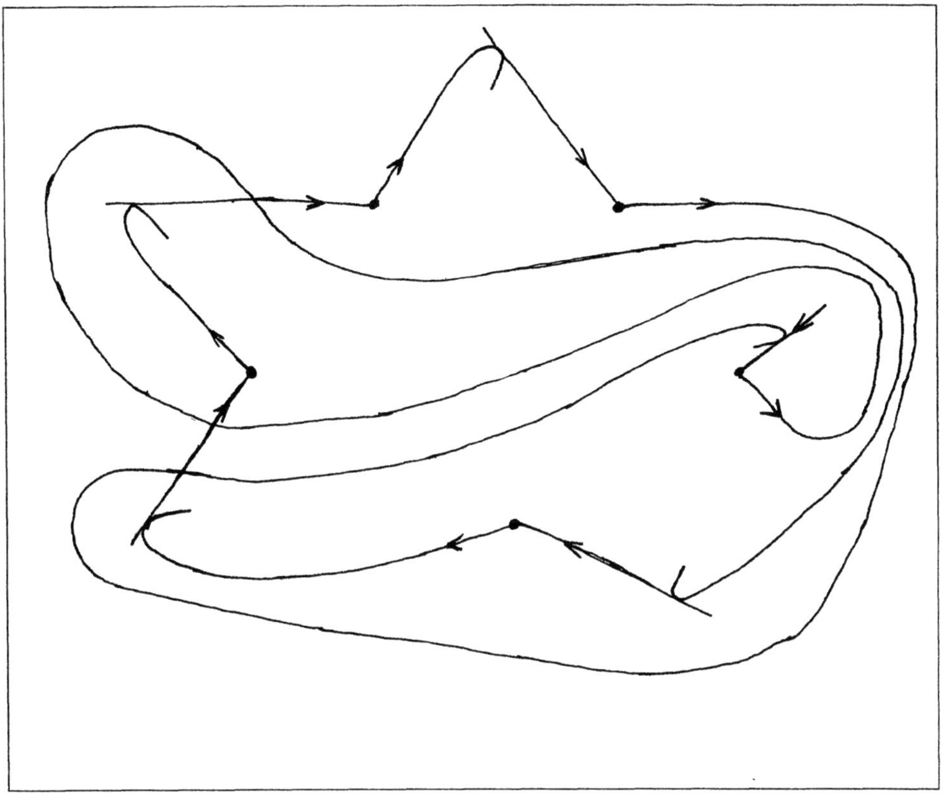

Figure 6: Manifold organization in the non cyclic periodic orbit xxyxy

H. G. Solari is a member of the Argentine Research Council in Science and Technology, CONICET.

This work has been partially supported by N.S.F. grant PHY88 – 43235.

References

[1] P. Grassberger and I. Procaccia, Phys. Rev. A **28**, 2591-2593 (1983).

[2] W. P. Thurston. Bull. AMS **19**, 417-432 (1988).

[3] P. Boyland. Braids and a topological method of proving positive topological entropy. Preprint.

[4] J. M. Gambaudo, S. van Strien and C. Tresser. Ann. Inst. H. Poincaré **49**, 335-356 (1989).

[5] B.G. Mindlin, H.G. Solari, M.A. Natiello, R. Gilmore and X.J. Hou, JNLS **1**, 147, (1991).

[6] F. Papoff, A. Fioretti, E. Arimondo, B. G. Mindlin, H. G. Solari and R. Gilmore, Phys. Rev. Lett. (1992), in press.

[7] H.G. Solari and R. Gilmore, Phys. Rev. A **37**, 3096, (1988)

[8] G. B. Mindlin , X.-J. Hou, H. G. Solari, R. Gilmore and N. B. Tufillaro, Phys. Rev. Lett. **64**, 2350-2353 (1990).

[9] E. Artin. Amm. Math. **48**, 101-126 (1947); **48**, 643-649 (1947)

[10] The American Scientist **xx** 113 (1951).

[11] M. A. Natiello and H. G. Solari. Preprint (1992).

[12] C. Grebogi, E. Ott and J. A. Yorke. Physica **D** **24**, 243, (1987).

Phase Bistability and Squeezing in a Two-photon Micromaser

M. Orszag, L. Roa and R. Ramírez
Facultad de Física, Universidad Católica de Chile,
Casilla 306, Santiago 22, Chile

Abstract

A quick tutorial review of the coherent and squeezed states is presented. We also derive the master equation for the two-photon micromaser, with atomic coherence. We find that the atomic coherence produces a bifurcation in the phase probability distribution and squeezing. The maximun squeezing is 36% and it occurs below threshold.

I. Introductory concepts and remarks

Heisenberg uncertainty principle predicts that it is not possible to be absolutely precise about the energy of a system, even the vacuum. This uncertainty manifests itself in so called vacuum fluctuations which is a sort of background quantum noise. These fluctuations impose a limit on the precision of optical measurements. For example phase fluctuations produce optical noise which smear out the interference patterns making very dificult a sensitive measurement.

Though at first sight the uncertainty principle seems to make imposible the decrease of this quantum noise, it has been recently shown that there are several ways of reducing or squeezing the fluctuations.

The reduction of the fluctuation of a given variable is possible at expense of increasing the fluctuation of the conjugate variable and vice-versa. For example increasing the fluctuations in amplitude would make it possible to reduce the phase fluctuations and would allow to send a very low power signal via the squeezed part of the light with little interference from quantum noise.

Squeezed light is one example of nonclassical light.[1] The possibility of nonclassical light was proposed by Roy Glauber in the 60's. The experimental realization of squeezed light was achieved in the Bell Labs in 1985.[2] The methods involves a non-linear process. Two red photons impinging on a crystal of lithium niobate combine to emit a blue photon of twice the frequency through a so called second harmonic generation. The emission of blue photons is more likely when the intensity of the red light is larger i.e. when the fluctuations make the amplitude large. Therefore the intensity fluctuations of the red light diminish at expense of phase fluctuations.

A single cavity mode electromagnetic field can be described by the conjugate electric and magnetic field operators:

E. Tirapegui and W. Zeller (eds.), Instabilities and Nonequilibrium Structures IV, 77–88.

$$E(z,t) = (2\omega^2/\epsilon_o V)^{1/2} q(t) \sin kz$$
$$H(z,t) = (2c^2\epsilon_o/V)^{1/2} p(t) \cos kz, \tag{1}$$

where $p(t)$ and $q(t)$ satisfy the equal-time conmutation relation $[p(t), q(t)] = -i\hbar$ and the uncertainty relation in the system ground state:

$$\langle(\Delta p)_o^2\rangle\langle(\Delta q)_o^2\rangle \geq \frac{\hbar\omega}{2}, \tag{2}$$

For some states, called "minimun uncertainty states" (MUS) the equality sign hold.

Introducing the destruction and creation operators:

$$a = (2\hbar\omega)^{-1/2}(\omega q + ip)$$
$$a^\dagger = (2\hbar\omega)^{-1/2}(\omega q - ip), \tag{3}$$

we can define a coherent state a right eigenstate of the destruction operator a:

$$a|\alpha\rangle = \alpha|\alpha\rangle, \tag{4}$$

where α is a complex amplitude $|\alpha|\exp(i\phi)$.

It can be shown that these states are MUS states and correspond to Gaussian wavepackets in position and momentum space:

$$\psi(q) = \langle q|\alpha\rangle = C\exp(-q^2/4\langle(\Delta q)_\alpha^2\rangle)$$
$$\psi(p) = \langle p|\alpha\rangle = C\exp(-p^2\langle(\Delta q)_\alpha^2\rangle/\hbar^2), \tag{5}$$

It is convenient to introduce the quadrature operators:

$$X = \frac{1}{2}(a + a^\dagger)$$
$$Y = \frac{1}{2i}(a - a^\dagger), \tag{6}$$

which satisfy the conmutation relation $[X, Y] = i/2$.

It is simple to show that their variances satisfy the uncertainty relation:

$$\langle(\Delta X)^2\rangle\langle(\Delta Y)^2\rangle \geq \frac{1}{16}, \tag{7}$$

For a coherent state their variances are equal:

$$\langle(\Delta X)^2_\alpha\rangle = \langle(\Delta Y)^2_\alpha\rangle = \frac{1}{4}, \tag{8}$$

In figure 1 we represent the uncertainties as an error circle centered in:

$$\langle(\Delta X + i\Delta Y)_\alpha\rangle = \alpha, \tag{9}$$

COHERENT STATES

A coherent state can be generated from the vacuum state by the action of the Glauber unitary displacement operator, [3]:

$$|\alpha\rangle = \hat{D}(\alpha)|0\rangle, \tag{10}$$

where

$$\hat{D}(\alpha) = \exp(\alpha a^\dagger - \alpha^* a), \tag{11}$$

acts as a translation operator on X and Y:

$$\hat{D}^{-1}(\alpha)X\hat{D}(\alpha) = X + \mathrm{Re}\alpha, \tag{12}$$

$$\hat{D}^{-1}(\alpha)Y\hat{D}(\alpha) = X + \mathrm{Im}\alpha, \tag{13}$$

Therefore the Glauber operator translates the vacuum error centered at the origin to a new position $(\mathrm{Re}\alpha, \mathrm{Im}\alpha)$

Hence a coherent state can be considered as a classical state with added vacuum fluctuation noise.

SQUEEZED STATES

A squeezed state can be genererated by the unitary squeeze operator, [4]:

$$\hat{S}(\zeta) = \exp(\frac{1}{2}\zeta^* a^2 - \frac{1}{2}\zeta a^{\dagger 2}), \tag{14}$$

where $\zeta = s\exp(i\theta)$.

This operator modifies the fluctuations of the quadratures:

$$\langle(\Delta X)^2_\alpha\rangle = \frac{1}{4}[\exp(-2s)\cos^2\frac{1}{2}\theta + \exp(2s)\sin^2\frac{1}{2}\theta]$$

$$\langle(\Delta Y)^2_\alpha\rangle = \frac{1}{4}[\exp(-2s)\sin^2\frac{1}{2}\theta + \exp(2s)\cos^2\frac{1}{2}\theta], \tag{15}$$

which, for $\theta = 0$ reduce to:

$$\langle(\Delta X)^2_\alpha\rangle = \frac{1}{4}\exp(-2s),$$

$$\langle(\Delta Y)^2_\alpha\rangle = \frac{1}{4}\exp(2s). \tag{16}$$

We notice that the variance of X is less than the vacuum value.

In the figure 2 we show an idealized squeezed state defined by:

$$|\alpha,\zeta\rangle = \hat{D}(\alpha)\hat{S}(\zeta)|0\rangle, \tag{17}$$

we notice that for $s > 0$ the X variance is compressed and the Y variance is expanded producing an elliptical error contour.

II. The two-photon micromaser. Bifurcation of the phase probability distribution and squeezing

In the present communication, we study the two photon micromaser [5-7], that consists in a low density three level atomic beam crossing a high Q microwave cavity [8,9].

The atoms crossing the cavity are initially prepared in a coherent superposition of the upper and lower level (Figure 3), that is:

$$|\psi(0)\rangle_{\rm at} = \beta_a|a\rangle + \beta_c|c\rangle. \tag{18}$$

The three level atom has a detuning Δ of the intermediate level with respect to the center of the transition (Figure 3). Also the states and have the same parity and one photon transitions are not allowed. Furthermore, we assume that the detuning is large ($\Delta/g \gg \sqrt{N}$) so that we have a truly two photon rather than a one photon cascade system. These two photons are resonant with the a-c transition.

The Hamiltonian of the system reads $H = H_o + V$,

$$\begin{aligned} H_o &= \sum_{i=a,b,c} E_i|i\rangle\langle i| + \hbar\omega(a^+a + \frac{1}{2}), \\ V &= \hbar g_{ab}[a|a\rangle\langle b| + |b\rangle\langle a|a^+] + \hbar g_{bc}[a|b\rangle\langle c| + |c\rangle\langle b|a^+]. \end{aligned} \tag{19}$$

Here a is the destruction operator of the field and g_{ab} and g_{bc} is the atom-radiation coupling constants for $a-b$ and $b-c$ transitions respectively. We have also used the rotating-wave approximation.

Introducing the interaction picture $V_I = \exp(iH_ot/\hbar)V\exp(-iH_ot/\hbar)$, the time evolution operator satisfies the differential equation $\dot{U}(t,t_o) = -i/\hbar V_I U(t,t_o)$. It is simple to find the propagator $U(\tau)$ under the approximation $\Delta/g \gg \sqrt{N}$. The result is [10]:

$$U(\tau) = \begin{pmatrix} 1 - a\hat{C}a^+ & 0 & -a\hat{C}a \\ 0 & e^{-i\phi(2\hat{n}+\delta)} & 0 \\ -a^+\hat{C}a^+ & 0 & 1 - a^+\hat{C}a \end{pmatrix}, \tag{20}$$

where $\hat{C} = (1 - e^{i\phi(2\hat{n}+\delta)})(a^+a + aa^+)^{-1}$, $\hat{n} = a^+a/2N_{\rm ex}$, $N_{\rm ex} \equiv Rt_{\rm cav} = t_{\rm cav}/t_{\rm at}$, $\phi \equiv 2N_{\rm ex}\tau g^2/\Delta$, $\delta \equiv 1/2N_{\rm ex}$. R is the atomic injection rate $1/t_{\rm at}$, $t_{\rm cav}$ is the cavity decay time, τ is the flight time of the atom through the cavity and $N_{\rm ex}$ represents the number of atoms going through the cavity during the cavity decay time. In equation (20) the 3x3 matrix elements correspond to the a-b-c sequence. Also we assume $g_{ab} = g_{bc} = g$.

Calling $\rho_{aa} \equiv |\beta_a|^2$, $\rho_{cc} \equiv |\beta_c|^2$ and $\rho_{ca} \equiv \beta_c^*\beta_a = \beta\exp(i\theta_{\rm at})$, we can write the master equation for the reduced density matrix for the field.

$$\dot{\rho}(t) = L\rho(t) + R[U(\tau)\rho(t)U^{+}(\tau) - \rho(t)], \tag{21}$$

where $L\rho$ represents the cavity losses. With zero thermal photons, this term can be written as:

$$L\rho = \frac{\omega}{2Q}(2a\rho a^{+} - a^{+}a\rho - \rho a^{+}a), \qquad \frac{\omega}{Q} = \frac{1}{t_{\text{cav}}}, \tag{22}$$

$U(\tau)\rho(t)U^{+}(\tau)$ is associated to the gain produced by an atom crossing the cavity during a time τ and t_{cav} is the cavity decay time. Replacing equations (20) and (22) in (21), we readily get for the ρ_{NM} matrix elements:

$$\begin{aligned}
\frac{\partial \rho_{NM}}{\partial t} &= \{\rho_{aa}[1 - \frac{N+1}{2N+3}(1 - e^{i\phi(2n+3\delta)})][1 - \frac{M+1}{2M+3}(1 - e^{i\phi(2m+3\delta)})] \\
&+ \rho_{cc}[1 - F(N)(1 - e^{i\phi(2n-\delta)})][1 - F(M)(1 - e^{-i\phi(2m-\delta)})] \\
&- (m+n+1)\}\rho_{NM} + 2\sqrt{(n+\delta)(m+\delta)}\rho_{N+1,M+1} \\
&+ \rho_{aa}G(N)G(M)(1 - e^{i\phi(2n-\delta)})(1 - e^{-i\phi(2m-\delta)})\rho_{N-2,M-2} \\
&+ \rho_{cc}G(N+2)G(M+2)(1 - e^{i\phi(2n+3\delta)})(1 - e^{-i\phi(2m+3\delta)})\rho_{N+3,M+3} \\
&- \rho_{ca}G(N+2)(1 - e^{i\phi(2n+3\delta)})(1 - \frac{M+1}{2M+3}e^{-i\phi(2n+3\delta)})\rho_{N+2,M} \\
&- \rho_{ca}G(M)(1 - e^{-i\phi(2m-\delta)})(1 - F(N)e^{i\phi(2n-\delta)})\rho_{N,M-2} \\
&- \rho_{ac}G(M+2)(1 - e^{-i\phi(2m+3\delta)})(1 - \frac{N+1}{2N+3}e^{i\phi(2n+3\delta)})\rho_{N,M+2} \\
&- \rho_{ac}G(N)(1 - e^{i\phi(2n-\delta)})(1 - F(M)e^{i\phi(2m-\delta)})\rho_{N-2,M},
\end{aligned} \tag{23}$$

where $F(l) = l/(2l-1)$ and $G(l) = \sqrt{l(l-1)}/(2l-1)$. We observe that in equation (23), the diagonal matrix elements ρ_{NN} are coupled with the off-diagonal elements $\rho_{N,N+d}$, with d even. This coupling occurs only when the atomic coherence ρ_{ca} is present.

In what follows, we present a numerical solution of equation (23) that will allows us to calculate the photon statistics, squeezing and phase distribution probabilities. In order to do this, we define a vector $\vec{\rho}$ whose components are $\rho_{N,N+d}$ with $d = 0, 2, 4\ldots$, and truncating the system to a finite dimension, the master equation (6) now can be cast into the form:

$$\dot{\vec{\rho}} = \tilde{A}\vec{\rho}, \tag{24}$$

where $\tilde{A}$ is a square matrix. We can now assume solutions of the form $\rho = \rho_o e^{\lambda t}$. The steady state solution will correspond to the zero eigenvalue case ($\lambda = 0$).

In all the following numerical results, we have taken $N_{\text{ex}} = 10$, $\theta_{\text{at}} = \pi$. In Figure 4 we show the steady state photon statistics versus N and ρ_{aa}. We notice that the threshold condition is modified as a function of ρ_{aa}. For $\rho_{aa} = 1$ we are above threshold and the maximum of the distribution corresponds to some non-zero value.

Now, if we take the ansatz $|\rho_{ac}| = \sqrt{\rho_{aa}\rho_{cc}}$, with $\rho_{aa} + \rho_{cc} = 1$, as we decrease ρ_{aa} we are both increasing ρ_{cc} and the atomic polarization $|\rho_{ac}|$. We have used this ansatz in all of our calculation. As we observe from the Figure 4, as ρ_{aa} decreases, the maximum of the distribution tends to shift towards the origin $N = 0$ and for $\rho_{aa} = 0.67$, the maximum is already at $N = 0$. Therefore, anywhere below this value, we are below threshold. This figure shows the dependence of the threshold on the atomic polarization.

Now we can calculate the steady state variances of x and y quadratures as a function of ρ_{aa} or the atomic polarization. We observe numerically that the steady state value of $\rho_{N,N+1}$ vanishes while $\rho_{N,N+2}$ goes to a non-zero value. This is due to the fact that photons are generated in correlated pairs and in the master equation (23) the photon statistics ρ_{NN} (which has a non vanishing steady state) is only coupled to $\rho_{N,N+2}$ or $\rho_{N,N-2}$

In Figure 5, we show the variance $(\Delta Y)^2 = (\Delta a_2)^2$ as a function of ρ_{aa}, for $\phi = .9\pi/2$, π, $3.2\pi/2$. The optimum squeezing of 36 % was observed for $\phi = 3.2\pi/2$ and at $\rho_{aa} = 0.2$. It is clear from Figure 4 that at this point we are below the threshold.

The question of a phase operator for the electromagnetic field is a log standing problem in Quantum Mechanics. Dirac [11] was the first to sugget a phase operator, although it was soon recognized that such an operator was not Hermitian.

Recently, sharp phase states have been defined [12,13] as:

$$|\varphi\rangle = \lim_{s\to\infty} |\varphi^{(s)}\rangle, \tag{25}$$

$$|\varphi^{(s)}\rangle = (s+1)^{-1/2} \sum_{m=0}^{s} e^{im\varphi}|m\rangle, \tag{26}$$

where $|m\rangle$ are the m-number state. The limit $s \to \infty$ in equation (25) has to be taken with caution and after the calculation have been performed *with finite s*.

By means of equation (25) we can define a phase probability distribution:

$$P(\varphi) = \langle \varphi(s)|\rho|\varphi(s)\rangle = \frac{1}{s+1}[1 + 2\sum_{r<r'}^{s} \mathrm{Real}(\rho_{r,r'} e^{-i\varphi(r-r')})], \qquad (27)$$

In Figure 6 we have calculated the steady state phase distribution $P(\varphi)$ as a function of the phase φ and ρ_{aa}, for $\phi = 3.2\pi/2$. As soon as ρ_{aa} has a value different from 1, a symmetric bifurcation in the $(\theta_o, \theta_o + 2\pi)$ interval is observed. This effect is reminiscent of the bifurcation found by Schleich *et al.* [14] for a strongly squeezed state and relatively low photon number. However, in the case of the two photon laser, the bifurcation is also present for the cases without squeezing and its physical origin is the phase locking effect produced by the atomic polarization.

Finally in Figure 7 we show the phase fluctuations $(\Delta\varphi)^2$ versus ρ_{aa} for $\phi = .9\pi/2$, π, $3.2\pi/2$. In the case $\phi = 3.2\pi/2$ we find, similarly to $(\Delta y)^2$, the maximum noise reduction. However $(\Delta Y)^2$ and $(\Delta\varphi)^2$ versus ρ_{aa} look very different due to the fact we are dealing with very low average photon numbers. As this number increases, these two quantities are related by $(\Delta\varphi)^2 = (\Delta Y)^2/\langle n\rangle$ [15]. We used $\theta_o = -\pi/2$ in both figures 6 and 7.

To summarize the results, we have found that the two photon laser exhibits a 36 % steady state squeezing below threshold and a bifurcation in the phase probability distribution, both effects present only when the atoms are prepared in a cohorent superposition of the lower and upper levels. Also the threshold is modified by the presence of the atomic polarization. The time dependent properties and phase diffusion of this system will be the subject of a future publication.

Acknowledgement

This work has been partially supported by Fondecyt grants No. 91-0631, 91-0628. One the authors (L.R.) acknowledges a Conicyt Doctoral Fellowship.

REFERENCES

[1] For a tutorial review of squeezed states see: D.F. Walls, Nature **306**, 141(1986)

[2] R.E. Slusher, L.W. Hollberg, B. Yurke, J.C. Mertz, J.F. Valley, Phys. Rev. Letters **55**, 2409(1985), reports the first observation of squeezed light.

[3] Glauber. R.J. Phys. Rev. **130**, 2529(1963), **131**, 2766 (1966)

[4] See, for example, H. Yuen, Phys. Rev. A13, 2226(1976)

[5] K.J. McNeil, D.F. Walls, J. Phys. **A8**, 104(1975)

[6] L.M. Narducci, W.W. Eidson, P. Furcinitti and D.C. Eteson, Phys. Rev. **A13**, 226(1976)

[7] L. Davidovich, J.M. Raimond, M. Brune and S. Haroche, Phys. Rev **A36**, 3771(1987)

[8] P. Goy, J.M. Raimond, M. Gross and S. Haroche, Phys. Rev. Lett. **50**, 1903(1983)

[9] D. Meschede, H. Walther and G. Müller, Phys. Rev. Lett **54**, 551(1985)

[10] M. Orszag, J.C. Retamal, Opt. Comm.**79**, 551(1990)

[11] P.A.M Dirac, Proc. R. Soc. London, Ser **A 114**, 243(1927)

[12] D.T. Pegg and S.M. Barnett, Europhys. Lett,**6** (1988) 483

[13] D.T. Pegg and S.M. Barnett, Phys. Rev. **A39**, 1665(1989)

[14] W. Schleich, R.J. Horowicz and S. Varro, Phys. Rev. **A40**, 7405(1989)

[15] J. Bergou, M. Orszag, M.O. Scully and K. Wodkiewicz, Phys. Rev. **A39**, 5136(1989)

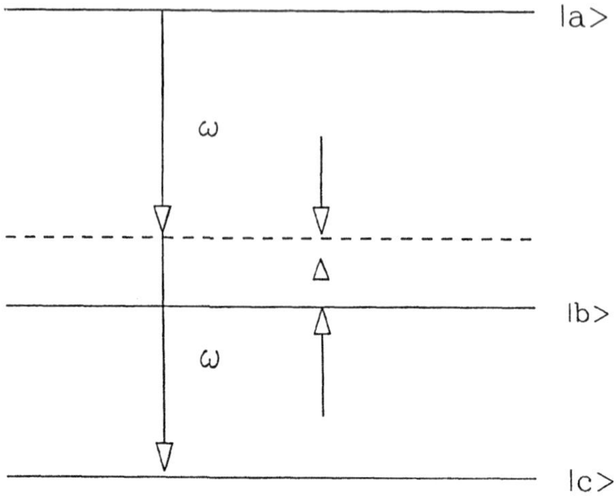

Figure 1. Graphical representation of a coherent state.

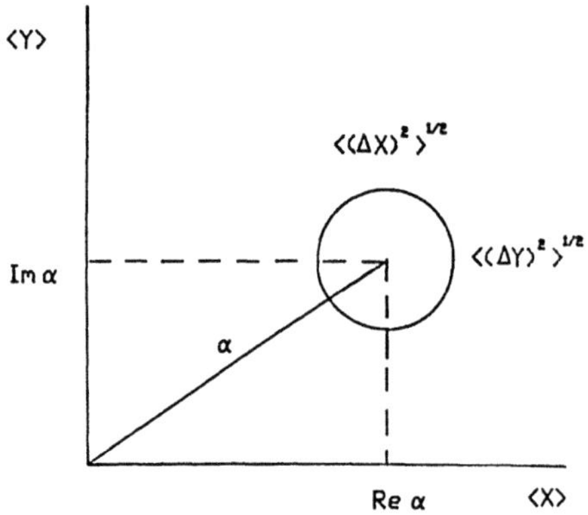

Figure 2. Graphical representation of a squeezed state.

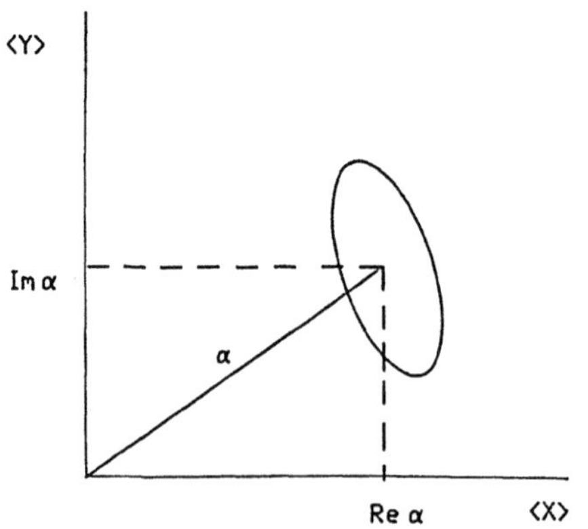

Figure 3. Energy levels relevant to the two photon laser with atomic polarization

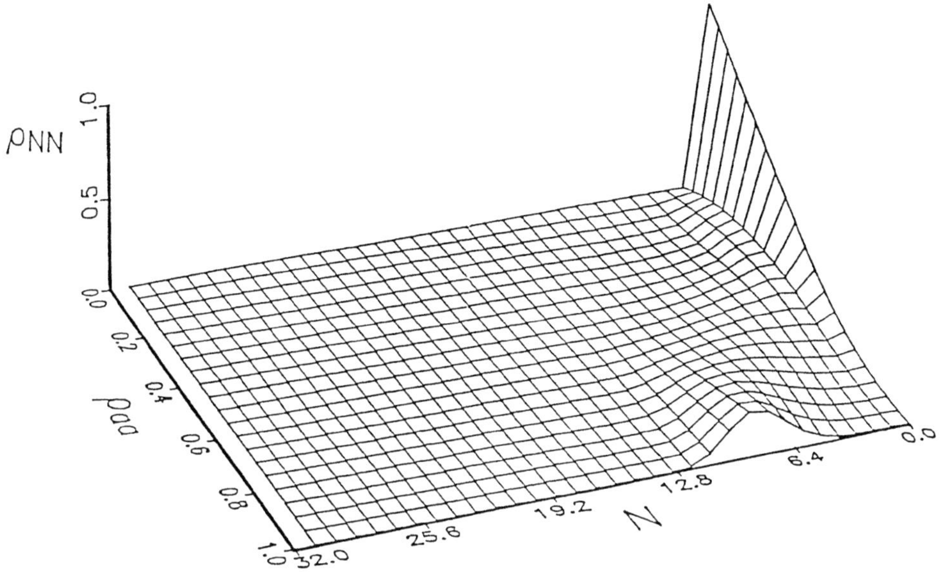

Figure 4. Steady state photon statistics versus N and ρ_{aa} for $N_{\text{ex}} = 10$, $\phi = 3.2\pi/2$, $\theta_{\text{at}} = \pi$.

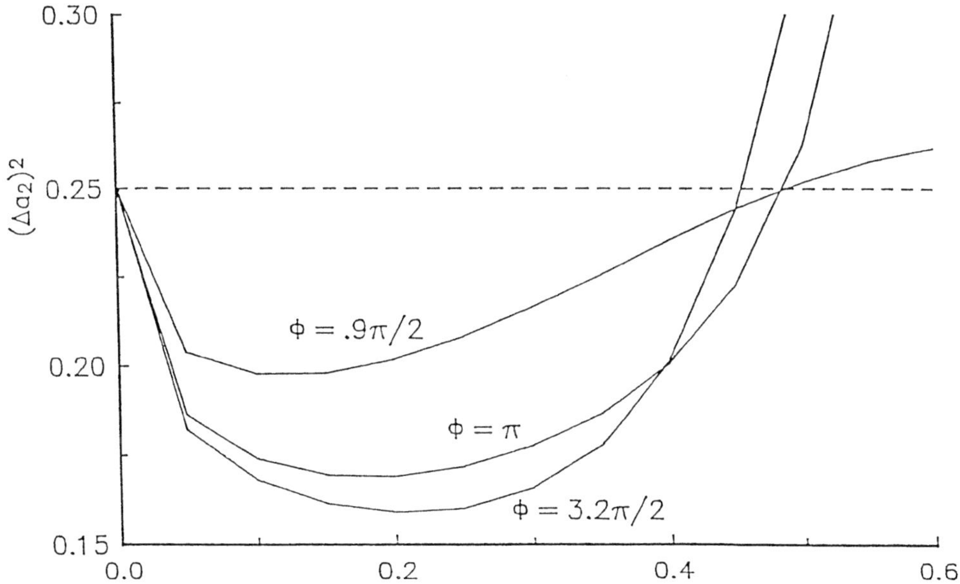

Figure 5. Steady state $(\Delta a_2)^2 = (\Delta Y)^2$ versus ρ_{aa} for $N_{\text{ex}} = 10$, $\theta_{\text{at}} = \pi$ and $\phi = 0.9\pi/2$, π and $3.2\pi/2$. The maximum squeezing (36 %) occurs for $\phi = 3.2\pi/2$ at $\rho_{aa} = 0.2$.

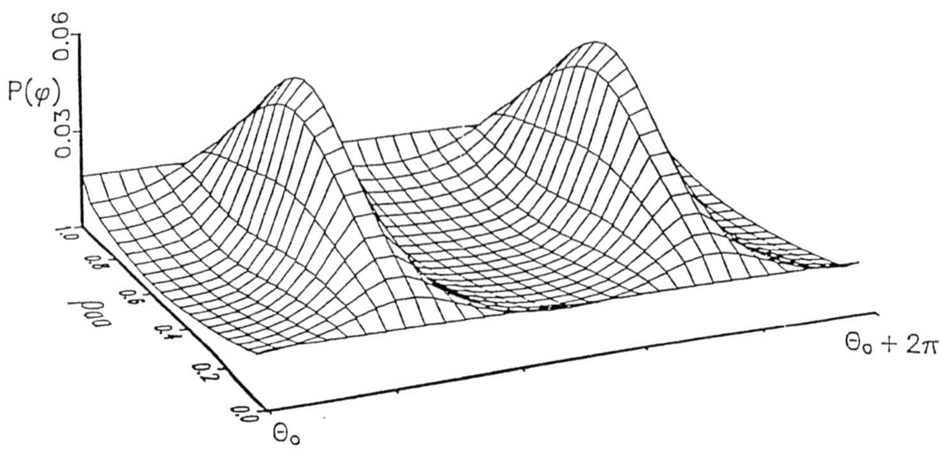

Figure 6. Steady state phase distribution versus ρ_{aa} and ϕ, for $N_{\rm ex} = 10$, $\phi = 3.2\pi/2$, $\theta_{\rm at} = \pi$. A symmetric bifurcation appears in the $(\theta_o, \theta_o{+}2\pi)$ phase interval. We used $\theta_o = -\pi/2$.

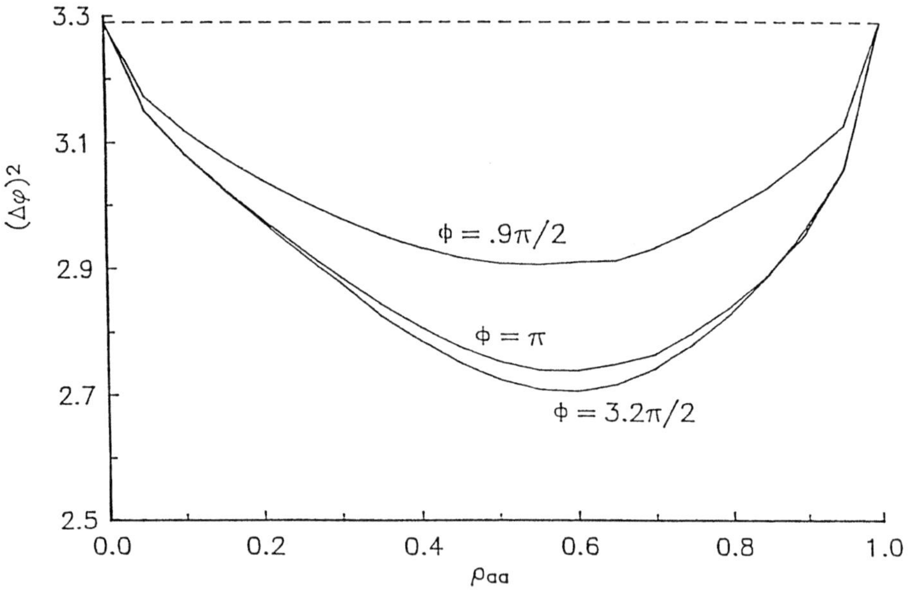

Figure 7. Steady state phase fluctuations $(\Delta\varphi)^2$ versus ρ_{aa} and for $N_{\rm ex} = 10$, $\theta_{\rm at} = \pi$ and $\phi = 0.9\pi/2$, π and $3.2\pi/2$. The maximum reduction takes place for $\phi = 3.2\pi/2$ at $\rho_{aa} = 0.6$. We used $\theta_o = -\pi/2$.

PHASE SPACE FLUCTUATIONS IN NEUTRAL FERMI LIQUIDS

B. Benhassine, M. Farine, E.S. Hernández[a], D. Idier,
B. Remaud and F. Sébille
Laboratoire de Physique Nucléaire UA IN2P3/CNRS et Université de Nantes N°57
2, rue de la Houssinière F-44072 NANTES Cedex
[a]Departamento de Física - Facultad de Ciencias Exactas y Naturales,
Universidad de Buenos Aires, 1428 Buenos Aires, Argentina

We study the dynamics of fluctuations on collective variables in neutral Fermi liquids. From theoretical approaches based on stochastic kinetic equations, the transport coefficients for collective observables can be formally deduced in the relaxation time approximation. Detailed numerical applications show that the dynamics of fluctuations can be extracted from noisy numerical simulations.

1. Introduction

Neutral Fermi liquids appear in nature either as extended systems like ^3He and neutron star matter or as droplets like atomic nuclei. Although drops of fermionic helium can, in principle, be produced at the laboratory, little is known about the phenomenology of such objects. By contrast, nuclei offer experimental situations whose interpretation often resort to the common source, the Landau theory of Fermi liquids, initially derived to describe the properties of the infinite system, where the majority of the physical insigths required for the proper statement and development of the problem can be found. In particular, the fluctuations in the nuclear dynamics constitute a field of current interest, stemming from the important amount of data presently coming out of medium energy heavy ions accelerators. It is commonly accepted that the time evolution of colliding heavy ions is mostly governed by a competition between the mean field and the many - body correlations; however, when the beam energy increases there are clear evidences that the nuclear system enters critical regimes where statements based on average one-body theories are meaningless.

Different approaches to theoretically treat the dynamics of fluctuations are linked to the so-called Boltzmann-Langevin approach which has been pioneered out by Bixon

E. Tirapegui and W. Zeller (eds.), Instabilities and Nonequilibrium Structures IV, 89–100.

and Zwanzig[1] for the hydrodynamical fluctuations of classical systems. Along this line of research, we derive a formalism to consistently get the dispersion of the collective variables out of the phase space fluctuations and show that such a formalism is amenable to numerical applications. We shall place ourselves in far-from-critical regimes where the concept of global relaxation time applies. As we shall see later, this approximation is less severe that it seems. Furthermore, we treat here fluctuations arising in the momentum space and restrict our study to regimes where the density fluctuations are damped and the system keeps its global homogeneity in r-space.

2. Stochastic phase space evolution

Very recently, Ayik and Grégoire[3] pioneered the inclusion, in the theoretical description, of a noise source that kicks the one-body distribution $\hat{f}(\mathbf{r},\mathbf{p},t)$ and forces it to fluctuate around an average Boltzmann - Uehling - Uhlenbeck (BUU) kinetic evolution $f(\mathbf{r},\mathbf{p},t)$. The final result of their derivation can be formalized in terms of a stochastic kinetic equation similar to the Boltzmann-Langevin one anticipated by Bixon and Zwanzig [1], that reads,

$$\frac{\partial \hat{f}}{\partial t}(\mathbf{r},\mathbf{p},t) = -(\mathbf{v}.\nabla_{\mathbf{r}} + \mathbf{F}.\nabla_{\mathbf{p}})\hat{f}(\mathbf{r},\mathbf{p},t) + K_f(\mathbf{r},\mathbf{p},t) + \delta K_{\underline{f}}(\mathbf{r},\mathbf{p},t); \tag{1}$$

where $\mathbf{F}$ is the total force including the spatial gradient of the mean field and K_f is the BUU collision kernel,

$$K_f(\mathbf{r},\mathbf{p},t) = \int d^3p_2 d^3p_3 d^3p_4 W_{1234}(\bar{f}_1\bar{f}_2 f_3 f_4 - f_1 f_2 \bar{f}_3 \bar{f}_4). \tag{2}$$

Here W_{1234} is the microscopic transition rate for the collision vertex $(\mathbf{p_1p_2}) \rightarrow (\mathbf{p_3p_4})$ and $\bar{f} = 1 - f$. The fluctuating collision term in (1) is related to the propagation of initial two-body correlations, and can be expressed as,

$$\delta K_f(\mathbf{r_1},\mathbf{p_1},t) = \int d^3r_2 d^3p_2 \{V(1,2), \mathcal{U}_{12}(t) f_2^{(c)}(1,2,0)\}, \tag{3}$$

where V(1,2) is the effective interaction between particles 1 and 2, $\{\}$ denotes a Poisson bracket, $\mathcal{U}_{12}(t)$ is the two-particle propagator in phase space and $f_2^{(c)}$ (1,2,0) is the irreducible two-body distribution,

$$f_2^{(c)}(1,2,t) = f_2(\mathbf{r}_1,\mathbf{p}_1,\mathbf{r}_2,\mathbf{p}_2,t) - f(\mathbf{r}_1,\mathbf{p}_1,t)f(\mathbf{r}_2,\mathbf{p}_2,t), \tag{4}$$

at t = 0.

If, in accordance with the usual treatments of the stochastic Langevin equation, one assumes a statistical ensemble of initial configurations where the correlations average out to zero, i.e., $< \delta K_f >= 0$, (hereafter, we use the symbol $< >$ to denote ensemble averages) and where the BUU distribution is the averaged fluctuating one, $f =< \hat{f} >$, it is possible to compute the correlation of the noise[3],

$$\sigma^2_{K_f}(\mathbf{r},\mathbf{p},t,\mathbf{r}',\mathbf{p}',t') =< \delta K_f(\mathbf{r},\mathbf{p},t)\delta K_f(\mathbf{r}',\mathbf{p}',t') >$$

$$= \sigma^2_{K_f}(\mathbf{r},\ \mathbf{p},\ t,\mathbf{r}',\ \mathbf{p}',\ t') = \delta(t-t')\delta(\mathbf{r}-\mathbf{r}')2D_f(\mathbf{p},\mathbf{p}'). \tag{5}$$

The diffusion matrix D_f can be expressed from the microscopic transition rate of (2) as derived by Ayik and Grégoire[3]. Furthermore, in the regime where fluctuations are limited to small amplitude excursions $\delta f = \hat{f} - f$, one can compute the evolution rule for the covariance of the distribution function,

$$\sigma^2_f(\mathbf{r},\ \mathbf{p},\ t,\mathbf{r}',\mathbf{p}',t) =< \delta f(\mathbf{r},\ \mathbf{p},\ t)\delta f(\mathbf{r}',\ \mathbf{p}',\ t) > . \tag{6}$$

In matriz notation the equation of motion is,

$$\frac{d}{dt}\sigma^2_f = A\sigma^2_f + \sigma^2_f\tilde{A} + 2D_f, \tag{7}$$

where A is a matrix that represents the linearized Boltzmann collision operator plus the streaming term[4].

A complementary viewpoint to deal with the dynamics of phase space fluctuations has been put forward by Randrup and Remaud[5], who consider the distribution f to be an Orstein-Uhlenbeck[4] process in phase space governed by a Fokker-Planck equation for the probability density $P(\hat{f})$. In such a frame, it is possible to extract the equations of motion for the mean value and the covariance of the stochastic $\hat{f}$, the first being the BUU equation and the second being identical to the one obtained by Ayik and Grégoire[3]. This is intrinsic to any Langevin vs Fokker-Planck description, and this derivation is more amenable to numerical simulations, since a detailed picture of both the dissipation and the diffusion mechanisms in terms of balance relationships is provided.

Finally, we would like to notice that an independent framework developed by Van Kampen[4] leads to both the BUU equation for f and to (7) when applied to fermion one-body distribution in phase space. The technique consists in assuming a master equation of motion (ME) for the occupancy of cells in phase space and expanding the ME in terms of a scale parameter ; the procedure gives naturally the evolution of averages and fluctuations of the stochastic $\hat{f}(\mathbf{r},\mathbf{p},t)$.

3. Brownian motion in collective space

We shall now analyze the features of the stochastic motion of the one-body macroscopic observables, especially, of multipole moments $Q_{lm}(f)$ of the fluctuating marginal distribution in momentum space,

$$\hat{f}(\mathbf{p},t) = \int d^3r \hat{f}(\mathbf{r},\mathbf{p},t) = \sum_{lm} \hat{q}_{lm}(t) Q_{lm}(\mathbf{p}), \tag{8}$$

since the time-dependent amplitudes $\hat{q}_{lm}(t) = \int d^3p \hat{f}(\mathbf{p},t) Q_{lm}(\mathbf{p})$ play the role of collective coordinates of the N-body system. Generally speaking, if Q($\mathbf{p}$) is an arbitrary one-body operator and q(t) its average with respect to the stochastic one body distribution $\hat{f}(\mathbf{r},\mathbf{p},t)$, the BUU-Langevin equation (1) induces a classical Langevin equation with multiplicative noise,

$$\dot{\hat{q}}(t) = v_Q + K_Q + \delta K_Q(t). \tag{9}$$

with

- a conservative velocity v_Q related to the free-flow in phase space,

$$v_Q = -\frac{1}{N} \int d^3r d^3p [\mathbf{F}.\nabla_p Q(\mathbf{p})] \hat{f}(\mathbf{r},\mathbf{p},t); \tag{10}$$

  (if the mean field is momentum-independent). This velocity vanishes for spatially homogeneous systems, i.e., $\mathbf{F} = 0$.

- a dissipative velocity,

$$K_Q = -\frac{1}{N}\int d^3r \int\int\int\int d^3p_1 d^3p_2 d^3p_3 d^3p_4 W_{1234}\frac{\Delta Q}{2} f_1 f_2 \bar{f}_3 \bar{f}_4, \tag{11}$$

where

$$\Delta Q = Q(\mathbf{p_1}) + Q(\mathbf{p_2}) - Q(\mathbf{p_3}) - Q(\mathbf{p_4}). \tag{12}$$

- a fluctuating velocity of the multiplicative type,

$$\delta K_Q(t) = \frac{1}{N}\int\int d^3r d^3p \delta K_f(\mathbf{r},\mathbf{p},t) Q(\mathbf{p}), \tag{13}$$

with vanishing ensemble average and correlation

$$\sigma^2_{K_Q}(t,t') =< \delta K_Q(t)\delta K_Q(t') >= 2D_Q(t)\delta(t-t'). \tag{14}$$

The time-dependent diffusion parameter is,

$$D_Q(t) = \frac{1}{N^2}\int d^3r \int\int d^3p d^3p' Q(\mathbf{p})Q(\mathbf{p}')D_f(\mathbf{p},\mathbf{p}')$$

$$= \frac{1}{N^2}\int d^3r \int\int\int\int d^3p_1 d^3p_2 d^3p_3 d^3p_4 W_{1234} f_1 f_2 \bar{f}_3 \bar{f}_4 (\frac{\Delta Q}{2})^2. \tag{15}$$

Now if the collision kernel of the BUU equation verifies the relaxation time approximation,

$$K_f(\mathbf{r},\mathbf{p},t) \sim -\frac{f(\mathbf{r},\mathbf{p},t) - f^{(0)}(\mathbf{r},\mathbf{p})}{\tau_f}, \tag{16}$$

the dissipative velocity reads,

$$K_Q(t) \sim -\frac{q(t) - q^{(0)}}{\tau_Q}, \tag{17}$$

with $q^{(0)}$ the asymptotic equilibrium value of the ensemble averaged $q = <\hat{q}>$ and $\tau_Q = \tau_f$, being

$$\frac{1}{\tau_Q} \sim \frac{1}{\tau_f} = \int\int\int d^3p_2 d^3p_3 d^3p_4 W_{1234} f_2^{(0)} \bar{f}_3^{(0)} \bar{f}_4^{(0)}. \tag{18}$$

On the other hand, consistently, the diffusion parameter D_Q should be close to

$$D_Q^{(0)} = \frac{1}{N^2}\int d^3r \int\int\int\int d^3p_1 d^3p_2 d^3p_3 d^3p_4 W_{1234} f_1^{(0)} f_2^{(0)} \bar{f}_3^{(0)} \bar{f}_4^{(0)} Q^2(\mathbf{p}_1)$$

$$= \frac{1}{\tau_f}\int\int d^3r d^3p_1 \frac{Q^2(\mathbf{p}_1)}{N^2} f_1^{(0)}. \tag{19}$$

Since the equilibrium distribution $f_1^{(0)}$ is isotropic in momentum space, we identify in Eq. (19) the equilibrium dispersion of the observable per particle

$$\sigma_Q^{2(0)} = \int\int d^3r d^3p [\frac{Q}{N}(\mathbf{p})]^2 f^{(0)}(\mathbf{p}) = D_Q^{(0)} \tau_f. \tag{20}$$

This expression is a characteristic result of classical Brownian motion theory. The evolution law for the fluctuation of q in the statistical ensemble, $\sigma_Q^2(t) =< [\delta q(t)]^2 >$ (being $\delta q(t) = \hat{q}(t) - q(t)$) is then,

$$\frac{d}{dt}\sigma_Q^2 = -\frac{2}{\tau_Q}\sigma_Q^2(t) + 2D_Q(t), \tag{21}$$

with solution

$$\sigma_Q^2(t) = \sigma_Q^2(0)e^{-2t/\tau_Q} + D_Q^{(0)}\tau_Q(1 - e^{-2t/\tau_Q}), \tag{22}$$

if D_Q is a constant identical to the asymptotic equilibrium value $D_Q^{(0)}$.

We then recover, for $t >> \tau_Q$, expression (20) with $\tau_Q = \tau_f$. Equation (22) represents the dynamics of the collective fluctuation in the close-to- equilibrium limit, where the ensemble average $q(t)$ undergoes as well an exponential decay with time scale τ_Q. A generalization involving a set of collective observables $Q_i(\mathbf{p})$ (i $= 1, ...$M) is straightforward. Thus, a numerical test of the phase space dynamics projected upon a small set of collective variables, may shed light on the validity of the assumptions involved in the simple Brownian motion model in collective space. Equation (15) is the key to the Brownian motion picture in collective space, provided the noise correlation in phase space is known.

4. Simulation and results

We now use the above formalism to implement dispersion calculations on the Landau-Vlasov test-particles model[6]. To reduce the numerical effort, we restrict our considerations to infinite systems with translational symmetry and focus on a characteristic collective variable of interest in collisions, namely the quadrupole moment. Our study is restricted to far-from-critical regimes, where the mean trajectory approximation is valid .

The collective diffusion coefficient (19) can be discretized in the frame of test-particle models. It reduces to a microscopic balance of the collective operator of interest during the binary collision process, which is simulated by the model. More precisely the diffusion coefficient is given by the summation over all collisions of the square of the operator $\Delta Q = Q_1 + Q_2 - Q_3 - Q_4$. Then, we can get the collective relaxation time from the evolution of the mean value.

The initial distribution in momentum space is chosen to reproduce, as closely as possible, the one of two colliding nucleus. We take in the center of mass frame, two Fermi spheres centered at $(0, 0, -\delta k/2)$ and $(0, 0, \delta k/2)$ where the z axis corresponds to the beam direction.

We have performed several simulations with different values of the initial shift δk, corresponding to different excitation energies of the system. As already found with such simulations[2], there is an obvious relaxation towards equilibrium and numerical fluctuations are introduced by the coarse graining of the system coordinates and the discrete character of the collision simulation. Notwithstanding the numerical noise, quantitative information on the system fluctuations can be worked out.

We consider the relaxation of the quadrupole moment operator $Q = 2p_z^2 - p_x^2 - p_y^2$ which represents a good signature of anisotropy. The exponential decay of this observable is obvious as seen in fig.1, which allows a rather precise estimate of the relaxation time τ_Q of (18). For this variable, the relaxation time approximation and the ensuing equilibrium dispersion law are then rather justified, even for the highest excited systems that we have considered.

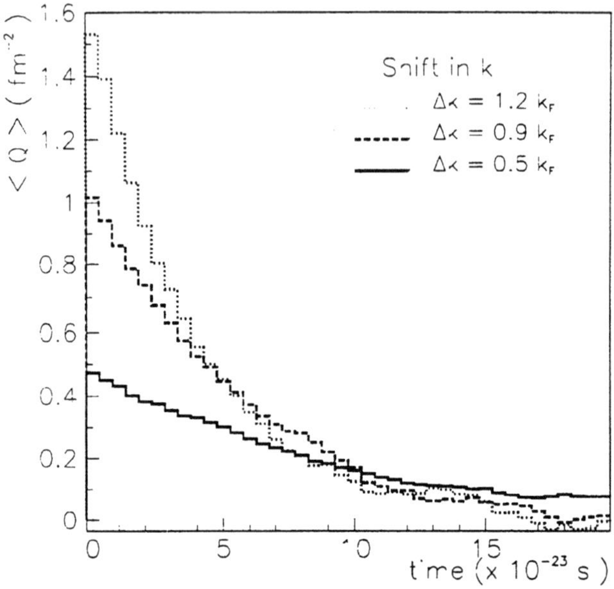

Fig. 1. Quadrupole moment relaxation for different values of the initial shift Δk of the two initial Fermi spheres.

Performing exponential fits on the time evolution curves of fig.1, we extract the relaxation times of the quadrupole moment; the results are plotted on fig.2 as functions of the final temperature. Those times are found to be of the same order than the relaxation times of the phase-space distribution ; this feature is a clear indication that the linear approximation in (17) is established even at high excitation energies. On the other hand, similar tests carried upon the transverse momentum flow exhibit identical qualitative and almost identical quantitative behavior, the small differences can be attributed to the numerical fluctuations (cf. Fig. 2). This indicates that the collective relaxation time is a quantity inherent to the system and depends only on its excitation energy.

We also compute the diffusion coefficients D_Q (fig. 3). Here, some care should be taken in presence of the numerical fluctuations. Due to the structure of the collision term, the numerical noise on the gain and loss coefficient tends to cancel out in the mean quantities ; for the diffusion process, we average the instantaneous diffusion coefficient on several time steps provided that this averaging time is much smaller than the global relaxation time. Even if some fluctuations remain, one clearly sees

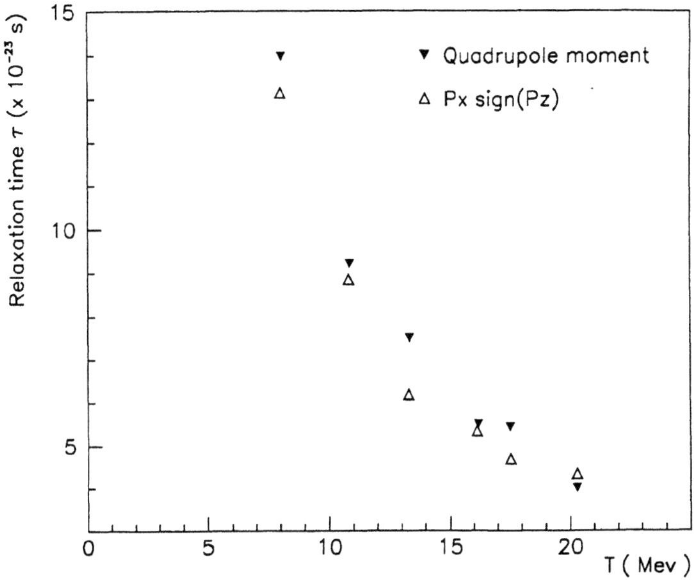

Fig. 2. The relaxation times of both quadrupole moment and transverse momentum.

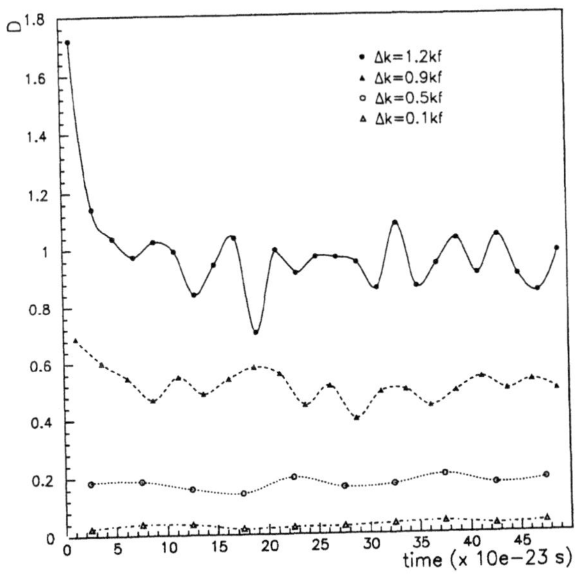

Fig. 3. Quadrupole diffusion coefficient as a function of time for different initial conditions.

that taking apart the first time steps of very excited systems ($\Delta k = 0.5$ and 1.3), the computed diffusion coefficient is fairly constant. We confirm for the whole phase space diffusion matrix, the conclusion of ref. 6 concerning its diagonal part : even if the system is very far from equilibrium, the fluctuations of the system keep close track of their equilibrium values ; this remark is important since it states that the leading parameter for fluctuations dynamics is the excitation energy regardless this energy is equilibrated or not.

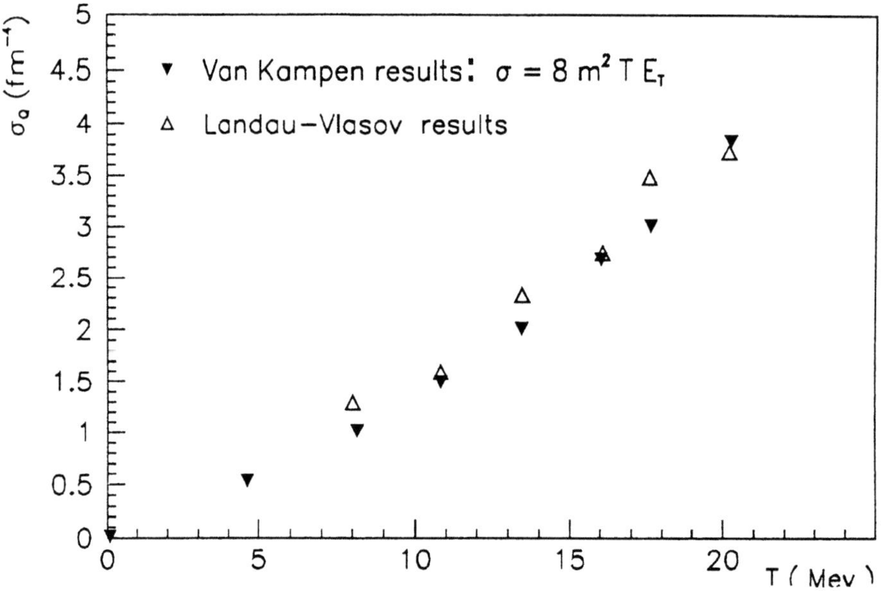

Fig. 4. Comparison of the quadrupole moment equilibrium dispersions in the numerical (Landau - Vlasov) and theoretical schemes.

The validity of the relaxation time approximation is again verified by the above results; then, we can calculate the expected equilibrium dispersions according to the method of Sect. 4 using the values of τ and D extracted from simulations. From the comparisons with the formal results arising from the application of Van Kampen's procedure[7] which are plotted in fig.4, one can check the accuracy of the Landau-Vlasov model. The equilibrium dispersions are reproduced with a margin of error less than 15%.

5. Conclusions

We have derived a consistent Brownian motion approach to the dynamics of collective variables in the linear regime, which allows a direct link of the transport coefficients in collective space to the microscopic two-body correlations induced by the residual interaction on the phase space distributions. We have then deduced the general equation (22) for the dynamics of collective fluctuations which tends to an equilibrium consistent with the general Brownian motion theory. The whole consistency of the framework is demonstrated by simulations based on pseudo-particle models which are numerical implementations of the Brownian approach. These simulations show that the inherent numerical fluctuations of the calculations do not prevent the extraction of transport coefficients. They show too that the equilibrium values reached in the Brownian approach are compatible with those derived at equilibrium with Van Kampen's approach. The fact that pseudo-particle models -here used in infinite matter- have been successfully applied to finite systems, paves the way to actual dynamical studies of fluctuations in systems like interacting heavy ions, keeping in mind that the excitation energy (either equilibrated or not) is the leading parameter for the fluctuations.

The authors thank J. Randrup and P. Chomaz for very fruitful discussions. One of us (E. S. H.) also wishes to ackowledge warm hospitality during her stay at the Laboratoire de Physique Nucléaire , Université de Nantes, and the CNRS od France for financial support.

References

1. M. Bixon and R. Zwanzig, Phys. Rev. 187 (1969) 267.
2. D. Idier, B. Benhassine, M. Farine, B. Remaud and F. Sébille, submitted to Nucl. Phys. A.
3. S. Ayik and C. Grégoire, Phys. Lett. B 212 (1988) 269 ; Nucl. Phys. A 513 (1990) 187.
4. N.G. Van Kampen, Stochastic Processes in Physics and Chemistry (North Holland, New York, 1981).

5. J. Randrup and B. Remaud, Nucl. Phys. A 514 (1990) 339.

6. C. Grégoire, B. Remaud, F. Sébille, L. Vinet and Y. Raffray, Nucl. Phys. A 465 (1987) 77.

7. B. Benhassine, M. Farine, E. S. Hernández, D. Idier, B. Remaud and F. Sébille, to be published.

POLAR DECOMPOSITION AND DENSE SIMILARITY TO UNITARY OPERATORS

N. Bertoglio
Facultad de Matemática
Pontificia Universidad Católica de Chile
Casilla 114-D, Santiago, Chile

S. Martínez and J. San Martín
Departamento de Ingeniería Matemática
Universidad de Chile
Casilla 170-3 Correo 3, Santiago, Chile

ABSTRACT. Prigogine and Misra have shown that any K-dynamical system is densely similar to a Markov process converging to equilibrium. In their construction the self-adjoint operator realizing the dense similarity is a function of the positive part of the polar form of the Markov operator. In this work we show that this property holds in a very general framework.

1. Introduction

A K-dynamical system is an abstract reversible dynamical system $(X, \mathcal{B}, \mu, T)$ such that there exists a σ-field $\mathcal{A}_0 \subset \mathcal{B}$ which is increasing, $T^{-1}\mathcal{A}_0 \subset \mathcal{A}_0$, and such that the increasing sequence $\mathcal{A}_n = T^n \mathcal{A}_0$ verifies:

$$\mathcal{A}_n \underset{n\to\infty}{\nearrow} \mathcal{A}_\infty = \mathcal{B}, \quad \mathcal{A}_n \underset{n\to-\infty}{\searrow} \mathcal{A}_{-\infty} = \mathcal{N} \text{ the trivial } \sigma\text{-field}$$

Denote by $E_n = \mathbb{E}^{\mathcal{A}_n}$ the mean expected value operator associated to $\mathcal{A}_n$ for $-\infty \leq n \leq \infty$ and by U_T the Koopman operator associated to T. The following imprimivity equalities are verified:

$$U_T^{-1} E_n U_T = E_{n+1} \text{ for } -\infty \leq n \leq \infty \tag{1.1}$$

Prigogine and collaborators (see references [1], [2]) have associated to a K-system the following Markov operator:

$$Q_T = (\sum_{n\in\mathbb{Z}} \nu_n E_n) U_T \tag{1.2}$$

E. Tirapegui and W. Zeller (eds.), Instabilities and Nonequilibrium Structures IV, 101–107.

where $(\nu_n : n \in \mathbb{Z})$ is a strictly positive probability vector on $\mathbb{Z}$. Notice that if we put $r_n = \sum_{k \geq n} \nu_k$ then the sequence $(r_n : n \in \mathbb{Z})$ is a strictly decreasing sequence with $r_{-\infty} = 1$ and $r_\infty = 0$. Now, Q_T can be set in the form:

$$Q_T = (\sum_{n \in \mathbb{Z}} r_n(E_n - E_{n-1}) + E_{-\infty})U_T \tag{1.3}$$

We shall use mainly this last form of Q_T.

The Markov operator Q_T has a densely defined inverse (which is not Markov)

$$Q_T^{-1} = U_T^{-1}(\sum_{n \in \mathbb{Z}} r_n^{-1}(E_n - E_{n-1}) + E_{-\infty}) \tag{1.4}$$

In fact the domain D of Q_T^{-1} is dense on $L^2(\mu)$ and from (1.3) and (1.4) we get $Q_T Q_T^{-1} = I_D$ the identity on D. The equality $Q_T^{-1} Q_T = I$ the identity on $L^2(\mu)$ follows from imprimivity relations (1.1).

The Markov operator Q_T converges to equilibrium, this means:

$$Q_T^{*n} f \underset{n \to \infty}{\overset{L^2(\mu)}{\longrightarrow}} E_{-\infty} f = \int f d\mu \text{ for any } f \in L^2(\mu) \tag{1.5}$$

In fact by using (1.1) it follows

$$Q_T^{*n} = U_T^{-n} \sum_{m \in \mathbb{Z}} (\prod_{k=0}^{n-1} r_{m+k})(E_m - E_{m-1}) + E_{-\infty}$$

Then $\|Q^{*n} f - E_{-\infty} f\|_2^2 = \sum_{m \in \mathbb{Z}} (\prod_{k=0}^{n-1} r_{m+k})^2 \|(E_m - E_{m-1})f\|^2$. Since $\prod_{k=0}^{n-1} r_{m+k} \underset{n \to \infty}{\longrightarrow} 0$, because $r_n \underset{n \to \infty}{\longrightarrow} 0$, we deduce (1.5).

The convergence to equilibrium is equivalent to the condition that the entropy functional converges to 0 for any density function of finite entropy. More precisely in [3] it was shown that (1.5) is equivalent to the property;

$$\int \phi(Q^{*n} f) d\mu \underset{n \to \infty}{\searrow} 0 \text{ for any } f \text{ verifying } f \geq 0, \int f d\mu = 1, \int \phi(f) d\mu < \infty$$

where $\phi(u) = u \log u$.

Now let us show that if the sequence $(r_n : n \in \mathbb{Z})$ verifies the property:

$$\sum_{n=1}^{\infty}(1-r_{-n})<\infty \tag{1.6}$$

then Q_T is densely similar to U_T via a Markov self-adjoint operator Λ.

We remind that dense similarity was introduced by Nagy and Foias (they called "quasi-similarity", [4] Chapter II p. 64): an operator Q defined on a Hilbert space H is said to be densely similar to a unitary operator U if there exists a self-adjoint operator Λ with densely defined inverse such that the following diagram commutes:

$$\begin{array}{ccccc} & & Q & & \\ & H & \longrightarrow & H & \\ \Lambda & \downarrow & & \downarrow & \Lambda \\ & H & \longrightarrow & H & \\ & & U & & \end{array}$$

i.e. $\Lambda Q = U\Lambda$ or equivalentely $Q^*\Lambda = \Lambda U^*$.

To show that Q_T is densely similar to U_T, Prigogine and collaborators introduce the following operator:

$$\Lambda = \sum_{n\in\mathbb{Z}} \lambda_n(E_n - E_{n-1}) + E_{-\infty} \text{ with } \lambda_n = \prod_{j=1}^{\infty} r_{n-j} \text{ for } n \in \mathbb{Z} \tag{1.7}$$

From the condition $\sum_{j=1}^{\infty}(1-r_{n-j})<\infty$ we deduce that λ_n belongs to $(0,1)$, then the self-adjoint operator has densely defined inverse. On the other hand λ_n is strictly decreasing and $\lambda_{-\infty} = \lim_{n\to-\infty}\lambda_n = 1$, $\lambda_\infty = \lim_{n\to\infty}\lambda_n = 0$. Then we can write $\Lambda = \sum_{n\in\mathbb{Z}}(\lambda_n - \lambda_{n+1})E_n$ with $\sum_{n\in\mathbb{Z}}(\lambda_n - \lambda_{n+1}) = 1$, so Λ is a Markov operator. Now let us show the commutativity relation:

$$\Lambda Q_T = U_T\Lambda \tag{1.8}$$

We have $\Lambda Q_T = \sum_{n\in\mathbb{Z}} \lambda_n r_n (E_n - E_{n-1})U_T + E_{-\infty} = U_T(\sum_{n\in\mathbb{Z}} \lambda_{n-1}r_{n-1}(E_n - E_{n-1}) + E_{-\infty})$. From definition (1.7) we find that $\lambda_{n-1}r_{n-1} = \lambda_n$ so $\Lambda Q_T = U_T\Lambda$ so (1.8) follows.

In the next section we shall prove that the above construction can be made in a more general framework, more precisely we shall prove that under very general conditions if the polar decomposition of an operator Q is written in the form $Q = \sqrt{QQ^*}U$ then it will be densely similar to U via a self-adjoint operator Λ which is a function of $\sqrt{QQ^*}$. We will also show that this is the case in the construction of Prigogine.

2. Main Results

The basic result of this section is the following one:

Theorem 1. Let Q be an operator with densely defined inverse and with polar decomposition $Q = \sqrt{QQ^*}U$. Assume $Q^*Q = G(QQ^*)$ with $G \geq 0$ increasing, denote $F(s) = \sqrt{G(s^2)}$ and $F^{(i)} = F \circ \cdots \circ F$ i-times. Let:

$$H(s) = \prod_{i=1}^{\infty} F^{(i)}(s)$$

Then

$$\Lambda Q = U\Lambda \text{ for } \Lambda = H(\sqrt{QQ^*})\ .$$

Moreover if $H(s) \in (0, \infty)$ for all $s \in [0, 1]$ we get that Q is densely similar to the unitary operator U. ■

The proof of this theorem will follow from the next two lemmas. The result established in the next one can be found in [5] Chapter VI, §2, section 7, but the proof is different.

Lemma 1. Let Q be an operator with densely defined inverse. Then $Q = UA = BU$ where $A = \sqrt{Q^*Q}, B = \sqrt{QQ^*}$ are positive operators and U is a unitary operator.

Proof. Consider the polar decompositions:

$$Q^* = AU_1,\ Q = BU_2$$

then $Q = U_1^* A, Q^* = U_2^* B$. The operators $A = \sqrt{Q^*Q}$ and $B = \sqrt{QQ^*}$ are positive and $U = U_1^*, V = U_2^*$ are unitary because Q has densely defined inverse. The operators U, V can be computed in the following way:

$$U = \lim_{n\to\infty} Qg_n(Q^*Q), \quad V = \lim_{n\to\infty} Q^*g_n(QQ^*)$$

with $g_n(u) = f_n(\sqrt{u})$ and $f_n(u) = \frac{1}{u}$ if $u \geq \frac{1}{n}$ and $\frac{1}{n}$ otherwise.

We have $Q(Q^*Q)^n = (QQ^*)^n Q$ for any $n \geq 0$, then for any polynomial p it is satisfied:

$$Qp(Q^*Q) = p(QQ^*)Q$$

Let g be any real Baire function, so $g = \lim_{n\to\infty} p_n$ with p_n a sequence of polynomials. We have $Qp_n(Q^*Q) = p_n(QQ^*)Q$ so:

$$0 =< (Qp_n(Q^*Q) - p_n(QQ^*)Q)x, y > \underset{n\to\infty}{\longrightarrow} < (Qg(Q^*Q) - g(QQ^*))x, y >$$

for any $x, y \in H$. We obtain that the equality

$$Qg(Q^*Q) = g(QQ^*)Q$$

holds for any real Baire function. Since $g(QQ^*)$ is a self-adjoint operator the above equality can be written as

$$Qg(Q^*Q) = (Q^*g(QQ^*))^*$$

Apply this equality to g_n and take the limit, to get:

$$U = \lim_{n\to\infty} Qg_n(Q^*Q) = \lim_{n\to\infty} (Q^*g_n(QQ^*))^* = V^*$$

Hence $Q = BV^* = BU = UA$, and the statement of the lemma holds. ■

Lemma 2. Assume Q has densely defined inverse and that the following relation holds:

$$Q^*Q = G(QQ^*) \text{ with } G \geq 0 \text{ increasing}$$

Then:

$$U^{-1}P_sU = P_{F^{-1}(s)}$$

where $F(s) = \sqrt{G(s^2)}, Q = \sqrt{QQ^*}U$ and (P_s) is the spectral family associated to $\sqrt{QQ^*}$.

Proof. Let $A = \sqrt{Q^*Q}, B = \sqrt{QQ^*}$. From hypotheses and the definition of F we get $A = F(B) = F(\int sdP_s) = \int F(s)dP_s$. Hence $A = \int sdP_{F^{-1}(s)}$.

On the other hand from Lemma 1 $UA = BU$, so $A = U^{-1}BU = U^{-1}(\int sdP_s)U$. Since $U^{-1}P_sU = U^*P_sU$ is a family of projections increasing with s, we get $A = \int sd(U^{-1}P_sU)$, then by uniqueness of the spectral family we deduce $P_{F^{-1}(s)} = U^{-1}P_sU$, as we wanted to show. ■

Now let us prove the theorem.

Proof of Theorem 1. We have $H(B)B = \int H(s)sdP_s$. On the other hand

$$UH(B)U^{-1} = \int H(s)d(UP_sU^{-1}) = \int H(s)dP_{F(s)} = \int H(F^{-1}(s))dP_s$$

Since H verifies

$$H(F^{-1}s) = H(s)s$$

we get $H(B)B = UH(B)U^{-1}$. Thus $H(B)BU = UH(B)$ or equivalently $H(B)Q = UH(B)$. Since the spectrum of QQ^* belongs to $[0,1]$ we get that if $H(s)$ is included in $(0,\infty)$ for $s \in [0,1]$, the self-adjoint operator $H(\sqrt{QQ^*})$ will have a densely defined inverse. ■

Let us show that the operator Q_T, defined in 1.2, verifies the construction made in Theorem 1. More precisely we shall prove:

Theorem 2. Let $(E_n; -\infty \leq n \leq \infty)$ be an increasing sequence of projections, with $E_\infty = I$, $(r_n : -\infty \leq n \leq \infty)$ be a strictly decreasing sequence verifying $r_{-\infty} = 1$, $r_\infty = 0$, $\sum_{k=1}^{\infty}(1-r_{-k}) < \infty$ and U a unitary operator satisfying the imprimitivity equalities $U^{-1}E_nU = E_{n+1}$ for $-\infty \leq n \leq \infty$. Take $Q = (\sum_{n\in\mathbb{Z}} r_n(E_n - E_{n-1}) + E_{-\infty})U$. Then $Q^*Q = G(QQ^*)$ with $G \geq 0$ increasing defined implicitly by $F(s) = \sqrt{G(s^2)} = r_{N(s)-1}$ where $N(s) = \inf\{n \in \mathbb{Z} : r_n \leq s\}$. The values $H(s) = \prod_{i=1}^{\infty} F^{(i)}(s)$ belongs to (0,1) and the operator Q is densely similar to U via $\Lambda = H(\sqrt{QQ^*})$ that is $\Lambda Q = U\Lambda$. Moreover, if $\sqrt{QQ^*} = \int sdP_s$, we have $\Lambda = \int_0^1 H(s)dP_s = \sum_{n\in\mathbb{Z}} \lambda_n(E_n - E_{n-1}) + E_{-\infty}$ with $\lambda_n = \prod_{j=1}^{\infty} r_{n-j}$.

In the proof of this Theorem we use the following result:

Lemma 3. Let $(E_n : -\infty \leq n \leq \infty)$ be an increasing sequence of projections with E_∞ the identity and $(r_n : -\infty \leq n \leq \infty)$ be a decreasing sequence with $r_{-\infty} = 1$, $r_\infty = 0$. Consider

$$A = \sum_{n\in\mathbb{Z}} r_n(E_n - E_{n-1}) + E_{-\infty}$$

Then:

$$A = \int_0^1 sdP_s$$

where $P_s = I - E_{N(s)-1}$ being $N(s) = \inf\{n \in \mathbb{Z} : r_n \leq s\}$. Moreover the spectral family has a discontinuity at $s = 1 : P_1 - P_{1^-} = E_{-\infty}$.

Proof. $I - A = \sum_{n\in\mathbb{Z}}(1 - r_n)(E_n - E_{n-1}) = \int_{-\infty}^{\infty} h(u)d\hat{E}_u$ with

$$h(u) = 1 - r_n \text{ if } u \in (n-1, n] \text{ and } \tilde{E}_u = E_n \text{ if } u \in [n-1, n)$$

The spectrum of A is contained on $[0,1]$ so we can write $A = \int_0^1 sdP_s$. Now:

$$1_{[1-s,1]}(I - A) = \int_0^1 1_{[1-s,1]}(1-u)dP_u = \int_0^s dP_u = P_s$$

On the other hand

$$1_{[1-s,1]}(I-A) = \int 1_{[1-s,1]}(h(u))d\tilde{E}_u = \int_{u:h(u)\geq 1-s} d\tilde{E}_u$$

$$= \sum_{n:r_n\leq s} (E_n - E_{n-1}) = I - E_{N(s)-1}$$

where $N(s) = \inf\{n : r_n \leq s\}$.

We have $N(1-) = -\infty$, so $P_1 - P_{1-} = I - (I - E_{-\infty}) = E_{-\infty}$. ■

Proof of Theorem 2.

We have $QQ^* = \sum_{n\in\mathbb{Z}} r_n^2(E_n - E_{n-1}) + E_{-\infty} = (\sum_{n\in\mathbb{Z}} r_n(E_n - E_{n-1}) + E_{-\infty})^2$. From the previous Lemma $QQ^* = \int_0^1 s^2 dP_s$ where $P_s = I - E_{N(s)-1}$. By using the imprimitivity equalities we get $Q^*Q = \sum_{n\in\mathbb{Z}} r_{n-1}^2(E_n - E_{n-1}) + E_{-\infty}$. Then it follows directly that $Q^*Q = G(QQ^*)$ with $F(s) = \sqrt{G(s^2)}$ verifying $F(s) = r_{N(s)-1}$.

Now, the hypotheses of Theorem 1 are satisfied and we obtain $\Lambda Q = U\Lambda$ with $\Lambda = H(\sqrt{QQ^*})$. Since $N(F(s)) = \inf\{n \in \mathbb{Z} : r_n \leq F(s)\} = \inf\{n \in \mathbb{N} : r_n \leq r_{N(s)-1}\} = N(s) - 1$, we obtain $F^{(2)}(s) = r_{N(s)-1}$ and in a recursive way we find $F^{(i)}(s) = r_{N(s)-i}$. Then $H(s) = \prod_{i=1}^{\infty} F^{(i)}(s) = \prod_{i=1}^{\infty} r_{N(s)-i} = \prod_{i=1}^{\infty}(1 - (1 - r_{N(s)-i}))$ which belongs to (0,1) because $\sum_{i=1}^{\infty}(1 - r_{n-i}) < \infty$ for any $n \in \mathbb{N}$, and the result follows. ■

Acknowledgments. This work was partially supported by FONDECYT 1208-91, 0881-92 and DTI, Universidad de Chile.

References

[1] B. Misra, I. Prigogine and M. Courbage, From deterministic dynamics to probabilistic descriptions, Physica A98 (1979) 1-26.

[2] S. Goldstein, M. Courbage and B. Misra, On intrinsic randomness of dynamical systems, J. Stat. Phys. 25 (1981) 111-126.

[3] S. Martínez, Non-equilibrium entropy on stationary Markov processes, Acta Appl. Math. 3 (1985) 221-238.

[4] B. Sz.-Nagy and C. Foias, Analyse Harmonique des Opérateurs de l'Espace de Hilbert, Masson et Cie (1967).

[5] T. Kato, Perturbation theory for linear operators, Springer Verlag 2nd. edition (1976).

SITE-EXCHANGE CELLULAR AUTOMATA

N. BOCCARA[1,2] and M. ROGER[1]
[1] DRECAM/SPEC, CE-Saclay, 91191 Gif-sur-Yvette Cedex, France
[2] Department of Physics, University of Illinois, Chicago,IL 60680, USA

ABSTRACT. We study the stationary density of nonzero sites of automata networks whose local rule consists of two subrules. The first one, applied synchronously, is a one-dimensional range-one cellular-automaton rule. The second, applied sequentially, mixes the site values. The mixing results from either a local or a nonlocal exchange of the site values.

1. Introduction

Automata networks are fully discrete dynamical systems [1]. Modeling populations of interacting individuals in theoretical epidemiogy in terms of automata networks appears to be promising [2]. Local interactions, responsible for the spread of the disease by contact, may be modeled by a probabilistic cellular-automaton rule, whereas the motion of the individuals, which is an essential factor, may be modeled by a site-exchange rule applied sequentially. Results obtained recently in [2] led us to study some general properties of automata networks whose local rule consists of two subrules. The first one is a synchronous cellular-automaton rule, and the second a sequential mixing rule. More specifically, we have considered one-dimensional range-one deterministic and probabilistic cellular automata and studied the stationary density of nonzero sites as a function of the probability p characterizing the first subrule (when it is probabilistic), and the degree of mixing m on which depends the second one.

One-dimensional range-one deterministic cellular-automaton rules are defined as follows. The state at time t is represented by a function $S_t: i \mapsto s(t,i)$ from the set of integers $\mathbf{Z}$ to the set of states $\{0,1\}$. The equation

$$s(t+1,i) = f\big(s(t,i-1), s(t,i), s(t,i+1)\big), \tag{1}$$

where f is a map from $\{0,1\}^3$ to $\{0,1\}$, determines the evolution of the cellular automaton whose local rule is f. $\mathcal{S} = \{0,1\}^{\mathbf{Z}}$ is the state space. An element of the state space is also called a configuration.

The probabilistic cellular-automaton rules considered in this paper are of the type

$$s(t+1,i) = X f\big(s(t,i-1), s(t,i), s(t,i+1)\big), \tag{2}$$

where X is a discrete random variable defined on $\{0,1\}$ such that

$$Pr(X=0) = 1-p \quad \text{and} \quad Pr(X=1) = p. \tag{3}$$

E. Tirapegui and W. Zeller (eds.), Instabilities and Nonequilibrium Structures IV, 109–118.

The parameter m that measures the degree of mixing is defined as follows. A site whose value is one is selected at random and swapped with another site value (either zero or one) also selected at random. The second site is either a neighbor of the first one (local site-exchange) or any site of the lattice (nonlocal site exchange). This operation is repeated $mc(m,t)N$ times, where N is the total number of sites and $c(m,t)$ the density of nonzero sites at time t for a degree of mixing equal to m. It is important to note that this mixing process—which models either short or long range moves in epidemiology—does not change the value of the density.

If the degree of mixing m tends to infinity, the value of the stationary density of nonzero sites $c(\infty,\infty)$ is correctly predicted by a mean-field-type approximation in which it is assumed that the probability, at time t, that a site value is equal to 1 is $c(m,t)$. This approximation is incorrect when m is not sufficiently large.

2. Deterministic Rules

We have studied Rules 18, 22 and 54 (rules are numbered following Wolfram [3]).

Rule 18

It is defined by the following map

$$f_{18}(x_1,x_2,x_3) = \begin{cases} 1, & \text{if } (x_1,x_2x_3) = (0,0,1) \text{ or } (1,0,0), \\ 0, & \text{otherwise.} \end{cases}$$

For this rule, configurations belonging to the attractor, contained in the limit set, consist of sequences of zeros of odd lengths separated by isolated ones. The average number of sequences of zeros of lengths $2n+1$ per site is equal to $1/2^{n+3}$. With respect to this background a sequence of two ones or a sequence of zeros of even length is a "defect" (Grassberger's kinks [4]). Since two sequences of zeros of odd lengths separated by two neighboring ones generate, at the next time step, a sequence of zeros of even length, configurations generated by Rule 18 may contain defects of one type only. During the evolution these defects move, and, when they meet, they annihilate pairwise. This process has been studied by Grassberger [4] who found that, starting from a random initial configuration, the density of defects decreases as $t^{-1/2}$. These defects may be viewed as particles. For Rule 18 all particles are identical, and a particle is its own antiparticle. This particularly simple picture is not general [5].

The evolution of the density of nonzero sites according to the mean-field approximation is determined by

$$c_{18}(\infty,t+1) = 2c_{18}(\infty,t)\big(1-c_{18}(\infty,t)\big)^2. \tag{4}$$

Therefore $c_{18}(\infty,\infty) = 1-1/\sqrt{2} = 0.29289$, whereas the stationary density of nonzero sites in the absence of mixing $c_{18}(0,\infty)$, which is exactly known [3], is equal to $\frac{1}{4}$.

In order to characterize the asymptotic behavior of $c_{18}(m,\infty)$ for very small and very

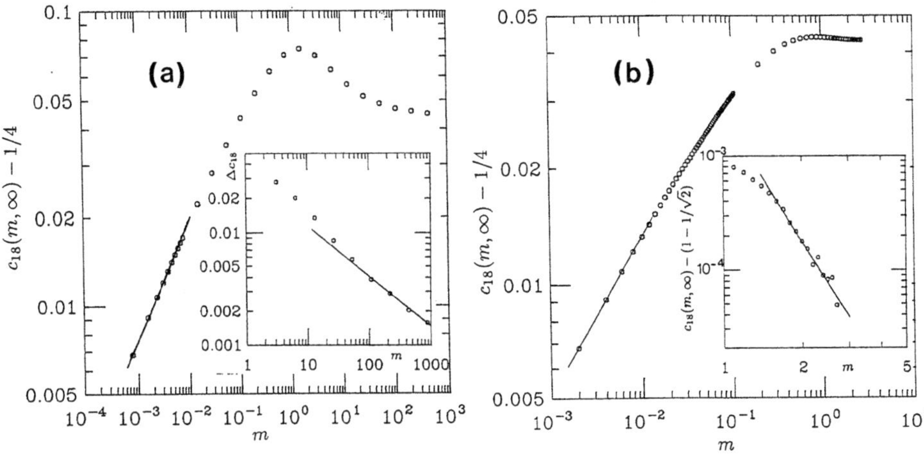

Figure 1. Rule 18. Asymptotic behaviors of $c_{18}(m, \infty)$. (a) local site exchange (b) nonlocal site exchange.

large m, we have defined the following exponents:

$$\alpha_0 = \lim_{m \to 0^+} \frac{\log\left(c_{18}(m,\infty) - c_{18}(0,\infty)\right)}{\log m}$$

$$\alpha_\infty = \lim_{m \to 0^+} \frac{\log\left(c_{18}(m,\infty) - c_{18}(\infty,\infty)\right)}{\log m}.$$

From our log-log plots we have obtained $\alpha_0 = 0.415 \pm 0.005$ (Fig. 1a) and $\alpha_\infty = 0.44 \pm 0.01$ (Fig.1a inset). These results may be understood as follows.

If starting from the attractor of the cellular automaton evolving according to Rule 18, we exchange a very small number of sites between two successive applications of Rule 18, we create defects at a rate which can be assumed to be proportional to m. Therefore, we may write

$$\lim_{t \to \infty} \frac{\partial c_{18}(m,t)}{\partial t} = O(m),$$

but, since for large t,

$$\lim_{m \to 0} c_{18}(m,t) = O(t^{-\gamma})$$

with $\gamma = \frac{1}{2}$, we have

$$t^{-\gamma-1} = O(m),$$

thus

$$\lim_{t \to \infty} c_{18}(m,t) = O(m^{\gamma/(\gamma+1)}).$$

This simple argument gives, at best, an order of magnitude. The assumption that defects are created at a rate proportional to m seems natural but might be inaccurate. Moreover, we assumed that the defects are created at random in the lattice. In the case of the local site exchange process, when we move a nonzero site to a neighboring zero site, we create two neighboring defects which have, therefore, a higher probability to annihilate than more distant defects. Hence, the exponent α_0 should be slightly greater than the value $\gamma/(\gamma+1) = \frac{1}{3}$, which is indeed the case.

When m is large, a given site has moved, on the average to a distance $\sqrt{m}$, thus, if the correlations created by the application of Rule 18 have a range that is small compared to this distance, the exponent α_∞ should, therefore, be close to $\frac{1}{2}$.

If the mixing process is nonlocal, only the argument explaining the behavior of $c_{18}(m,\infty)$ for small m remains valid. The motion of nonzero sites not being diffusive, the behavior of $c_{18}(m,\infty)$ for large m is, however, no more valid. We have obtained (Figure 1b) $\alpha_0 = 0.425 \pm 0.005$ and $\alpha_\infty = 3.7 \pm 0.1$.

Rule 54

Rule 54 is defined by

$$f_{54}(x_1,x_2,x_3) = \begin{cases} 1, & \text{if } (x_1,x_2x_3) = (0,0,1), (0,1,0), (1,0,0) \text{ or } (1,0,1). \\ 0, & \text{otherwise.} \end{cases}$$

As for Rule 18, the evolution toward the attractor may be viewed as annihilation processes of interacting particlelike structures. The background is periodic in space and time, both periods being equal to 4. Three types of particles may be distinguished (Figures 2a, b, c, d). Two of them are nonpropagating and periodic in time. Their periods are equal to 4. They may be generated by sequences of zeros whose lengths are greater that 3. We denote them by g_e and g_o according to whether they consist of sequences of zeros of even or odd length. There exists also a propagating particle w, which may be generated by three zeros following three ones or the converse. This particle may propagate to the right or to the left. Its velocity is equal to 1. These particles have a rather rich variety of interactions [3]. Their number decreases very slowly. Simulations done on a lattice of 10^4 sites for a number of time steps larger than 10^8 show that the number of g_e-particles tends to zero as $t^{-\gamma}$. The exponent γ determined over two decades from 3×10^6 to 3×10^8 time steps is approximately equal to 0.15. For small m, the argument given for Rule 18 is still valid and leads to the value 0.13 for the exponent α_0. For small and large values of m, the arguments given for Rule 18 are still valid. But, for small m, since the number of defects decreases very slowly, m has to be very small in order to create defects at a sufficiently low rate. We have found $\alpha_0 = 0.11 \pm 0.01$ (Fig. 3a), to be compared to $\gamma/(\gamma+1) = 0.13$, and $\alpha_\infty = 0.53 \pm 0.01$ (Fig. 3a inset). For nonlocal site exchange we have found $\alpha_0 = 0.19 \pm 0.01$, notably higher than $\gamma/(\gamma+1)$, and $\alpha_\infty = 5.3 \pm 0.1$ (Fig. 3b).

Rule 22

This rule is defined by the map

$$f_{22}(x_1,x_2,x_3) = \begin{cases} 1, & \text{if } (x_1,x_2x_3) = (0,0,1), (0,1,0) \text{ or } (1,0,0) \\ 0, & \text{otherwise.} \end{cases}$$

In this case the evolution toward the attractor cannot be viewed as annihilation processes of interacting particlelike structures. For large t the density of nonzero sites tends to its

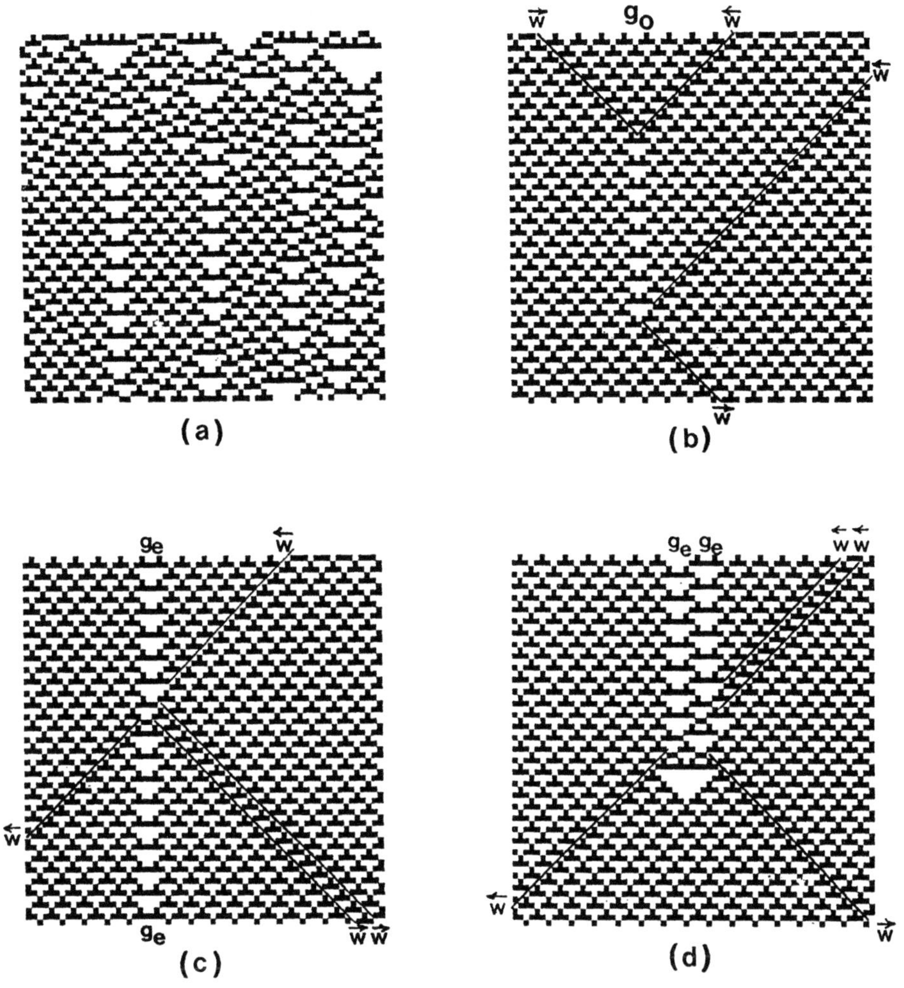

Figure 2. (a) spatio-temporal pattern of a cellular automaton evolving according to Rule 54 starting from a random initial state. (b-d) few examples of interactions between particles (see text).

stationary value exponentially, that is, the exponent γ is infinite. Hence, for small , m, the argument given for Rule 18 leads to $\alpha_0 = 1$, that is, $c_{22}(m, \infty)$ behaves linearly for small m, which is indeed the case (Figure 4a). For large m, we have found $\alpha_\infty = 0.55 \pm 0.01$ in agreement with the argument given for Rule 18. For nonlocal site-exchange, we have found $\alpha_0 = 0.86 \pm 0.01$, which is rather different from the linear behavior found for the local site-exchange process, and $\alpha_\infty = 4.4 \pm 0.1$ (Figure 4b).

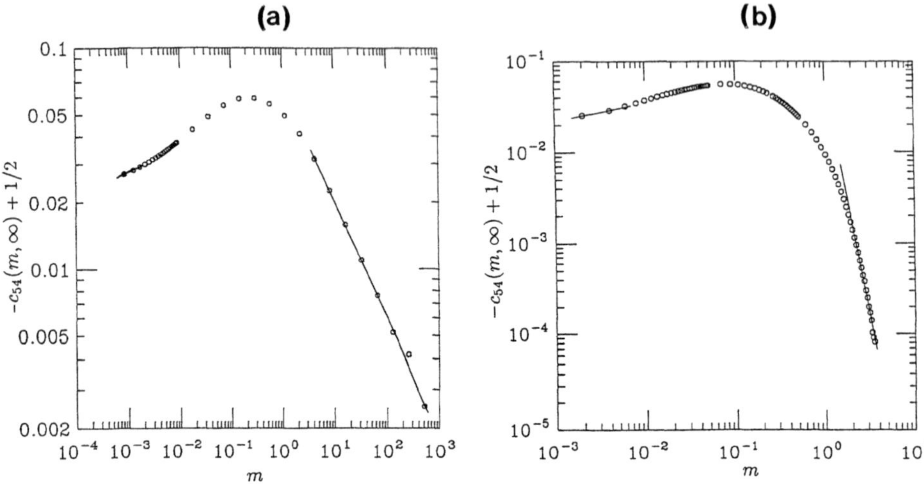

Figure 3. Rule 54. Asymptotic behaviors of $c_{54}(m,\infty)$. (a) local site exchange (b) nonlocal site exchange.

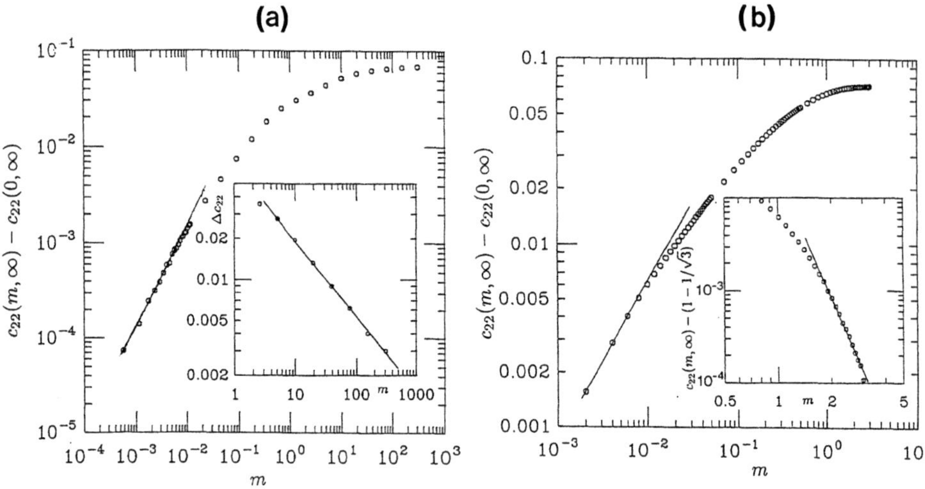

Figure 4. Rule 22. Asymptotic behaviors of $c_{22}(m,\infty)$. (a) local site-exchange (b) nonlocal site-exchange.

3. Probabilistic Rules

We have studied only one rule, namely Rule 18. For a fixed value of the parameter m, the cellular automaton exhibits, at $p = p_c(m)$, a transcritical bifurcation similar to a second-order phase transition whose order parameter is nonnegative. The behavior of the stationary density of nonzero sites $c_{18}(m,p,\infty)$ in the neighborhood of the bifurcation point may be characterized by a critical exponent $\beta(m)$ defined by

$$\beta(m) = \lim_{p \to p_c(m)+0} \frac{\log c_{18}(m,p,\infty)}{\log(p - p_c(m))}, \tag{5}$$

Within the mean-field approximation, we have

$$c_{18}(\infty,p,t+1) = 2pc_{18}(\infty,p,t)\big(1 - c_{18}(\infty,p,t)\big)^2. \tag{6}$$

Therefore

$$p_c(\infty) = \tfrac{1}{2} \quad \text{and} \quad \beta(\infty) = 1. \tag{7}$$

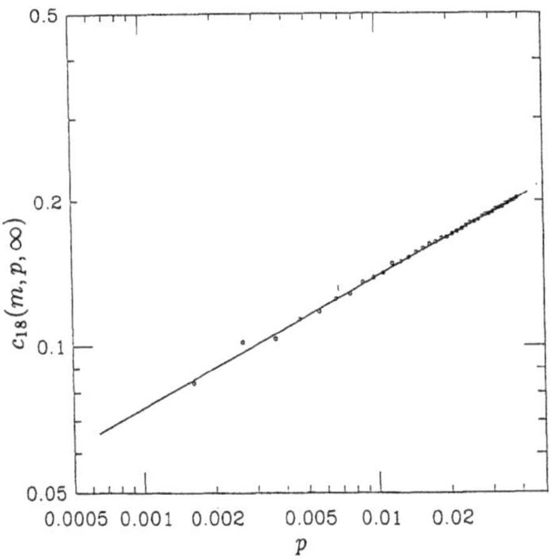

Figure 5. Typical variation of $c_{18}(m,p,\infty)$ as a function of p for $m = 0.41$ (local site-exchange).

Figure 5 shows a typical variation of the stationary density $c_{18}(m,p,\infty)$ as a function of p for a fixed value of m in the case of short range moves (local site-exchange). $p_c(m)$ is not equal to $\frac{1}{2}$ and the exponent $\beta(m)$ is clearly different from 1. The variations of $p_c(m)$ and β as fonctions of m are represented in Figure 6. Two regimes, reminiscent of the crossover phenomenon found in phase transitions theory [6], may be distinguished. In the small m regime, i.e., for $m \lesssim 10$, $p_c(m)$ and particularly $\beta(m)$ have their $m = 0$ values. In the large m regime, i.e., for $m \gtrsim 10$, $p_c(m)$ and $\beta(m)$ have their mean-field values.

The value we have found for $\beta(0)$ is the same as for directed percolation [7]. This result is in favor of the universality of critical behavior for one-dimensional cellular automata with one absorbing state, the universality class being characterized by the space dimensionality and, probably, the dependence of the mean-field map for small densities [8].

In the case of long range moves (nonlocal site-exchange), the variations of $p_c(m)$ and β as fonctions of m, represented in Figure 7, are strikingly different. While local and nonlocal site-exchange mixing processes lead to similar asymptotic behaviors for the stationary

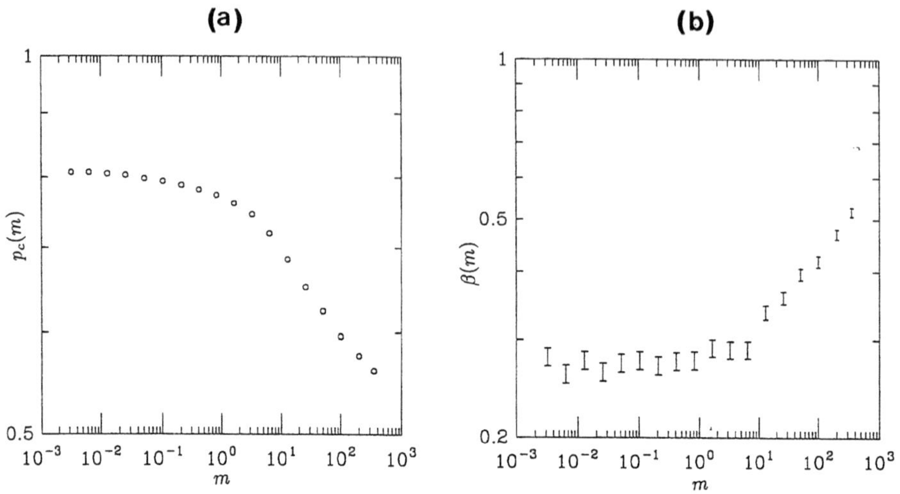

Figure 6. p_c (a) and β (b) as functions of m. Local site-exchange.

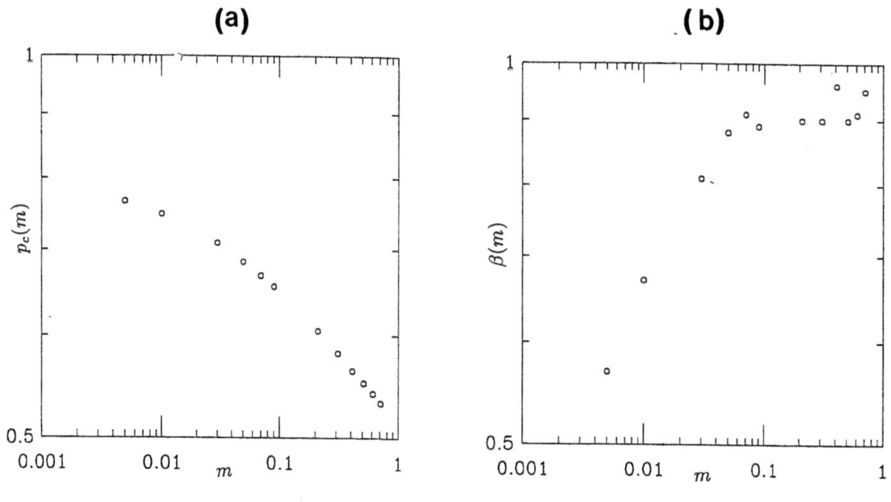

Figure 7. p_c (a) and β (b) as functions of m. Nonlocal site-exchange.

density $c(m,p,\infty)$ for small values of m when p is close to one, this is no more the case when p is close to p_c. When the stationary density is very small, the perturbation of the spatio-temporal pattern is much smaller for short-range moves than for long-range moves.

For $m=0$ the derivatives $\partial p_c/\partial m$ and $\partial \beta/\partial m$ are infinite. The behaviors of $p_c(m)$ and $\beta(m)$ for small m may, therefore, be characterized by the exponents α_p and α_β defined by

$$\alpha_p = \lim_{m\to 0+} \frac{\log\left(p_c(m) - p_c(0)\right)}{\log m} \tag{8a}$$

$$\alpha_\beta = \lim_{m \to 0^+} \frac{\log\left(\beta(m) - \beta(0)\right)}{\log m} \tag{8b}$$

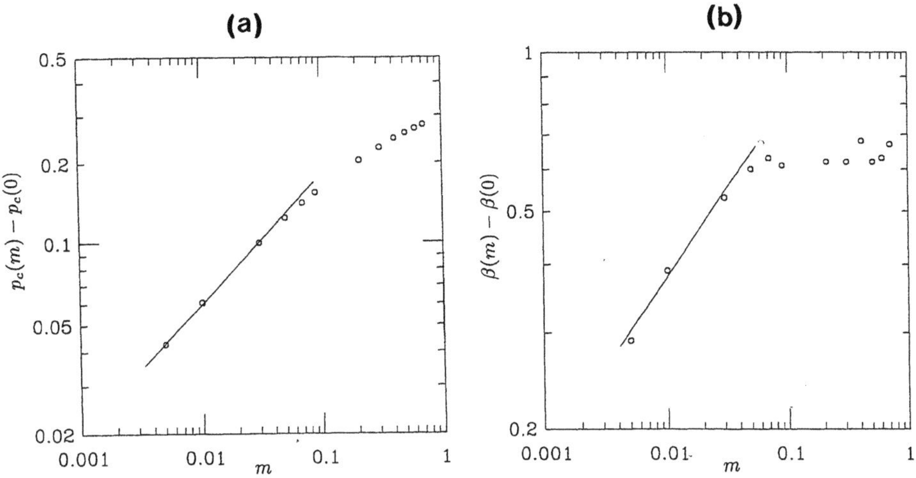

Figure 8. Log-log plots of (a) $p_c(m) - p_c(0)$ and (b) $\beta(m) - \beta(0)$ as function of m.

From our simulations, we have found (Figures 8a and b)

$$\alpha_p = 0.48 \pm 0.01 \quad \text{and} \quad \alpha_\beta = 0.33 \pm 0.02. \tag{9}$$

Notice that, to determine the exponent α_β, it is very important to determine $\beta(m)$ with a sufficient precision, that is we need to measure the percolation threshold $p_c(m)$ with great precision.

4. Conclusion

Motivated by recent results obtained in theoretical epidemiogy, using models formulated in terms of automata networks, we have studied some general properties of automata networks whose local rule consists of two subrules. The first one, applied synchronously, is a one-dimensional range-one cellular-automaton rule, and the second, applied sequentially, mixes the site values. The mixing results from either a local or a nonlocal exchange of the site values.

When the degree of mixing m tends to infinity, the evolution of the automata network is correctly predicted by a mean-field-type approximation which assumes that the probability, at time t, that a site value is equal to 1 is equal to the density of nonzero sites at time t. This approximation is not exact when $m \neq \infty$.

For three deterministic and one probabilistic one-dimensional range-one cellular automata, we have studied the variation of the stationary density of nonzero sites as a function of the degree of mixing.

When the cellular automaton is deterministic, we have determined the asymptotic behaviors for small and large degrees of mixing, and given some simple arguments to explain our results.

When the cellular automaton is probabilistic, it exhibits a transcritical bifurcation similar to a second-order phase transition whose order parameter is nonnegative. The behavior of the stationary density of nonzero sites in the neighborhood of the bifurcation point may be characterized by a critical exponent which has been determined as a function of the degree of mixing. According to whether the mixing results from local or nonlocal exchange of site values, the behaviors of the bifurcation point and the critical exponent are different.

Acknowledgment

One of the authors (NB) is very grateful to the organizers of the workshop for their very kind hospitality during his stay in Chile.

References

1 Goles, E. and Martínez, S. (1990) *Neural and Automata Networks. Dynamical Behavior and Applications*, Kluwer, Dordrecht.
2 Boccara, N. and Cheong, K. (1992) *Automata Network SIR Models for the Spread of Infectious Diseases in Populations of Moving Individuals*, J. Phys. A: Math. Gene. **25**, 1-25.
3 Wolfram, S. (1983) *Statistical Mechanics of Cellular Automata*, Rev. Mod. Phys. **55**, 601-44.
4 Grassberger, P. (1983) *Discussion of Random Walks in Domain Walls, Mainly in* $k = 2$, $r = 1$ *Cellular-Automaton Rule 18*, Phys. Rev. **A 28**, 3666-7.
5 Boccara, N., Nasser, J. and Roger, M. (1991) *Particlelike Structures and their Interactions in Spatiotemporal Patterns Generated by One-Dimensional Deterministic Cellular-Automaton Rules*, Phys. Rev. **A 44**, 866-75.
6 Boccara, N. (1976) *Symétries Brisées*, Hermann, Paris.
7 Bease, J. (1977) *Series Expansions for the Directed-Bond Percolation Problem*, J. Phys. C: Solid State Phys. **10** 917-924.
8 Bidaux, R., Boccara, N. and Chaté, H. (1989) *Order of the Transition Versus Space Dimensionality in a Family of Cellular Automata*, Phys. Rev. **A 39**, 3094-3105.

STEADY STATE SEGREGATION IN DIFFUSION-LIMITED BIMOLECULAR REACTIONS: EFFECT OF STRONG SPACE DISORDER AND A GALANIN APPROACH

HORACIO S. WIO*
Centro Atómico Bariloche
8400-S.C.Bariloche, ARGENTINA

Abstract

The problem of bimolecular reactions $A + B \rightarrow C$ is discussed from two different points of view. First, analyzing the effect of strong space diosorder within an Effective Medium Approximation for the continuum. Second we analyzed a simple model of the Galanin stochastic form, by means of a path integral approach.

1. INTRODUCTION

The study of diffusion-reaction systems has shown an increased interest in the last years. A short sample of recent reviews on this and related subjects is shown in Ref. [1]. As is well known there are two time scales that characterize these systems: the diffusion time (the time that elapses between collisions of reacting particles) and the reaction time (the time that particles take to react when in proximity). If the reaction time is much larger than the diffusion time, the process is reaction limited. In such a case the kinetics is described by classical rate equations. But in recent years the interest has been focused on the less tractable diffusion-limited processes, where the reaction time scale is negligible compared with the diffusion time [1]. Applications of diffusion-limited reactions include ionic recombination, electron-hole recombination, surface catalysis, polymers in solution, to indicate just a few. An aspect pointed out almost a decade ago by Zeldovich and coworkers was the possibility of macroscopic segregation [7,8]. Among other problems, authors have been concerned with the influence of dimensionality, initial conditions, sources, conservation laws, etc [1,2,4,5]. Also the form of the kinetic or rate equations has received some attention [3,6,10,11].

In this contribution we present two different aspects for the treatment of diffusion-limited systems, for the case $A + B \rightarrow C$ (where $C = B$ or 0). The first one refers to

*Member of CONICET, ARGENTINA.

E. Tirapegui and W. Zeller (eds.), Instabilities and Nonequilibrium Structures IV, 119–132.

the effect of spatially correlated disorder on the diffusive properties of the mentioned systems and, consequently, its effects on the macroscopic segregation characteristics. The second one is related to a simple description based on a Feinberg-Galanin model $A + B \to B$, similar to one used in neutron diffusion with absorption, and to a path-integral approach to it.

2. STRONG SPATIAL DISORDER

Here we address the effect of spatial disorder on the diffusive properties of the above mentioned diffusion-limited systems. As previously discussed, there are basically two mechanisms, reaction and diffusion, competing in the evolution of the system. The reaction process alone, in our case annihilation, already produces segregation. The diffusion process leads to homogenization of species and in this way reduces segregation. However, when diffusion stirring is inefficient, as in low dimensional systems, macroscopic segregation occurs. In this framework the effect of disorder can be understood by considering changes in the diffusive behaviour of particles. A direct method of analysis can be achieved by considering an Effective Medium Aproximation (EMA). Such an approximation provides with an equivalent time dependent diffusion coefficient calculated from the statistical properties of the random disordered field. Then we can use standard analysis of the reaction diffusion equation with an effective diffusion coefficient which takes into account the effect of disorder.

The EMA has been used only for discrete systems. In this case, and based on operator projection techniques and ordered cumulants, a systematic way of obtaining an EMA and perturbations around it has been developed in Ref. [14]. Although it is usually thought that EMA is only valid for discrete models, it is possible to generalize such an approximation to a continuous medium. This generalization has been made in the case of weak disorder [16] and can be extrapolated in general to strong disorder [17,18]. We will apply here this new procedure discussing only the one-dimensional problems, the higher dimensional situation will not be considered here.

From Ref. [16], it is clear that the effect of weak disorder will not alter the conclusions of Refs. [1,5], as it turns out that the resulting effective diffusion constant only implies a change in its magnitude but not a time dependence. We then concentrate on the most interesting case of strong disorder [14,15,16,17]. We follow the approach of Refs. [1,5]. This means we assume the validity of the reaction diffusion equations for the sum and difference variables: $\rho(x,t) = 1/2[\rho_A(x,t) + \rho_B(x,t)]$ and $\gamma(x,t) = 1/2[\rho_A(x,t) - \rho_B(x,t)]$ where ρ_A and ρ_B are the local densities of the A and B reactants respectively. The equations are:

$$\dot{\gamma}(x,t) = \partial_x \eta(x) \partial_x \gamma(x,t) + R_\gamma(x,t) \tag{1a}$$

$$\dot{\rho}(x,t) = \partial_x \eta(x) \partial_x \rho(x,t) - k\left[\rho^2 - \gamma^2\right] + R_\rho(x,t), \tag{1b}$$

where R_γ and R_ρ are the associated source terms, k the rate coefficient and the diffusion coefficient $\eta(x)$ is a random field. Applying the continuous version of the EMA [16] to (1.a) we arrive at:

$$\dot{\overline{\gamma}}(x,t) = \int_0^t D(t-t')\partial_x^2\overline{\gamma}(x,t')\,dt' + R_\gamma(x,t) \tag{2}$$

where the overbar indicates average over field configurations. The effective time dependent diffusion coefficient is given, in the Laplace transformed space, $t \rightarrow s$, (hereafter Laplace-transformed functions are written as the same functions with argument s instead of t), by the self consistent condition [17,18]:

$$\langle \frac{\eta - D(s)}{1 + \frac{1}{D(s)}(\eta - D(s))(1 - l\,G_1(0,s))} \rangle = 0 \tag{3}$$

where $\langle - \rangle$ indicates average over field configurations, l is the correlation length of the field $\eta(x)$, and G_1 is the one point Green function in the ordered case. This condition is equivalent to the EMA condition in a discrete medium with statistical independent sites separated by a distance l. This is a very useful condition because in this way it is easy to generalize known results from the equivalent discrete medium.

As shown in Ref. [5] one way to obtain information regarding the degree of segregation is through the analysis of the correlation function $C(x-x',t) = \langle \gamma(x,t)\gamma(x',t) \rangle$. From the knowledge of the initial correlation $\langle \gamma(x,0)\gamma(x',0) \rangle$ and correlation of sources $\langle R_\gamma(x,t)R_\gamma(x',t) \rangle$ we can obtain $C(x-x',t)$ as:

$$C(x-x',t) = \int dx_1 \int dx_2 \overline{G_2}(x,x',t/x_1,x_2,0)\langle \gamma(x_1,0)\,\gamma(x_2,0) \rangle +$$

$$\int_0^t dt' \int dx_1 \int dx_2 \overline{G_2}(x,x',t/x_1,x_2,t')\langle R_\gamma(x_1,t')\,R_\gamma(x_2,t') \rangle \tag{4}$$

$\langle - \rangle$ means average over all possible randomness. We have assumed that disorder is independent of initial conditions and sources. $\overline{G_2}(x,x',t/x_1,x_2,0)$ is the disorder averaged two-point Green function. Here we make the following ansatz (In fact it is exact, see Ref. [18]) for such a Green function:

$$\overline{G_2}(x,x',t/x_1,x_2,0) \approx \overline{G_1}(x,t/x_1,0)\,\overline{G_1}(x',t/x_2,0), \tag{5}$$

where $\overline{G_1}(x',t/x,0)$ is the one point averaged Green function, solution of (2) with $R_\gamma = 0$ and $\delta(x-x')$ as initial conditions. A thorough discussion of this Ansatz and other related aspects will be given elsewhere [18]. In the following we consider two kinds of strong disorder given respectively by a percolative and a generalised Poisson models of disorder. In both cases we study the influence of initial random conditions, a more general presentation including random source terms may be found in Refs. [17,18].

We first analyse percolative disorder. In this case the random field η is defined by a two-level process with values $\eta = \Delta$ and $\eta = 0$ with probability α and $(1-\alpha)$ respectively. For the sake of simplicity we take an exponential correlated field with a correlation length l :

$$\langle \eta(x)\eta(x')\rangle = \Delta^2 \exp(- \mid x - x' \mid /l) \tag{6}$$

The field can be thought of as regions that are exponentially distributed in which diffusion is possible or not possible. This field has the strongest disorder. The exact stationary solution of (1a) could be trivially obtained for a given realization, and other exact calculations are also possible. In particular it is not difficult to obtain the longtime limit of the equivalent diffusion coefficient defined, in terms of the Laplace variable s, as $D = \frac{1}{2}s^2\langle x^2(s)\rangle$. It reads

$$D(s) = \frac{1}{6}\alpha l^2 s. \tag{7}$$

This exact result can be compared with the one obtained with the EMA condition (3). As in the discrete case the EMA deals with the correct time dependence but does not lead to a correct coefficient [14].

Taking $D(s)$ from (8) in (2) and Fourier transforming we obtain for $\overline{G_1}$

$$lim_{t\to\infty}\overline{G_1}(k,t/0) \sim \frac{1}{1+D_1k^2} \tag{8}$$

with $D_1 = \frac{\alpha l^2}{6}$. Then Eq(5) gives (neglecting sources):

$$\langle\gamma(k,t)\gamma(k',t)\rangle \approx \overline{G_1}(k,t/0)\overline{G_1}(k',t/0)\langle\gamma(k,0)\gamma(k',0)\rangle \tag{9}$$

As in Refs. [1,5] we can choose uncorrelated or correlated initial conditions. In the first case

$$\langle\gamma(k,0)\gamma(k',0)\rangle_u = \frac{N}{2}[\delta_{k+k',0} - \delta_{k,0} - \delta_{k',0}]$$

which leads to

$$lim_{t\to\infty}\Big\langle\gamma(x,t)\gamma(0,t)\Big\rangle_u \approx \frac{n}{2D_1^{1/2}}\Big[(1+\frac{\mid x\mid}{D_1^{1/2}}\exp(-\frac{\mid x\mid}{D_1^{1/2}})\Big] \tag{10}$$

rendering

$$lim_{t\to\infty}\Big\langle\gamma(x,t)^2\Big\rangle_u = \frac{n}{2D_1^{1/2}} \tag{11}$$

We remark that here the correlations are first defined in a finite volume V ,where N particles are distributed,and the infinite volume limit is taken at the end of calculations when the Fourier transformed correlation is antitransformed. The density n appearing in (10) and (11) is N/V in the above defined limit.

From the stationary solution of the Eq.(1b) we arrive at $\langle\rho_{st}(x)^2\rangle = \langle\gamma_{st}(x)^2\rangle$. Then the segregation parameter introduced in Refs. [1,5] turns out to be

$$lim_{t\to\infty} S(t) = \frac{\langle\gamma^2\rangle}{\langle\rho^2\rangle} = 1 \tag{12}$$

The result (12) clearly indicates that the subdiffusive behaviour due to the percolative disorder increases the segregation with respect to the ordered case. In this last case segregation is only reached in the extinction limit. On the other hand if we consider a correlated initial condition, with c the geminate correlated placement [5], we have for the initial correlations

$$\Big\langle\gamma(k,0)\gamma(k',0)\Big\rangle = \frac{N}{2}\delta_{k+k',0}(1 - cos(kc)) \tag{13}$$

This leads to

$$C_c(x,t) = C_u(x,t) - \frac{1}{2}[C_u(x-c,t) + C_u(x+c,t)] \tag{14}$$

so that

$$\Big\langle\gamma(t)^2\Big\rangle_c = C_c(0,t) \approx_{t\to\infty} \frac{n}{2D_1^{1/2}}\Big[1 - (1 + \frac{c}{D_1^{1/2}})\exp(-c/D_1^{1/2})\Big] \tag{15}$$

indicating a reduction of the segregation due to the initial correlation. We find a kind of competition between the effective diffusion length $D_1^{1/2}$ and the initial geminate placement c. If $c >> D_1^{1/2}$ correlated particles are in different diffusive regions and this is equivalent to the uncorrelated situation, so we recover the previous result (11). In the opposite case, $c < D_1^{1/2}$, the probability of extinction of the correlated particles is larger and consequently the number of segregated particles is reduced.

From (10) and (14) it is possible to do an estimation of a segregation length. In the uncorrelated case it is clear that a length scale given by $l_s = D_1^{1/2}$ exists. In the correlated case a possible estimation can be given by the length in which the correlation is zero. When $D_1^{1/2} > c$ this length is given by

$$l_s = \frac{c[\exp(c/D_1^{1/2}) - \exp(-c/D_1^{1/2})]}{\exp(c/D_1^{1/2}) + \exp(-c/D_1^{1/2}) - 2} \tag{16}$$

which tends to $D_1^{1/2}$ if $D_1^{1/2} >> c$. In the case $c > D_1^{1/2}$ it is evident that the correlations in the correlated and uncorrelated initial condition cases are very similar.

Now we consider Poisson disorder. In this case the random field $\eta(x)$ is given by a spatial distribution of random pulses separated by an exponentially distributed random distance of mean λ. For the sake of simplicity we take exponential pulses with a height ω and decay parameter $1/l$, ω being a random variable exponentially distributed with mean ω_0. Strong disorder appears when $\lambda l < 1$, so that when the mean distance between pulses is greater than its width [12,13], Poisson strong disorder is weaker than percolative disorder because the space remains connected. This corresponds to a disorder of class c [14]. Diffusive regions (exponential pulses) are only connected by an exponentially small diffusion coefficient. This is the cause of the subdiffusive behaviour found with this kind of disorder. Applying the EMA condition (3) to this case we obtain an equivalent diffusion coefficient for small s (long time) given by

$$D(s) = D_2 s^{\nu}, \tag{17}$$

where

$$\nu = \frac{1-\lambda l}{1+\lambda l}, \tag{18}$$

and

$$D_2 = \Big(\frac{1}{2\Gamma(1-\lambda l)}\Big)^{2/(1-\lambda l)} (2\omega_0)^{\frac{2\lambda l}{1+\lambda l}} l^{\frac{2(1-2\lambda l)}{(1+\lambda l)}} . \tag{19}$$

With this coefficient in Eq.(2) we obtain

$$\langle \gamma^2(t) \rangle_u \approx \frac{n\Omega(\frac{\nu+1}{2})}{4D_2^{1/2}\Gamma(\frac{\nu+1}{2})} t^{\frac{-(1-\nu)}{2}} \tag{20}$$

for the uncorrelated initial condition and

$$\langle \gamma^2(t) \rangle_c \approx \frac{nc^2\Omega(\nu)}{8D_2^{3/2}\Gamma(\frac{3\nu-1}{2})} t^{\frac{-3(1-\nu)}{2}} \quad \nu > 1/3 \tag{21}$$

for the correlated case. Ω is the function:

$$\Omega(z) = \frac{1}{2\pi i}\int_{r-i\infty}^{r+i\infty} \frac{[u(1-u)]^{-z}}{(1-u)^{(-(\nu+1)/2} + u^{(1-\nu)/2}} du \ .$$

For the segregation parameter an analysis similar to the one of the percolative case gives for the long time limit $S \to 1$, so the segregation is reached in the limit of extinction. The difference with the ordered case is that segregation and extinction are slowed down due to the subdiffusive behaviour induced by the disorder. This slowing down is more intense, when disorder is stronger, i.e when $\lambda l \to 0$.

3. FEINBERG-GALANIN MODEL

The model we want to discuss in this section was originally used in neutron diffusion problems, where it is known as the Feinberg-Galanin model [19]. In the following we present the model, a standard method of solution using stochastic techniques as well as a path-integral approach [21]. The latter offers the possibility of extending the results of this model from the case $A + B \to B$ to the $A + B \to 0$ one.

3.1. The Model

For sake of completeness, we repeat here a few of the results of Ref[19]. We consider the case of a one-dimensional diffusion equation for the processes $A+B \to B$ modelled by the Galanin theory [19]

$$\partial_t N(x,t) = D\partial_x{}^2 N(x,t) - \gamma\delta(x - \epsilon(t))N(x,t) + S_0(x,t) \tag{22}$$

where $N(x,t)$ is the density of the A reactant, γ is the Galanin constant, $\epsilon(t)$ is the random position of the B reactant (that here plays the role of the absorber), and $S_0(x,t)$ is an independent source of A particles that here we will consider as zero. The only hypothesis that is necessary to take into account is the Markov character of the processes $\epsilon(t)$ [20]. As we are interested in calculating $\langle N(x,t)\rangle$, that is the averaged particle density, calling $n(x,t) = \langle N(x,t)\rangle$ and taking averages in equation (22), we obtain

$$\partial_t n(x,t) = D\partial_x{}^2 n(x,t) - \gamma A(x,t) + S_0(x,t) \tag{23}$$

where $A(x,t) = \langle\delta(x - \epsilon(t))N(x,t)\rangle$. The effect of the reactant B on A is completely specified by $A(x,t)$. If this quantity is known it is simple to evaluate $n(x,t)$. Taking the integral form of Eq. (22):

$$\begin{aligned} N(x,t) &= \int_{-\infty}^{+\infty} dx' K^{(0)}(x,t|x',0)N(x',0) + \int_0^t dt' \int_{-\infty}^{+\infty} dx' K^{(0)}(x,t|x',t')S_0(x',t') \\ &\quad -\gamma \int_0^t dt' \int_{-\infty}^{+\infty} dx' K^{(0)}(x,t|x',t')\delta(x' - \epsilon(t'))N(x',t') \end{aligned} \tag{24}$$

In the following we adopt $S_0 = 0$, and

$$S(x,t) = N(x,0)\delta(t)\ . \tag{25}$$

Iterating now Eq. (24), multiplying by $\delta(x - \epsilon(t))$ and taking averages, we find:

$$\begin{aligned} A(x,t) &= \langle\delta(x - \epsilon(t))N(x,t)\rangle \\ &= \int_0^t dt' \int_{-\infty}^{+\infty} dx' K^{(0)}(x,t|x',t')\langle\delta(x - \epsilon(t))\rangle S(x',t') \\ &\quad - \gamma \int_0^t dt' \int_{-\infty}^{+\infty} dx' \int_0^{t'} dt'' \int_{-\infty}^{+\infty} dx'' \langle\delta(x - \epsilon(t))\delta(x' - \epsilon(t'))\rangle \\ &\quad \times K^{(0)}(x,t|x',t')K^{(0)}(x',t'|x'',t'')S(x'',t'') + \dots\ . \end{aligned} \tag{26}$$

As ϵ is a Markov process we have

$$\begin{aligned} \langle \delta(x-\epsilon(t)) \rangle &= \int_{-\infty}^{+\infty} \delta(x-\epsilon) P(\epsilon, t) d\epsilon \\ &= P(x,t) = \int_{-\infty}^{+\infty} \omega(x,t|x_0,0) P(x_0) dx_0 \,, \end{aligned} \tag{27}$$

and

$$\begin{aligned} \langle \delta(x-\epsilon(t))\delta(x'-\epsilon(t')) \rangle &= P(x,t;x',t') \\ &= \omega(x,t|x',t') \int_{-\infty}^{+\infty} \omega(x,t|x_0,0) P(x_0) dx_0 \,, \end{aligned} \tag{28}$$

where we have used the standard notation for the joint and the initial probability distributions, and $\omega(x,t|x',t')$ is the conditional probability for the process ϵ. Substitution of these results into Eq. (26) leads to a closed equation for $A(x,t)$:

$$\begin{aligned} A(x,t) &= \int_0^t dt' \int_{-\infty}^{+\infty} dx' K^{(0)}(x,t|x',t') S(x',t') \int_{-\infty}^{+\infty} \omega(x,t|x_0,0) P(x_0) dx_0 \\ &\quad -\gamma \int_0^t dt' \int_{-\infty}^{+\infty} dx' K^{(0)}(x,t|x',t') \omega(x,t|x',t') A(x',t') \,. \end{aligned} \tag{29}$$

With this result, we are in principle able to solve Eq. (23) for $n(x,t)$.

3.2. Point of View of Path Integrals

The problem posed by Eq. (22) is equivalent to a Schrödinger equation (in imaginary time) with a δ potential [22]. As is well known [21,22], the path-integral representation of its solution could be written as

$$N(x_f,t_f) = \int_{-\infty}^{+\infty} dx_0 K(x_f,t_f|x_0,0) N(x_0,0) \tag{30}$$

where

$$K(x_f,t_f|x_0,0) = \iint \mathcal{D}[x]\, e^{-S[x]} \tag{31}$$

with $\mathcal{D}[x]$ the usual measure. Following Ref. [22], the action $S[x]$ has the form

$$S[x] = S^{(0)}[x] + S^{(1)}[x] \tag{32a}$$

$$S^{(0)}[x] = \frac{1}{4D} \int_0^{t_f} \dot{x}^2(t) \,, dt \tag{32b}$$

$$S^{(1)}[x] = \int_0^t dt \gamma \delta(x(t)-\epsilon(t)) \,. \tag{32c}$$

Expanding the exponent in Eq. (31) in powers of γ

$$e^{-S[x]} = e^{-S^{(0)}[x]} \left\{ 1 + \sum_{n=1}^{\infty} \frac{(-1)^n}{n!} (S^{(1)}[x])^n \right\} . \tag{33}$$

Eq. (31) can be written as

$$K(x_f,t_f|x_0,0) = K^{(0)}(x_f,t_f|x_0,0) + K^{(1)}(x_f,t_f|x_0,0)\,, \tag{34a}$$

$$K^{(0)}(x_f,t_f|x_0,0) = \iint \mathcal{D}[x]\, e^{-S^{(0)}[x]}\,, \tag{34b}$$

$$K^{(1)}(x_f,t_f|x_0,0) = \sum_{n=1}^{\infty}(-\gamma)^n G^{(n)}(x_f,t_f|x_0,0)\,. \tag{34c}$$

where

$$G^{(n)}(x_f,t_f|x_0,0) = \frac{1}{n!}\iint \mathcal{D}[x]\,(S^{(1)}[x])^n e^{-S^{(0)}[x]} \tag{35}$$

with

$$(S^{(1)}[x])^n = \int_0^{tf} dt_n \int_0^{tf} dt_{n-1}\dots\int_0^{tf} dt_1 \prod_{k=1}^{n}\delta(x(t_k)-\epsilon(t_k))\,. \tag{36a}$$

As is very well known, ordering in time ($t_1 < t_2 < \cdots < t_n$), and changing integration limits, we may rewrite Eq. (36a) as

$$(S^{(1)}[x])^n = n!\int_0^{tf} dt_n \int_0^{tn} dt_{n-1}\dots\int_0^{t2} dt_1 \prod_{k=1}^{n}\delta(x(t_k)-\epsilon(t_k))\,. \tag{36b}$$

Now, with the help of the δ's, we make a partition of the time interval in $G^{(n)}$, which gives

$$\begin{aligned} G^{(n)}(x_f,t_f|x_0,0) &= \int_0^{tf} dt_n \dots \int_0^{t3} dt_2 \int_0^{t2} dt_1 K^{(0)}(x_f,t_f|\epsilon_n,t_n) \\ &\times \left[\prod_{k=1}^{n} K^{(0)}(\epsilon_k,t_k|\epsilon_{k-1},t_{k-1})\right] K^{(0)}(\epsilon_1,t_1|x_0,0)\,. \end{aligned} \tag{37}$$

At this point we must resort to an average over the processes ϵ. Such averaging, following ideas similar to the ones in the previous section, lead us to

$$n(x,t) = \int_{-\infty}^{+\infty} dx_0 \langle K(x_f,t_f|x_0,0)\rangle N(x_0,0) \tag{38}$$

with

$$\langle K(x_f,t_f|x_0,0)\rangle = \langle K^{(0)}(x_f,t_f|x_0,0)\rangle + \langle K^{(1)}(x_f,t_f|x_0,0)\rangle\,, \tag{39a}$$

$$\langle K^{(0)}\rangle = K^{(0)}(x_f,t_f;x_0,0)\,, \tag{39b}$$

$$\langle K^{(1)}\rangle = \sum_{n=1}^{\infty}(\gamma)^n \langle G^{(n)}(x_f,t_f|x_0,0)\rangle\,. \tag{39c}$$

Considering Eq. (37) and the results of Ref. [19] we find

$$\begin{aligned} \langle G^{(n)}(x_f,t_f|x_0,0)\rangle &= \int_{-\infty}^{+\infty} d\epsilon_n \dots \int_{-\infty}^{+\infty} d\epsilon_1 \int_0^{t_f} dt_n \dots \int_0^{t_2} dt_1 \\ &\times K^{(0)}(x_f,t_f|\epsilon_n,t_n)\left[\prod_{k=1}^{n} K^{(0)}(\epsilon_k,t_k|\epsilon_{k-1},t_{k-1})\right] \\ &\times K^{(0)}(\epsilon_1,t_1|x_0,0)P(x_0,0;\epsilon_1,t_1\dots\epsilon_n,t_n,x_f,t_f)\,. \end{aligned} \tag{40}$$

Now, from Eq. (28) and Ref. [19], we have for the joint probability distribution

$$P(x_0, 0; \epsilon_1, t_1 \ldots \epsilon_n, t_n, x_f, t_f)$$
$$= \omega(x_f, t_f|\epsilon_n, t_n)\left[\prod_{k=1}^{n} \omega(\epsilon_k, t_k|\epsilon_{k-1}, t_{k-1})\right]\omega(\epsilon_1, t_1|x_0, 0) \,. \qquad (41)$$

If we also integrate over the initial distribution of reactant A, i.e. $N(x_0, 0)$, we obtain

$$\langle\!\langle G^{(n)}(x_f, t_f|x_0, 0)\rangle\!\rangle = \int_{-\infty}^{+\infty} dx_0 \langle G^{(n)}(x_f, t_f|x_0, 0)\rangle N(x_0, 0)$$
$$= \int_{-\infty}^{+\infty} d\epsilon_n \ldots \int_{-\infty}^{+\infty} d\epsilon_1 \int_0^{t_f} dt_n \ldots \int_0^{t_2} dt_1 K^{(0)}(x_f, t_f|\epsilon_n, t_n)\omega(x_f, t_f|\epsilon_n, t_n)$$
$$\times \left[\prod_{k=2}^{n} K^{(0)}(\epsilon_k, t_k|\epsilon_{k-1}, t_{k-1}) \,, \omega(\epsilon_k, t_k|\epsilon_{k-1}, t_{k-1})\right]$$
$$\times \int_{-\infty}^{+\infty} dx_0 \int_{-\infty}^{+\infty} dy_0 K^{(0)}(\epsilon_1, t_1|x_0, 0)\omega(\epsilon_1, t_1|y_0, 0)N(x_0, 0)P(y_0) \,. \qquad (42)$$

This result, together with Eqs. (34), is completely coincident (making a term by term comparison) with the one we would obtain by solving Eq. (23) via a path-integral method. However, there are more direct ways of obtaining the path-integral solution to (23), but the above procedure seems to be the most adequate one in order to allow us to consider the case $A + B \to 0$. In such a case, we need to solve a coupled set of equations of the form (22), implying that $\omega(x, t|x', t')$ will not be fixed externally but must be selfconsistently obtained solving this system (i.e., it would not necessarily result to be Gaussian).

3.3. Results for the Reaction $A + B \to B$

As we said before, we consider the reaction example $A + B \to B$, where A plays the role of the absorbed element and B of the absorbent. Both A and B have a diffusive behaviour. We will consider the following situations : in the first case B is concentrated in the origin and A is uniformly distributed; in the second case, both A and B are uniformly distributed. We have for the Green's function and for the conditional probability

$$K^{(0)}(x, t|x', t') = \frac{1}{\sqrt{4\pi D(t - t')}} \exp -\frac{(x - x')^2}{4D(t - t')} \,, \qquad (43)$$

and

$$\omega(x, t|x', t') = \frac{1}{\sqrt{4\pi d\,(t - t')}} \exp -\frac{(x - x')^2}{4d(t - t')} \,. \qquad (44)$$

The initial condition for A is $S(x',t') = N_0\delta(t')$ for both cases. For B we have

$$P_1(x_0,t_0) = P_0\delta(x_0)\ , \qquad P_2(x_0,t_0) = P_0\ . \tag{45}$$

Replacing this in Eq. (29), and transforming Fourier-Laplace in space and time respectively, we obtain

$$A_1(k,p) = \frac{N_0 P_0}{\sqrt{2\pi}(p+k^2 d)} \frac{\sqrt{p+k^2\beta}}{\sqrt{p+k^2\beta}+\gamma/\sqrt{8\pi(D+d)}} \tag{46}$$

and

$$A_2(k,p) = \frac{N_0 P_0 \delta(k)\sqrt{2\pi}}{p} \frac{\sqrt{p+k^2\beta}}{\sqrt{p+k^2\beta}+\gamma/\sqrt{8\pi(D+d)}} \tag{47}$$

where $\beta = Dd/(D+d)$.

The integral form of the Eq. (23) for the density n is

$$n = N_0 - \gamma \int_0^t dt' L^{-1}[A(0,p),p,t']\ , \tag{48}$$

where L^{-1} stands for the inverse Laplace transform. In the last equation we took $k=0$ in order to obtain the total density. As we want to obtain the relative density we adopt $\Delta n = n/n_0$ where n_0 is the value that we obtain from n making $\gamma = 0$ (i.e., without absorption). Then, we have

$$\Delta n = 1 - \sqrt{32\pi}\alpha P_0 (D+d)\gamma^{-1}\left(2\sqrt{\tau/\pi} + e^{\tau}\,\mathrm{Erfc}\,(\sqrt{\tau}) - 1\right)\ , \tag{49}$$

where $\tau = \gamma^2 t/[8\pi(D+d)]$. In the first case $\alpha = 1$ while in the second $\alpha = 2\pi$. It is worth to stress the fact that these solutions do not have the usual exponential or inverse power law decreasing behavior [1], instead they vanish at a finite time.Finally let us consider the following case. Assume the reactants B do not diffuse, but have a fixed spatial distribution according to $P_B(x) = N_B(1+\cos(k_0 x))$. The solution of the problem, where now we must take $\omega(x,t|x',t') = \delta(x-x')$, have the characterstic that, $n(x,t)$, after a short initial transient, reaches a form (decaying in time) with a spatial modulation of wavenumber k_0, that corresponds to the *conjugate* of $P_B(x)$. Such result is the signature of segregation (originated by fluctuations), and when treating, within a similiar approach, the case $A + B \to 0$, we have found (in a low order approximation) a similar result [24].

4. FINAL REMARKS

Summarizing the effect of disorder on the time evolution and segregation of the diffusing particles, we have seen that: When disorder is weak only a quantitative change of the diffusion coefficient is possible and no important changes with respect to the ordered situation appear. When the disorder is so strong that it produces subdiffusive behaviour but not strong enough to disconnect diffusive regions, we obtain a slowing down of the reaction process and also of the segregation (Poisson disorder). In the percolative case, where disorder is so strong that produces isolated diffusive regions, extinction is not possible and a segregated steady state appears. The effect of disorder in the case with sources is similar. We observe an infinite growth if disorder is able to isolate correlated sources. On the contrary, if it is not so the behaviour is like in ordered media. As a final remark, strong or intermediate disorder produces changes in the exponents of Eqs. 12, 15, 20 and 21, an important result in a problem where exponents have been thought to be universal. On the other hand, weak disorder changes coefficients but not exponents.

Through the analysis of a simple stochastic model of the Galanin type [19], we have here presented a discussion of the diffusion-limited bimolecular reaction $A + B \rightarrow B$. This was done by using a path integral approach. Models of this kind could be also extended to other bimolecular reactions (i.e. $A + B \rightarrow 0$, $A + A \rightarrow 0$, etc.). It is in those cases that the path integral approach could show all its power, as this formalism provides several alternatives of approximation schemes adequate to treat different situations.It is worth remarking here that the problem of diffusion in a random media has been discussed recently within a functional integral framework [23]. However, the ideas developed here are related but of a different kind. Within this path integral formalism it is also possible to treat higher dimensional cases. All this points will be discussed elsewhere [24]. Another aspect of interest is the possibility of the ϵ processes not being Markovian. Such a case would imply, in the approach of Ref. [19] an almost unsurmountable dificulty. However, the possibility of treating non- Markovian problems is well known within the path integral schemes [25].

Acknowledgements

The research work presented here is the result of collaboration with M. A. Rodriguez, L. Pesquera, C. B. Briozzo, F. G. Nicolini, G. Abramson. Fruitful discussions with D. H. Zanette, C. Borzi, D. ben Avraham, A. Blumen, P. Alemany are greatly acknowledged. Support from : CONICET, Argentina; the Ministerio de Educación y Ciencia, Spain, through its Programa de Cooperación con Iberoamérica; and Fundación Antorchas, Argentina, are also acknowledged.

References

[1] Recent reviews are: A. Blumen, J. Klafter and G. Zumofen, in *Optical Spectroscopy of Glasses*, edited by I. Zschokke (D. Reidel Pub. Co., 1986); V. Kuzovkov and E. Kotomin, Rep. Prog. Phys. **51**, 1479 (1988); A. S. Mikhailov, Phys. Rep. **184**, 307 (1989); D. ben-Avraham, M. A. Burschka and Ch. R. Doering, J. Stat. Phys. **60**, 695 (1990); K. Lindenberg, B. J. West and R. Kopelman, in it Noise and Chaos in

Nonlinear Dynamical Systems, edited by F. Moss, L. A. Lugiato and W. Schleich, (Cambridge U. P., Cambridge, 1990).

[2] L. W. Anacker and R. Kopelman, Phys. Rev. Lett. **58**, 289 (1987).

[3] S. Kanno, Prog. Theoret. Phys. **79**, 721 (1988).

[4] D. ben-Avraham and Ch. R. Doering, Phys. Rev. **A37**, 5007 (1988).

[5] K. Lindemberg, B.J. West and R. Kopelman, Phys. Rev. Lett. **60**, 1777 (1988).B. J. West, R. Kopelman and K. Lindemberg, J. Stat. Phys. **54**, 1429 (1989).

[6] E. Clement, L.M. Sanders and R. Kopelman, Phys. Rev. **A39**, 6455 (1989).

[7] A. A. Orchinnikov and Y. B. Zeldovich, Chem. Phys. **28**, 215 (1978).

[8] Y. B. Zeldovich and A. S. Mikhailov, Sov. Phys. Usp. **30**, 977 (1987).

[9] V. Kuzovkov and E. Kotomin, Rep. Prog. Phys. **51**, 1479 (1988).

[10] A. S. Mikhailov, Phys. Rep. **184**, 307 (1989).

[11] Ch. R. Doering and D. ben-Avraham, Phys. Rev. **A38**, 3035 (1988). D. ben-Avraham and Ch. R. Doering, Phys. Rev. **A39**, 6436 (1989).

[12] S. Alexander, J. Bernasconi, W. R. Schneider and R. Orbach, Rev. Mod. Phys. **53**, 175 (1981). J. W. Haus and K. W. Kehr, Phys. Rep. **150** , 263 (1987). S. Havlin and D. ben-Avraham, Adv. Phys. **36** , 695 (1987).

[13] S. Havlin, M. Schwartz, R. Blumberg, A. Bunde and H. E. Stanley, Phys. Rev. **A40**, 1717 (1989).

[14] M. A. Rodriguez, E. Hernandez-Garcia, L. Pesquera and M. San Miguel, Phys. Rev. **B40**, 4212 (1989). E. Hernandez-Garcia, M. A. Rodriguez, L. Pesquera and M. San Miguel, Phys. Rev. **B42** , 10653 (1990).

[15] J. Heinrichs, Phys. Rev. Lett. **52** , 1262 (1984). G. Nicolis and V. J. Altares, J. Phys. Chem. **93** , 2861 (1989).

[16] A. Valle, M. A. Rodriguez and L. Pesquera, Phys. Rev. **A43** , 948 (1991).

[17] H. S. Wio, M. A. Rodriguez, C. B. Briozzo and L. Pesquera , Phys. Rev. **A44** , R813 (1991).

[18] M. A. Rodriguez, H.S.Wio, C. B. Briozzo and L. Pesquera , to be submitted for publication.

[19] I. Martinez and M. A. Rodriguez, Ann. Nucl. Energy **12**, 113 (1985); M. M. R. Williams, *Random Processes in Nuclear Reactors* (Pergamon P., Oxford, 1974).

[20] N. G. van Kampen, *Stochastic Processes in Physics and Chemistry* (North Holland, Amsterdam, 1981); C. W. Gardiner, *Handbook of Stochastic Methods* (Springer-Verlag, Berlin, 1983).

[21] R. P. Feynman and A. R. Hibbs, *Quantum Mechanics and Path Integrals* (McGraw-Hill, New York, 1965); L. S. Schulman, *Techniques and Applications of Path Integrals* (Wiley, New York, 1981); F. Langouche, D. Roekaerts and E. Tirapegui, *Functional Integration and Semiclassical Expansions* (D. Reidel Pub. Co., Dordrecht, 1982); M.Freidlin, *Functional Integration and Partial Differential Equations* (Princeton U. P., New Jersey, 1985).

[22] S. V. Lawande and K. V. Bhagwat, Phys. Lett. A **131**, 8 (1988), B. Gaveau and L. S. Schulman, J. Phys. A **19** 1833 (1986).

[23] L. C. Botelho and E. P. da Silva, J. Phys. A **23**, 1829 (1990); R.Tao, Phys. Rev. A **43**, 5284 (1991).

[24] H. S. Wio, F. Gonzalez Nicolini, M. A. Rodriguez and G. Abramson, to be submitted for publication.

[25] H. S. Wio, P. Colet, M. San Miguel, L. Pesquera and M. A. Rodriguez, Phys. Rev. A **40**, 7312 (1989).

Schrödinger's Cat and Squeezing

M. Orszag, R. Ramirez, J.C. Retamal, L. Roa
Facultad de Física, Universidad Católica de Chile,
Casilla 306, Santiago 22, Chile

Abstract. For particular values of the atomic flight time through the two photon micromaser cavity, if the atom is polarized, the field evolves to a pure state made of a superposition of even-photon number states or odd-photon number states. As the upper limit of the trap micromaser, these states approach a perfectly squeezed state.

Recently, with sophisticated experimental techniques, it seems plausible to generate macroscopic quantum superpositions, the so-called "Schrödinger's cat". Particularly interesting candidates, in quantum optics, are the one and two photon micromaser. One photon micromasers have been studied both experimentally [1,2] and theoretically [3,4]. Of particular interest is the lossless case, where the trapping states were found which are pure $|n\rangle$ states or tangent or cotangent states, depending on whether the atoms are injected in the upper state or in a coherent superposition of the two states, respectively [3,4]. In the present paper, we analyze the two photon micromaser consisting of a single mode lossless cavity with a low injection of three level atoms in which a way, as to have, on the average, only the atom inside the cavity, at a given time. We consider a three level atom, as per figure 1, assuming that the atoms are all identically prepared, in a coherent superposition of the upper and lower states, that is [6]:

$$|\psi\rangle_{\rm at} = \alpha|a\rangle + \gamma|c\rangle, \tag{1}$$

In the absence of electromagnetic losses and for large detuning, such that, the time evolution operator of the system is given by [6]:

$$U(\tau) = \begin{pmatrix} 1 - a\hat{C}a^{+} & 0 & -a\hat{C}a \\ 0 & e^{-i\phi(2\hat{N}+1)} & 0 \\ -a^{+}\hat{C}a^{+} & 0 & 1 - a^{+}\hat{C}a \end{pmatrix}, \tag{2}$$

where $\hat{C} = (1 - e^{-i\phi(2\hat{N}+1)})(a^{+}a + aa^{+})^{-1}$, $\phi \equiv \tau g^2/\Delta$, Δ being the detuning (fig. 1), g the coupling constant and the flight time of the atom through the cavity. Whether, in steady state, one generated or not a Schrödinger's cat (pure state)

E. Tirapegui and W. Zeller (eds.), Instabilities and Nonequilibrium Structures IV, 133–139.

can be measured via the entropy, defined as (after l atoms have passed through the cavity):

$$S_l = -\mathrm{Tr}\rho_f^{(l)} \log \rho_f^{(l)}, \tag{3}$$

If $\lim S_l = 0$, we have generated a pure state. [If stands for the field clamity operates] In order to investigate this point, as well as the squeezing of the field, we study the evolution of an arbitrary state, that is

$$\sum_N S_N|N\rangle(\alpha|a\rangle + \gamma|c\rangle) \rightarrow \sum_N S_N[|a_n\rangle|a\rangle + |c_n\rangle|c\rangle], \tag{4}$$

where

$$|a_n\rangle = \alpha[1 - 2i\frac{N+1}{2N+3}e^{-i\Omega_{N+1}} \sin \Omega_{N+1}]|N\rangle - 2i\gamma G(N)e^{-i\Omega_{N-1}} \sin \Omega_{N-1}|N-2\rangle$$

and

$$|c_n\rangle = -2i\alpha G(N+2)e^{-i\Omega_{N+1}} \sin \Omega_{N+1}|N+2\rangle + \gamma[1-2i\frac{N}{2N-1}e^{-i\Omega_{N-1}} \sin \Omega_{N-1}]|N\rangle$$

Where $\Omega_l = (2l+1)\phi/2$ and $G(l) = \sqrt{l(l-1)}/(2l-1)$. From the eq.(4), we immediatly see the trapping states.

When

$$\Omega_{N_d-1} = \frac{1}{2}(2N_d - 1)\phi = q\pi, \qquad q \text{ integer} \tag{5}$$

we have a downward trapping in $|N = N_d\rangle$ and when

$$\Omega_{N_u+1} = \frac{1}{2}(2N_u + 3)\phi = p\pi, \qquad p \text{ integer}, \tag{6}$$

we have an upward trapping in $|N = N_u\rangle$ and the Fock space is divided in isolated blocks, so that if our initial field distribution is within a given block, we have a closed dynamics, leading to a "Schrödinger's cat" (pure state). Once the steady state is a superposition of number states restricted between a lower and upper bound, N_d and N_u respectively, then when one extra atom crosses the cavity, the atom-field system would be modified only by some global phase factors and a different atomic superposition, that is:

$$\sum_{N=N_d}^{N_u} S_n|N\rangle(\alpha|a\rangle + \gamma|c\rangle) \rightarrow e^{i\phi'} \sum_{N=N_d}^{N_u} S_n|N\rangle(\alpha'|a\rangle + \gamma'|c\rangle), \tag{7}$$

where ϕ', α' and γ' are independent of N.

By comparing the right-hand-side of equations (4) and (7), we readily get:

$$\alpha = \alpha' e^{i\phi'}$$
$$\gamma = \gamma' e^{i\phi'}$$

$$S_{N+2} = -\frac{\alpha}{\gamma}\sqrt{\frac{N+1}{N+2}}S_N. \tag{8}$$

The last equality in eq (8), gives us a recursion relation that generates even or odd superposition of $|N >$ states:

$$|f\rangle_{\text{even}} = S_o \sum_{N=N_d/2}^{N_u/2} (-1)^N (\frac{\alpha}{\gamma})^N \sqrt{\frac{(2N-1)!!}{(2N)!!}} |2N\rangle, \tag{9-a}$$

$$|f\rangle_{\text{odd}} = S_1 \sum_{N=(N_d-1)/2}^{(N_u-1)/2} (-1)^N (\frac{\alpha}{\gamma})^N \sqrt{\frac{(2N)!!}{(2N+1)!!}} |2N+1\rangle, \tag{9-b}$$

where S_o and S_1 are normalization constants. Defining the usual hermitian quadratures for the field:

$$a_1 = \frac{1}{2}(a + a^+)$$
$$a_2 = \frac{1}{2i}(a - a^+) \tag{10}$$

and making use of $|f >_{\text{even}}$, with N_d=0, we readily get:

$$(\Delta a_2)^2 = \frac{1}{4}\Big[1 + \frac{2}{M}\sum_{N=0}^{N_u/2}(2N)|\frac{\alpha}{\gamma}|^{2N}\frac{(2N-1)!!}{2N!!}\{1 - |\frac{\gamma}{\alpha}|\}\Big] \tag{11}$$

where:

$$M = \sum_{N=0}^{N_u/2} |\frac{\alpha}{\gamma}|^{2N}\frac{(2N-1)!!}{2N!!} \tag{12}$$

and we have chosen a relative phase between α and γ equal to π.

An approximate calculation of M indicates that the minimum value of $(\Delta a_2)^2$ occurs in the neighborhood of $x = |\alpha/\gamma| = 1$ (from below) for large N_u. This is further verified in figure 2, where in the case of large N_u (1000), the minimum for $\langle(\Delta a_2)^2\rangle$ correspondes to a nearly perfect squeezed state.

Finally, in figure 3 we show $(\Delta a_2)^2$ and S (entropy) versus the number of atoms l crossing the cavity, for a pure initial vacuum state of the field. We notice that after 4500 atoms have passed through the cavity, $S = 0$ and therefore the steady state of the field is a pure state (even state) and the above arguments are correct. This result was obtained as follows: The master equation for the reduced field density, after the atoms pass through the cavity, is given by $\rho_f^{(k)} = \text{Tr}_{\text{at}}(U\rho_f^{(k-1)}\rho_{\text{at}}U^\dagger)$ that eventually (after a large number of atoms went through the cavity) leads to a steady state that is, in the present example, an even state, if the initial state of the field is a superposition of even $|n\rangle$ states in the corresponding block.

We finish this work with some comments

(1) Although a model with a lossless cavity might seems academic, actual experiments in micromasers have been performed with extremely high Q-values and so, this model is quite realistic.

(2) We observed numerically that for $|\alpha/\gamma| < 1$, the maximum of the photon number distribution occurs for $N = 0$. However, this maximum moves abruptly to a non-zero value as we cross the $|\alpha/\gamma| = 1$ line. Therefore $|\alpha/\gamma| = 1$ is a critical point. As a matter of fact, for $N_u \to \infty$, $(\Delta a_2)^2$ suffers a sudden jump, at $x = 1$ from 0 to 0.25.

(3) One may ask the question of how to generate detect one of these states. To generate an even state, we have to prepare the field initially within the corresponding block. In order to do that, we could send a beam of atoms excited in the upper state with an appropiate speed, as to generate a pure n-state $|N_i\rangle$. A second beam will have atomic polarization and an interaction time that will correspond to a pair of integers N_d and N_u such that $N_d \leq N_i \leq N_u$.

As to the experimental verification of these results, as in the one photon case, it requires a very low temperature such that the number of thermal photons is much less than one. This can be achieved experimentally [1,2]. On the other hand, one would expect, as in the one photon case, these states to be robust to cavity damping [7] and small velocity fluctuations in the beam [8].

(4) From the trapping condition, Eq. (6), we see that $\phi/2\pi = \tau g^2/2\pi\Delta = p/(2N_o+3)$ implies that is $\phi/2\pi$ a rational number. The limit where the large squeezing occurs (N_u large) does not imply that the coupling constant or τ has to be very small or Δ very large. It may simply imply that $\phi/2\pi$ approaches an irrational number, with both p and N_u large. This suggests a connection between this problem and number theory and chaos.

References

[1] D. Meschede, H. Walther, and G. Müller, Phys. Rev. Lett 54, 551 (1985)

[2] G. Rempe, H. Walther and N. Klein, Phys. Rev. Lett. 58,353 (1987)

[3] P. Filipowicz, J. Javanainen, and P. Meystre, J. Opt. Soc. Am. **B3**, 906 (1986)

[4] P. Filipowicz, J. Javanainen, and P. Meystre, Phys. Rev. **A34**, 3077 (1986)

[5] J.J. Slosser and P. Meystre, Phys. Rev. A41,3867 (1990)

[6] M. Orszag, R. Ramírez, J.C. Retamal, L. Roa, Phys. Rev. **A45**, 6717(1992)

[7] J.J Slosser, P. Meystre and E. Wright, Opt. Lett. **15**, 233(1990)

[8] M. Orszag, J.C. Retamal, C. Saavedra, Phys. Rev. **A45**, 2118(1992)

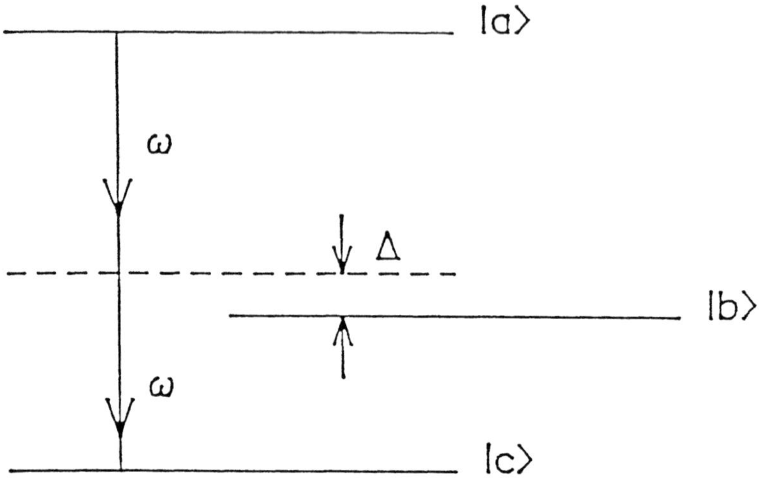

Figure 1.- Energy levels relevant to the two-photon micromaser with atomic polarization.

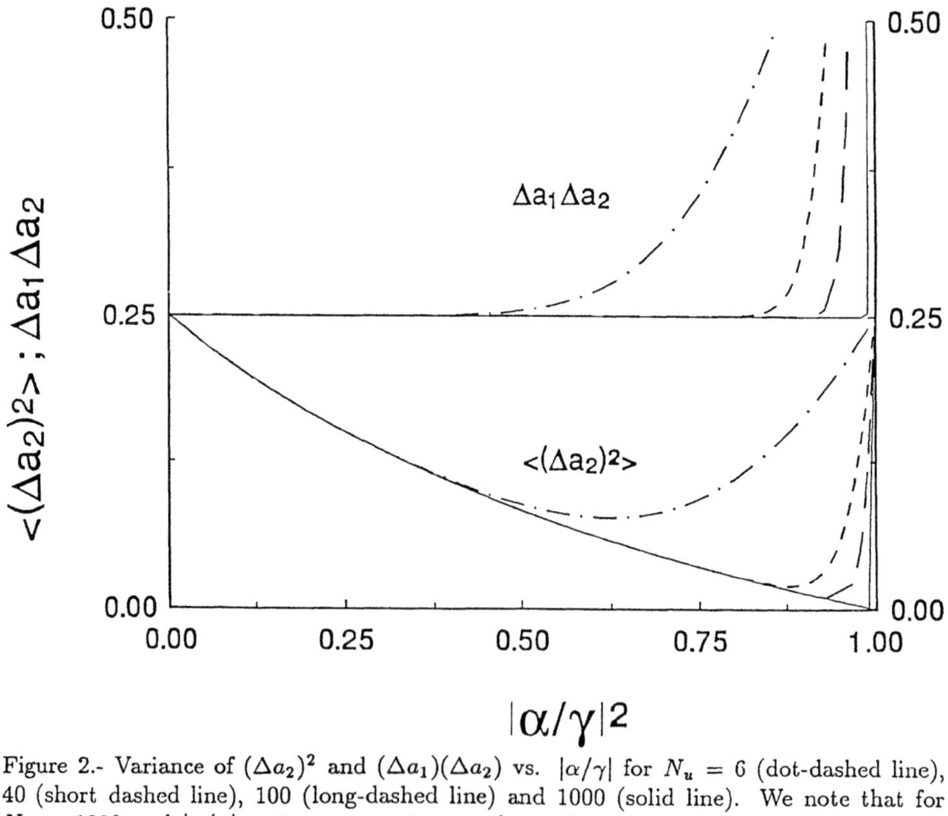

Figure 2.- Variance of $(\Delta a_2)^2$ and $(\Delta a_1)(\Delta a_2)$ vs. $|\alpha/\gamma|$ for $N_u = 6$ (dot-dashed line), 40 (short dashed line), 100 (long-dashed line) and 1000 (solid line). We note that for $N_u = 1000$ and $|\alpha/\gamma| = 1 - \epsilon$ we get a nearly perfect squeezed state, which is not a minimum uncertainty state.

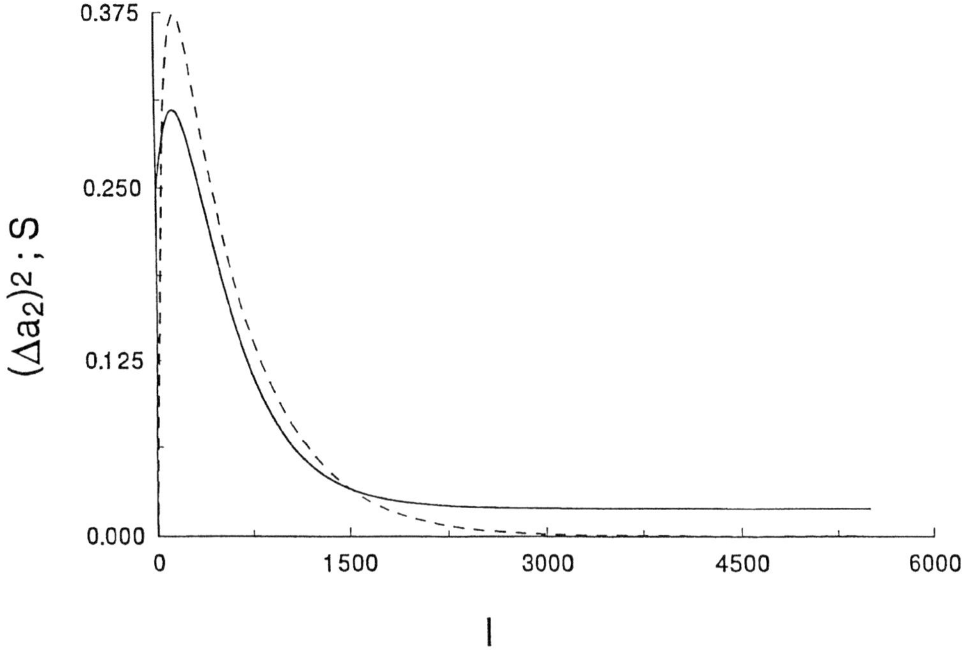

Figure 3.- $(\Delta a_2)^2$ (solid line) and S (dashed line) vs. the number of injected atoms l. For $l \geq 4500$, S vanishes and $(\Delta a_2)^2 = 0.0194$. The initial state of the field is the vacuum state with $N_d = 0$, $N_u = 40$ and $|\alpha/\gamma| = 0.88$.

SEQUENTIAL ITERATION FOR EXTREMAL AUTOMATA

Eric Goles and Gonzalo Hernández
Departamento de Ingeniería Matemática
Universidad de Chile
Casilla 170-3 Correo 3, Santiago, Chile
email: egoles@uchcecvm.bitnet

ABSTRACT. We study the sequential iteration of an automata class defined by local rules evolving only to local extreme values. Under the symmetry assumption of the cellular space we determine a Lyapunov operator driving the automaton dynamics. This operator allows us to characterize the steady state as a set of fixed points and to give bounds for the transient phase.

1. Introduction

Let $G = (V, E)$ be an undirected, connected, finite graph with $|V| = n$. To each vertex $i \in V$ we assing values $x_i \in Q = \{q_1, ..., q_n\}$. We define a local transition function in each vertex, $f_i : Q^{|V_i|} \to Q$, where V_i is the neighborhood of vertex i. We will say that f_i is an Extremal Rule (ER) iff: $f_i(x_j : j \in V_i) \in \{m_i, x_i, M_i\}$ and $x_i \notin]m_i, M_i[\Rightarrow f_i(x_j : j \in V_i) = x_i$, where $m_i = \min\{x_j, j \in V_i\}$ and $M_i = \max_j\{x_i : j \in V_i\}$. That is to say, vertex i, under function f_i, may change only to a local extreme value. Furthermore, if it is a local minimum it remains unchanged.

These rules where introduced in [KB, GK] as image restauration procedures. Later, the convergence to fixed points for the synchrous iteration (all the sites are updated at the same time) was establishes for FES rules; i.e. $x_i \in]w_i, M_i[\Rightarrow f_i(x_j : j \in V_i) \in \{m_i, M_i\}$, ($f_i(x_j : j \in V_i) = x_i$ otherwise). Clearly FES is a subclass of ER. Moreover, in [GK] a $0(n^2)$ bound for convergence to fixed points was established for the class, called NE rules; i.e.

$$f_i(x_j : j \in V_i) = \begin{cases} m_i & \text{if } x_i - m_i < M_i - x_i \text{ and } x_i \in]m_i, M_i[\\ M_i & \text{if } M_i - x_i > x_i - m_i \text{ and } x_i \in]m_i, M_i[\\ x_i & \text{otherwise} \end{cases}$$

In this paper we study the sequential iteration of ER rules, i.e. the vertices of G are updated one by one in a prescribed order. Without loss of generality one may assume that the iteration order is $\{1\}, \{2\}, ..., \{n\}$. For this model we introduce Lyapunov operators driving the sequential dynamics which allows us to prove, for the ER class we prove the convergence to fixed points and we give a general, rather bad, bound for the transient

E. Tirapegui and W. Zeller (eds.), Instabilities and Nonequilibrium Structures IV, 141–148.

time. Moreover, for some particular ER rules we exhibit polynomial bounds and also examples where the bound is reached.

2. Lyapunov Functionals

A Lyapunov functional is a local monotonous aditive quantity H depending of the configuration on graph G. Further information about these operators, in the framework of automata networks, can be seen in [Go,GM]. We have the following result:

Proposition 1. Let $G = (V, E)$, be an undirected, conected graph with $|V| = n$ and suppose a set of ER local functions. The quantity:

$$H(x) = -\frac{1}{2}\sum_{i=1}^{n}\sum_{j\in V_i} a^{|x_i-x_j|},$$

$a > d = \max|V_i|$, is a Lyapunov functional for the sequential iteration; i.e: if $x' = (x'_1, ..., x'_n)$, where $x'_i = x_j \quad \forall j \neq i,\ x'_i = f(x_j : j \in V_i)$, then $\Delta H = H(x') - H(x)$ ≤ 0 and the inequality is strict if and only if $x'_i \neq x_i$.

Proof. Suppose that vertex i is updated. One gets

$$\Delta H = -\frac{1}{2}\sum_{i=1}^{n}\sum_{j\in V_i} a^{|x'_i-x'_j|} + \frac{1}{2}\sum_{i=1}^{n}\sum_{j\in V_i} a^{|x_i-x_j|}$$

¿From definition on $x' = (x_1, ..., x_{i-1}, x'_i, x_{i+1}, ..., x_n)$ and symmetry of $|y_i - y_j|$:

$$\Delta H = -\sum_{j\in V_i} a^{|x'_i-x_j|} + \sum_{j\in V_i} a^{|x_i-x_j|}$$

Clearly, $x'_i = x_i \Rightarrow \Delta H = 0$. The case $x'_i \neq x_i \Rightarrow m_i < x_i < M_i$ and $x'_i \in \{m_i, M_i\}$. Furthermore,

$$|x_i - x_j| \leq M_i - m_i - 1 \quad \forall j \in V_i$$

Since $x'_i \in \{m_i, M_i\}$, in the first term of ΔH appears the expression $a^{|M_i-m_i|}$, hence:

$$\begin{aligned}\Delta H &\leq -a^{M_i-m_i} + da^{M_i-m_i-1}\\ &\leq a^{M_i-m_i}(\frac{d}{a} - 1) = (d-a)a^{M_i-m_i-1}\end{aligned}$$

since $d < a \Rightarrow \Delta H < 0$. ■

Corollary 1. The sequential iteration on the class ER converges to fixed points.

Proof. Suppose the system is in steady state, i.e. $x(0), x(1), ..., x(T-1)$ is a cycle with period $T \geq 1$ If there exists $t \in [0, T-1]$ such that $x(t) \neq x(t-1)$ then $H(x(t)) < H(x(t-1))$, hence $H(x(0)) < H(x(0))$, which implies $T = 1$. ■

Previous Lyapunov functional permits to obtain a bound of the transient time. Given a set of extremal rules $X = (f_i)_{i=1}^n$ we define the maximum transient time

$$\tau_X(G) = \max_{t \geq 0}\{t / x(t) \neq x(t-1)\}$$

Proposition 2. For any set X of Extreme Rules on a graph $G = (V, E)$.

$$\tau_X(G) \leq |E|(d+1)^{q-2}$$

Proof. Clearly, one may choice $a = d+1$, so $|\Delta H| \geq d+1$. On the other hand:

$$-a^{|q-1|}|E| \leq -\frac{1}{2}\sum_i \sum_{j \in V_i} a^{|q-1|} \leq H(x) \leq -\frac{1}{2}\sum_i d_i = -|E|,$$

so

$$-|E|(d+1)^{q-1} \leq H(x) \leq -|E|$$

hence

$$\tau_X(G) \leq \frac{|E|((d+1)^{q-1}-1)}{d+1} \leq |E|(d+1)^{q-2} \quad ■$$

Remark. Clearly the number of configurations is q^n. So, by considering $q = n$ (i.e. $Q = \{1, 2, ..., n\}$) one gets n^n admisible configurations for the system. On the other hand $d \leq n-1$ so our bound is $0(|E|n^{n-2}) = 0(n^n)$ which is not significativily better than the total number of configurations. It is not difficult to see that by coding states $\{1, ..., n\}$ in base two one may obtain a better bound $\tau_X(G) \sim 0(n|E|n^{\frac{Ln(q-1)}{Ln 2}}) \sim 0(n^{\frac{Ln(n)}{Ln 2}})$. In order two obtain more realistic bounds we shall study some particular rules in next paragraph.

3. Polinomial Transient Bounds

Let us first consider the Median Extremal Rule (MER), defined as follows:

$$f_i(x_j : j \in V_i) = \begin{cases} x_i & \text{if } x_i \notin]m_i, M_i[\\ M_i & \text{if } |\{j \in V_i / x_i \geq x_j\}| > |\{j \in V_i / x_i < x_j\}| \\ m_i & \text{if } |\{j \in V_i / x_i \geq x_j\}| < |\{j \in V_i / x_i < x_j\}| \\ x_i & \text{in tie cases.} \end{cases}$$

For instance,

2		5			2		4	
3	4	5	$\to 4$ (tie-case)		3	4	5	$\to 7$
1		7			1		7	

Proposition 3. $H(x) = -\frac{1}{2}\sum_{i=1}^{n}\sum_{j\in V_i}|x_i - x_j|$ is a Lyapunov functional for the sequential iteration of MER

Proof. Suppose first that we update the i-th vertex and $x_i' = M_i$, so

$$\begin{aligned}
\Delta H &= -\sum_{j\in V_i}|M_i - x_j| + \sum_{j\in V_i}|x_i - x_j| = \\
&= -d_iM_i + \sum_{j\in V_i} x_j + \sum_{x_i\geq x_j}(x_i - x_j) + \sum_{x_i<x_j}(x_j - x_i) \\
&= -d_iM_i + 2\sum_{x_i<x_j} x_j + x_i|\{j\in V_i/x_i\geq x_j\}| - x_i|\{j\in V_i/x_i<x_j\}|
\end{aligned}$$

Let us call $C(x_i \geq x_j) = |\{j \in V_i/x_i \geq x_j\}|$ and $C(x_i < x_j) = |\{j\in V_i/x_i < x_j\}|$, hence

$$\begin{aligned}
&= -(C(x_i\geq x_j) + C(x_i<x_j))M_i + 2\sum_{x_i<x_j} x_j + x_i(C(x_i\geq x_j) - C(x_i<x_j)) \\
&\leq -(C(x_i\geq x_j) + C(x_i<x_j))M_i + 2M_iC(x_i<x_j) + x_i(C(x_i\geq x_j) - C(x_i<x_j)) \\
&\leq M_i(C(x_i<x_j) - C(x_i\geq x_j)) + x_i(C(x_i\geq x_j) - C(x_i<x_j)) \\
&\leq (M_i - x_i)(C(x_i<x_j) - C(x_i\geq x_j))
\end{aligned}$$

since $x_i' = M_i \neq x_i, C(x_i < x_j) < C(x_i \geq x_j)$, so $\Delta H < 0$.

If $x_i' = m_i \neq x_i$: in a similar way one gets

$$\begin{aligned}
\Delta H &= -\sum_{j\in V_i} x_j + d_im_i + \sum_{x_i\geq x_j}(x_i - x_j) + \sum_{x_i<x_j}(x_j - x_i) \\
&= -2\sum_{x_i\geq x_j} x_j + (C(x_i<x_j) + C(x_i\geq x_j))m_i + x_i(C(x_i\geq x_j) - C(x_i<x_j)) \\
&\leq m_i(C(x_i<x_j) - C(x_i\geq x_j)) + x_i(C(x_i\geq x_j) - C(x_i<x_j)) \\
&\leq (x_i - m_i)C(x_i\geq x_j) - C(x_i<x_j)) < 0. \quad \blacksquare
\end{aligned}$$

Corollary 2. The sequential iteration on MER converges in $0(q|E|)$ steps to fixed points.

Minimun Rule (MIR): $\forall i = 1, ..., n$

$$x_i \notin]m_i, M_i[\Rightarrow f_i(x_j : j \in V_i) = x_i \quad \text{and} \quad x_i' = f_i(x_j : j \in V_i) \in \{x_i, m_i\}$$

Thus, in these rules all the nodes may only change to the local maximun (MAR) or, the local minimun (MIR). Clearly, these rules belong to ER.

Proposition 4. For the sequential update, the quantities

$$H_M(x) = -\sum_{i=1}^{n} x_i \quad \text{and} \quad H_m(x) = \sum_{i=1}^{n} x_i$$

are Lyapunov functional for MAR and MIR, respectively.

Proof. If vertex i is updated one gets: $\Delta H_M = -x_i' + x_i$ or $\Delta H_m = x_i' - x_i$, which implies directly the result. ■

Corollary 3. The sequential iteration of MAR or MIR converges in $0(q \cdot n)$ steps to fixed points.

Proof. We consider only the MAR case, the other one is analogous. Clearly, $|\Delta H| \geq 1$ and $-q \cdot n \leq H_M(x) \leq 0$, so $\tau_X(G) \leq qn$. ■

Remark. In this case we obtain $\tau_X(G) = 0(n^2)$. Moreover, consider the following automaton:

Take a chain $G = (\{1, ..., 7\}, V)$, where V are the nearest interactions edges and the node 7 is neighbor of all the rest. Let us consider the following set of MAR:

Nodes 1,2,3,7 evolve to the maximun of its neighbors if there exists a neighbor with odd state.

Nodes 4,5,6 evolve to the maximun of its neighbors if there exists a neighbor with even state.

By taking the initial configuration (8,6,2,3,5,7,0), one gets for the sequential iteration applied from left to right the dynamics exhibited in figure 1.

It is easy to see that the information takes a linear time to reach the node 6 from the node 2 or 3. In the opposite direction the behavior is analogous.

Clearly, by adding a new node between the nodes 6 and 7, with initial state 9, and the same rule of these nodes, the transient length increase in $O(n)$ steps.

Iterating this process, in both sides of the lattice, one obtains a $O(n^2)$ transient lenght.

Proof. Clearly $|\Delta H| \geq 1$ and

$$-(q-1)|E| \leq -\frac{1}{2}\sum_i (q-1)d_i \leq H(x) \leq 0$$

so, $\tau_X(G) \leq (q-1)|E|$. ■

Remark. Since $|E| \leq n^2$ and $q \leq n$ one gets $\tau_X(G) = 0(n^3)$.

Let us now consider the Majority Rule (MR) defined by:

$$f_i(x_j : j \in V_i) = \begin{cases} x_i & \text{if } x_i \notin]m_i, M_i[\\ M_i & \text{if } C(x_j = M_i) > C(x_j = x_i) \\ m_i & \text{if } C(x_j = m_i) > C(x_j = x_i) \\ x_i & \text{otherwise} \end{cases}$$

where, $C(x_j = a) = |\{j \in V_i / x_j = a\}|$

That is to say, site i changes if and only if there exists only one extreme more represented than itself.

Proposition 5. $H(x) = -\frac{1}{2}\sum_{i=1}^{n}\sum_{j \in V_i} \delta(x_i, x_j)$ is a Lyapunov functional for the sequential iteration on MR, where δ is the Kroenecker function, i.e.:

$$\delta(u,v) = \begin{cases} 1 & \text{if } u = v \\ 0 & \text{otherwise.} \end{cases}$$

Proof. Clearly $\Delta H = -C(x_j = x_i') + C(x_j = x_i)$, which implies the result. ■

Corollary 4. The sequential iteration on MR converges in $0(|E|)$ steps to fixed points.

Proof. Clearly $|\Delta H| \geq 1$ and $-|E| \leq -\frac{1}{2}\sum_i d_i \leq H(x) \leq 0$, so $\tau_X(G) \leq |E|$. ■

Remark. In this case we obtain $\tau_X(G) = 0(n^2)$.

A very simple class of Extremal Rules is established when all the nodes in the graph may only evolve to the maximun (or the minimum) that appears in its neighborhood. We associate to this rules Lyapunov functionals which gives a quadratic bound for the transient length. Furthermore, we exhibit an example which reaches the bound.

Consider the following rules:

Maximun Rule (MAR): $\forall i = 1, ..., n$

$$x_i \notin]m_i, M_i[\Rightarrow f_i(x_j : j \in V_i) = x_i \quad \text{and} \quad x_i' = f_i(x_j : j \in V_i) \in \{x_i, M_i\}$$

8	6	2	3	5	7	0
8	6	2	3	5	7	0
8	6	2	3	5	7	0
8	6	6	3	5	7	0
8	6	6	6	5	7	0
8	6	6	6	7	7	0
8	6	6	6	7	7	0
8	6	6	6	7	7	0
8	6	6	6	7	7	0
8	6	6	6	7	7	0
8	6	6	6	7	7	0
8	6	6	7	7	7	0
8	6	6	7	7	7	0
8	6	6	7	7	7	0
8	6	6	7	7	7	0
8	6	6	7	7	7	0
8	6	6	7	7	7	0
8	6	7	7	7	7	0
8	6	7	7	7	7	0
8	6	7	7	7	7	0
8	6	7	7	7	7	0
8	6	7	7	7	7	0
8	6	7	7	7	7	0
8	7	7	7	7	7	0
8	7	7	7	7	7	0
8	7	7	7	7	7	0
8	7	7	7	7	7	0
8	7	7	7	7	7	0
8	7	7	7	7	7	0
8	7	7	7	7	7	0
8	8	7	7	7	7	0
8	8	8	7	7	7	0
8	8	8	8	7	7	0
8	8	8	8	8	7	0
8	8	8	8	8	8	0

Figure 1. Quadratic transient in a finite chain for MAR

Conclusions

In this paper we studied a wide class of local rules, called Extremal Rules, which generalize FES [GK]. We demostrate the convergence only to fixed points, by associating a

Lyapunov functional to the sequential iteration. This approach also gives an exponential bound for the transient time of these rules.

Finally, in section 3, we present particular cases of ER, which has polynomial bounds for the transient time. Also me exhibit a case (MAR or MIR rules) where a quadratic bound is reached.

The problem to determine polynomial bounds for the wide class of FES seems hard. Numerical simulation allows us to conjecture the existence of a general polynomial bound.

Acknowledgments

Partially supported by FONDECYT 91-1211, DTI. and PG04192 U. de Chile and C-10003, Fundación Andes.

References

[GK] W. Goddard and D.J. Kleitman, Convergence of a Transformation on a Weighted Graph.

[GM] E. Goles and S. Martinez, Neural and Automata Networks, Kluwer Academic Publishers, 1990.

[Go] E. Goles, Lyapunov Functionals Associated to Automata Networks, in Automata Networks in Computer Science, F. Fogelman, Y. Robert, M. Tchuente (eds), Manchester University Press, 1987, 58-81.

[KB] H.P. Kramer and J.B. Bruckner, Iterations of a Nonlinear Transformation for Enhancement of Digital Images, Pattern Recognition 7 (1975), 53-58.

A stochastic approach towards the study of the one-dimensional Landau Ginzburg distribution

C. Van den Broeck [1]
U.C.S.D., La Jolla, CA 92093, U.S.A.
M. Malek Mansour
U.L.B. Campus Plaine, 1050 Bruxelles, Belgium

Abstract
We present a new approach that is inspired by the investigation of a stochastic model for reaction diffusion systems, and allows to obtain analytic results for the one-dimensional Landau Ginzburg potential. In the low noise limit, properties of the Landau Ginzburg system are linked to the solution of a simple functional equation. Some surprising, non-universal features of the model are derived.

1 Introduction

The Landau Ginzburg potential appears in many different physical problems, both in equilibrium and nonequilibrium context [2, 3]. Our original motivation was the study of nonequilibrium transitions in reaction diffusion systems [4]. In the following, we describe how a stochastic description of these systems led us to formulate a new analytic approach to investigate the properties of systems described by the Landau Ginzburg potential. Even though this analysis is restricted to the one-dimensional case, it is of interest since universality has been shown to break down in this case [5]. Also, the present approach is an alternative to the transfer matrix method in that it is specifically geared to the calculation of local properties (e.g. the local probability density), rather than global features (e.g. the partition function).

E. Tirapegui and W. Zeller (eds.), Instabilities and Nonequilibrium Structures IV, 149–159.

Consider first the following so called Schlogl reaction, taking place in a well stirred tank reactor [6]:

$$A \rightleftarrows X \;\; ; \;\; 2X + B \rightleftarrows 3X \tag{1}$$

The concentrations of A and B are kept constant. By a suitable parametrization and scaling of the variables, the concentration of species X, represented by the intensive variable x, obeys a kinetic equation of the following form:

$$\partial_t x = -U'(x) \tag{2}$$

with

$$U(x) = \lambda x^2/2 + \mu x^4/4 \tag{3}$$

where λ and μ are control variables, with $\mu > 0$. The interest of this chemical model is that the above equations have a generic form for all systems undergoing a so-called cusp bifurcation: for $\lambda > 0$, the potential U possesses a unique, globally stable minimum at $x^{st} = 0$, whereas for $\lambda < 0$, this state looses its stability in favor of two new, locally stable states located at $x_1^{st} = \sqrt{-\lambda/\mu}$ and $x_2^{st} = -\sqrt{-\lambda/\mu}$.
In order to take into account the intrinsic fluctuations, associated to the discreteness of the chemical process, one usually introduces a Gaussian white random "force" f:

$$\partial_t x = -U'(x) + f(t) \tag{4}$$

with

$$< f(t) >= 0 \quad < f(t)f(t') >= \epsilon\delta(t-t') \tag{5}$$

where ϵ represents the strength of the (thermal) noise. The validity of this phenomenological approach is open to discussion [4, 7], but we take it for granted to make the connection with properties of the Landau Ginzburg potential. As a consequence of the introduction of the white random force f(t), x(t) becomes a Markovian stochastic process. Its probability density $P(x,t)$ obeys the following Fokker Planck equation [7]:

$$\partial_t P(x,t) = \partial_x[U'(x)P(x,t)] + \frac{\epsilon}{2}\partial_x^2 P(x,t) \tag{6}$$

The steady state solution is given by the following Boltmann type of distribution:

$$P^{st}(x) \sim exp[-\frac{2}{\epsilon}U(x)] \tag{7}$$

We note that for $\lambda > 0$, and in the limit $\epsilon \longrightarrow 0$, only fluctuations of the order $x \sim \epsilon^{1/2}$ will arise with finite probability. In this limit, the nonlinear term in the potential is negligible, and the probability density is essentially Gaussian. On the other hand, one observes an amplification of the fluctuations at the bifurcation point, $\lambda = 0$, of the type $x \sim \epsilon^{1/4}$, whereas the probability density is no longer Gaussian. Finally, for $\lambda < 0$, the probability density possesses two peaks, located at the deterministic states x_1^{st} and x_2^{st}, around each of which Gaussian fluctuations of the order $\delta x \sim \epsilon^{1/2}$ will be observed.
The problem that we address here is how this picture is modified when a linear chain of such systems are coupled together by nearest neighbor harmonic interactions. In the chemical context, such a constraint is realized by a diffusive coupling of the chemical reactors. The time evolution of the system is now described in terms of the concentrations x_i in the various cells $i, i = -\infty, +\infty$:

$$\partial_t x_i = -U'(x_i) + \frac{D}{2}(x_{i+1} + x_{i-1} - 2x_i) + f_i(t) \tag{8}$$

with

$$< f_i(t) >= 0 < f_i(t) f_j(t') >= \epsilon \delta_{i,j} \delta(t - t') \tag{9}$$

where D represents the hopping rate between the cells. Note that we have not included the noise term that is associated to this dissipative transport process, even though the fluctuation dissipation theorem stipulates that there has to be such a term. In a real chemical system, this noise term is present; it is required to obtain the correct equilibrium distribution function. However, it turns out [4] that the noise associated to the hopping process can be neglected in the close vicinity of the critical point $\lambda = 0$, which is the situation that is of major interest to us here. Furthermore, the connection with the Landau Ginzburg potential arises precisely in the absence of this term, as we now proceed to show. The multi-variate Fokker Planck equation, equivalent to the set of Eqs. (8)-(9), reads:

$$\partial_t P(\{x_i\}, t) = \sum_{k=-\infty}^{+\infty} \partial_{x_k} \{[\partial_{x_k} U(x_k) - \frac{D}{2}(x_{k+1} + x_{k-1} - 2x_k)] P(\{x_i\}, t)\} + \frac{\epsilon}{2} \partial_{x_k}^2 P(\{x_i\}, t) \tag{10}$$

By inspection, one verifies that the steady state probablitity has the following form:

$$P^{st}(\{x_i\}) \sim exp\{-\frac{2}{\epsilon}\sum_{k=-\infty}^{+\infty}[U(x_k)+\frac{D}{4}(x_k-x_{k+1})^2]\} \tag{11}$$

i.e. it features the discretized form of the one-dimensional Landau Ginzburg probability density!
The traditional approach in statistical mechanics is to perform the integral of the r.h.s. over the variables x_i (using e.g. the transfer matrix approach) to obtain the partition function Z. Our approach will differ in two respects. First, we will use the set of equations determining $P^{st}(\{x_i\})$, rather than the explicit form of $P^{st}(\{x_i\})$ itself. Secondly, we focus on the calculation of local properties of the system, such as the reduced one-cell probability density, $P_1^{st}(x_k)$, which is a relevant object in the context of the above described reaction diffusion system. The main steps of the calculation are given in section 2, while a discussion of the results is presented in sections 3 and 4.

2 Calculation of reduced probability densities

We first note that the stationary probability density $P^{st}(\{x_i\})$ obeys the following set of partial differential equations:

$$\frac{\epsilon}{2}\frac{\partial P^{st}(\{x_i\})}{\partial x_k} = [-\frac{\partial U(x_k)}{\partial x_k}+\frac{D}{2}(x_{k+1}+x_{k-1}-2x_k)]P^{st}(\{x_i\}) \ \forall k \tag{12}$$

This can be seen by inspection, or by noting that $P^{st}(\{x_i\})$ is the stationary solution of the Fokker Planck equation (10).

Furthermore, the stationary probability density $P^{st}(\{x_i\})$ possesses the property of spatial Markovianity. More precisely, the conditional probabilities P_C depend only on the value of x that is specified in the nearest cell, e.g.:

$$P_C(x_k|x_m, x_{m+1}, ...) = P_C(x_k|x_m) \ \forall m > k \tag{13}$$

This property also implies that P_C obeys the following Chapman Kolmogorov equation:

$$P_C(x_i|x_l) = \int dx_k P_C(x_i|x_k)P_C(x_k|x_l) \ \forall i < k < l \tag{14}$$

Since any conditional probability can thus be written as a superposition of two-cell conditional probabilities we conclude that all the information on the statistical properties are contained in the one and two-cell probabilities $P_1(x_k)$ and $P_2(x_k, x_{k+1})$. Furthermore, these densities can be obtained from the solution of the set of equations (12). For example, by integrating Eq.(12) over all variables except x_k and x_{k+1}, and doing the same for the corresponding equation obtained by replacing x_k by x_{k+1}, one obtains two partial differential equations for $P(x_k, x_{k+1})$, which are easily solved. One finds:

$$\begin{aligned} P_2(x_k, x_{k+1}) &= Z_2^{-1} exp\{\frac{2}{\epsilon}[-U(x_k) - U(x_{k+1}) - \frac{D}{4}(x_{k+1} - x_k)^2 \\ &+\frac{D}{2}\int_{x_k} dx'_k[E(x'_k) - x'_k] + \frac{D}{2}\int_{x_{k+1}} dx'_{k+1}[E(x'_{k+1}) - x'_{k+1}]\} \end{aligned} \qquad (15)$$

Z_2 is a normalization constant. We have also introduced the conditional average $E(x)$, which will play a central role in our approach. It is defined by:

$$E(x) =< x_k \mid x_{k+1} = x >= \int dx_k \, x_k \, P_C(x_k \mid x_{k+1} = x) \qquad (16)$$

We note in passing that this average does not depend on the cell under consideration or on the choice of the neighbor, since the statistical properties of the system possess translational and left right invariance.
By a procedure, similar to that leading to Eq.(15), one finds that:

$$P_1(x_k) = Z_1^{-1} exp\{\frac{2}{\epsilon}(-U(x_k) + D\int_{x_k} dx'_k[E(x'_k) - x'_k])\} \qquad (17)$$

where Z_1 is a normalization constant.

Note that we have made good on our promise to calculate the local properties: the previous equation features the effective local potential being the sum of the bare potential U plus a term proportional to the coupling coefficient D. The evaluation of the local statistical properties has thus been reduced to the calculation of the conditional average $E(x)$. The latter can be determined by the solution of the integro-differential equation that is obtained by inserting the results of Eqs.(15) and (17) in the definition of $E(x)$, cf. Eq. (16). In this way, we have obtained, at least in principle, an exact solution of the problem.

In general, the self-consistent equation determining $E(x)$ is extremely complicated. A major simplification however takes place when considering the low-noise limit $\epsilon \to 0$. By applying steepest descent to Eq.(16), we find that:

$$E(x) = x^* \tag{18}$$

where x^* is the (supposedly unique) value of x_k that maximizes the expression proportional to ϵ^{-1} in the exponent of P_C, i.e. it is a solution of the following implicit equation:

$$-U'(x^*) + \frac{D}{2}(x - x^*) + \frac{D}{2}[E(x^*) - x^*] = 0 \tag{19}$$

By combining Eqs.(18) and (19), the following simple functional equation for $E(x)$ results:

$$E(x) + E^{-1}(x) = 2x + \frac{2}{D}U'(x) \tag{20}$$

where E^{-1} stand for the inverse function. Eq.(20) is one of our central results. Together with Eqs.(15) and (17) and spatial Markovianity, it completely determines the statistical properties of the system.

3 Application to the Landau Ginzburg potential

For the simple case of a harmonic potential $U(x) = \lambda x^2/2$, Eq.(20) can be solved exactly:

$$E(x) = \alpha x \tag{21}$$

with:

$$\alpha = \frac{\lambda}{D} + 1 - \sqrt{(\frac{\lambda}{D} + 1)^2 - 1} \tag{22}$$

The one-cell probability density takes on the following Gaussian form:

$$P_1(x) \sim exp\{-\frac{1}{\epsilon}[\lambda + D(\alpha - 1)]x^2\}. \tag{23}$$

with fluctuations of the order $x \sim \epsilon^{1/2}$. These results are actually valid for all values of ϵ, and can be derived directly from the density $P^{st}(\{x_i\})$ by invoking the well-known properties of a multi-Gaussian density.

Let us now focus on the "standard" Landau Ginzburg potential $U(x) = \frac{\lambda}{2}x^2 + \frac{\mu}{4}x^4$. When λ switches from a positive to a negative value, the local potential undergoes a cusp bifurcation, with the loss of stability of the reference state $x = 0$, and the arisal of two new stable states at $x = \pm\sqrt{-\frac{\lambda}{\mu}}$. The question is how this local instability is affected by the coupling of each cell to its neigbors. Is there a macroscopic symmetry breaking with the divergence of the correlation length? What is the form of the effective local potential? To answer these questions, we first note that, in view of the $x \rightarrow -x$ symmetry of the system, one has that $E(x) = -E(-x)$, i.e. $E(x)$ is an uneven function of x. Furthermore, although it seems to be impossible to find the exact analytic solution to Eq.(20), we remember that this result was derived by steepest descent in the low noise limit $\epsilon \rightarrow 0$, so that it suffices to find a solution in this same limit. We will restrict ourselves here to non-negative values $\lambda \geq 0$. In this case, the fluctuation of x around x=0 are small, i.e. of order ϵ^a, where $a > 0$ has yet to be determined. Using this property, it is clear that for $\lambda > 0$, the linear term in the r.h.s. of Eq.(20) is dominant, and the Gaussian result prevails, cf. Eqs.(21)-(23). The fluctuations are thus of the order $x \sim \epsilon^{1/2}$ ($a = 1/2$). It is revealing to evaluate the correction to this result by calculating the coefficients of the Taylor expansion of $E(x)$ around x=0. One finds:

$$E(x) = \alpha x + \frac{\mu}{D}[1 - \sqrt{\frac{D}{8\lambda}}]x^3 + O(\frac{x^5}{\lambda^{3/2}}) \tag{24}$$

It is clear from Eq.(24) that all the coefficients of the uneven terms in x, starting with x^3, diverge when $\lambda \rightarrow 0$, indicating that E(x) is no longer analytic in x at the critical point $\lambda = 0$. By a resummation of the most divergent terms, or by applying a singular perturbation calculation to Eq.(20), the following result is found for $E(x)$ in the vicinity of the critical point $\lambda = 0$:

$$E(x) = x - \frac{x}{\sqrt{D}}\sqrt{2\lambda + \mu x^2} + \frac{1}{D}(\lambda x + \mu x^3) + ... \tag{25}$$

In particular, at the critical point $\lambda = 0$, one finds:

$$E(x) = x - \sqrt{\frac{\mu}{D}}x|x| + ... \tag{26}$$

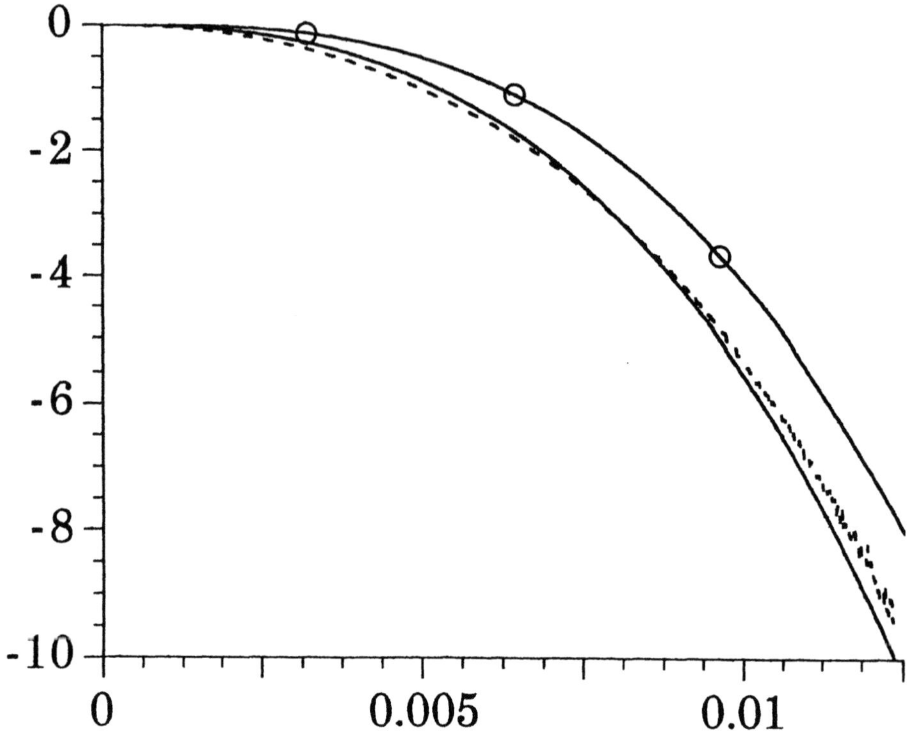

Figure 1: $ln[P_1(x)/P_1(0)]$ versus x for $\epsilon = 10^{-7}$, $D = 0.1$, $\mu = 3$ at the critical point $\lambda = 0$. The dashed line represents the results of the simulations; the solid line, marked by circles, corresponds to the cubic density, cf. Eq.(27), whereas the full line includes a correction term of the form of $x^2\epsilon^{-1/2}$.

Inserting this result into the one-cell probability density yields:

$$P_1(x) \sim exp(-\frac{2}{3\epsilon}\sqrt{D\mu}|x|^3) \tag{27}$$

The local fluctuations thus scale as $x \sim \epsilon^{1/3}$. We have performed extensive numerical simulations to verify these remarkable results. It turns out that one has to include first order corrections, which are of order $\epsilon^{1/6}$ [8], to obtain agreement between theory and simulations, cf. Figure 1.

4 The spatial correlation function

The amplification of the local fluctuations and the non-analyticity of $E(x)$ raises the question as to the existence of a phase transition at $\lambda = 0$ in the limit $\epsilon \to 0$ (which may be of the equilibrium or nonequilibrium type, depending on the physical content of the problem under consideration). To investigate this aspect in more detail, we turn to the study of the spatial correlation function. This quantity can again be linked to properties of the function $E(x)$. To see this, we note that, in the steepest descent approximation, the following result holds:

$$\begin{aligned} < f(x_k) \mid x_{k+1} = x > &= \int dx_k \, f(x_k) \, P_C(x_k \mid x_{k+1} = x) \qquad (28) \\ &= f(x*) \\ &= f(E(x)) \end{aligned}$$

On the other hand, one finds, using the Chapman Kolmogorov relation:

$$\begin{aligned} < x_k \mid x_{k+2} = x > &= \int dx_{k+1} \int dx_k \, x_k \, P_C(x_k \mid x_{k+1}) \, P_C(x_{k+1} \mid x_{k+2}) \\ &= < E(x_{k+1}) \mid x_{k+2} = x > \\ &= E(E(x)) \qquad (29) \end{aligned}$$

where, in the last step, we have applied Eq.(28) for $f = E$. By repeating a similar argument, one proves that

$$< x_k \mid x_{k+r} = x >= E(E(...E(x)...)) \qquad (30)$$

is the r-fold iterate of the function E, evaluated at x. Note that the spatial correlation function $< x_k \, x_{k+r} >$ is obtained from the above given conditional average, by multiplication of this result with $xP_1(x)$, followed by an integration over x.

It is helpfull to rephrase Eq.(30) in terms of the properties of the one-dimensional map $(x, E(x))$. Whenever this map has a unique stable fixed point at $x = 0$, the approach to this fixed point upon iteration of the map describes the way in which the conditional average decays to zero with the distance r. For example, in the linear case, cf. Eqs.(21) and (22), the approach of the fixes point $x = 0$ is exponentially fast. As a result, the conditional average, and hence the correlation function are also decaying in an exponential fashion, $< x_k \mid x_{k+r} = x >= \alpha^r\, x$. Note the classical divergence of the correlation length $l_c = -1/ln\alpha$ as $l_c \sim \lambda^{-1/2}$ for $\lambda \to 0$. On the other hand, at the critical point $\lambda = 0$, the function $E(x)$ is tangent to the first bisector, cf. Eq.(26), and the convergence to the fixed point obeys an inverse power law:

$$< x_k \mid x_{k+r} = x >= E(E(...E(x)...)) \approx \frac{1}{\sqrt{\frac{\mu}{D}}\, r + \frac{1}{x}} \tag{31}$$

implying that the spatial correlations decay as $1/r$.

5 Conclusion

In summary we found that, in the limit $\epsilon \to 0$, the "critical point" $\lambda = 0$ is characterized by a nonanalyticity of the local probability densities, by an amplification of the local fluctuations $x \sim \epsilon^{1/3}$, and by the arisal of long-range spatial correlations, decaying as $1/distance$. The latter behavior corresponds to a marginal situation as far as phase transitions are concerned, since the integral of the correlation function has a weak, logarithmic divergence.
We have obtained these results using a new calculation procedure, which links the local statistical properties directly to the solution of the functional equation for the conditional average E(x), cf. Eq. (20). We expect that the technique presented here could be applied to other types of lattices (e.g. the Bethe lattice), other types of potentials, as well as to continuum problems.

References

[1] Permanent address: L.U.C., B-3590 Diepenbeek, Belgium.

[2] see e.g. S.K. Ma, **Modern theory of critical phenomena**, Benjamin, New-York (1976).

[3] D.J. Scalapino, M. Sears and R.A. Ferrell, Phys. Rev **B6**, 3409 (1972).

[4] M. Malek Mansour, C. Van den Broeck, G. Nicolis and J.W. Turner, Ann. Phys. **131**, 283 (1981).

[5] G.A.Baker, Phys. Rev. Lett. **60**, 1844 (1988); J. Stat. Phys. **55**, 505 (1989).

[6] F. Schlogl, Z. Phys. **253**, 147 (1972).

[7] N. G. Van Kampen, **Stochastic Processes in Physics and Chemistry**, North-Holland, Amsterdam (1983).

[8] M. Malek Mansour and C. Van den Broeck, A new analytic approach for the study of the one-dimensional Landau Ginzburg probability distribution, preprint.

EFFECTIVE POTENTIAL FOR STOCHASTIC PROCESSES

H. CALISTO
Facultad de Física
Pontificia Universidad Católica de Chile
Casilla 306, Santiago 22, Chile

E. CERDA and E. TIRAPEGUI
Facultad de Ciencias Físicas y Matemáticas
Universidad de Chile
Casilla 487-3, Santiago, Chile

ABSTRACT

The notion and possible uses of the effective potential are considered for Markov processes associated to nonlinear systems in the presence of noise. We calculate this function up to first order in a parameter measuring the intensity of the noise with a method which can be generalized to more complicated cases. Some interpretations of the potential are discussed.

An important problem in nonlinear systems is the search of robust effects of noise. An example is the phenomena of noise induced transitions[1] which up to now has been defined by qualitative changes in the stationary probability of the stochastic process describing the system. This criteria has given interesting results but it is limited by the knowledge of the stationary probability for which no systematic approximation is known in the general case. Due to this we propose here the use of the effective potential to study these effects. This function has the property of being a minimum at the mean value of the process and one can always calculate it in perturbation theory using the intensity of the noise as an expansion parameter. The weak noise behavior should then be exhibited by the potential. Although in the case of the finite number of state variables the mean value is unique the existence of metastable states with large scape times induces local minimas in the approximate potential as we have checked in several calculations we present here. Let us start by introducing the essential concepts.

We consider a stochastic process $q(t) \in \mathrm{R}$, $t \in [t_0, T]$. Let the functional $\omega[q(\cdot)]$ be the probability of the realization (trajectory) $q(\cdot)$. We define through a functional integral the generating functional

$$Z[J(\cdot)] = \int \mathcal{D}q\, \omega[q(\cdot)] \exp\left[\frac{1}{\eta}\int_{t_0}^{T} dt J(t) q(t)\right] \tag{1}$$

We introduce the standard discretization we shall use: $t_j = t_0 + j\varepsilon$; $j = 0, \ldots, N+1$; $t_{N+1} = T$; $q(t_j) = q_j$; $J(t_j) = J_j$. Then $\omega[q(\cdot)]$ becomes a function $\omega(q_0, \ldots, q_{N+1})$

E. Tirapegui and W. Zeller (eds.), Instabilities and Nonequilibrium Structures IV, 161–173.

and the discretized version defining (1) will be

$$Z[J(\cdot)] = \int \prod_{i=0}^{N+1} dq_i \, \omega(q_0, \ldots, q_{N+1}) \exp\left[\frac{\varepsilon}{\eta} \sum_{\ell=0}^{N+1} J_\ell q_\ell\right] \tag{2}$$

The parameter η measures the intensity of the noise and will appear in a natural way in all the cases we study here. One has $Z[J(\cdot) = 0] = 1$ and the expansion

$$Z[J(\cdot)] = \sum_{n \geq 0} \frac{\eta^n}{n!} \int_{t_0}^{T} d\tau_1 \ldots d\tau_n \, \langle q(\tau_1) \cdots q(\tau_n)\rangle \, J(\tau_1) \ldots J(\tau_n) \tag{3}$$

where $\langle q(\tau_1) \cdots q(\tau_n)\rangle$ are correlations functions of the stochastic process $q(t)$. The functional $W[J(\cdot)] = \eta \ln Z[J(\cdot)]$, $W[J(\cdot) = 0] = 0$ generates the cumulants $\langle\langle q(\tau_1) \cdots q(\tau_n)\rangle\rangle$ and one has

$$W[J(\cdot)] = \sum_{n \geq 1} \frac{\eta^{n-1}}{n!} \int_{t_0}^{T} d\tau_1 \ldots d\tau_n \, \langle\langle q(\tau_1) \cdots q(\tau_n)\rangle\rangle \, J(\tau_1) \ldots J(\tau_n) \tag{4}$$

One defines next the Legendre transform $\Gamma[Q(\cdot)]$ of $W[J(\cdot)]$ by[2]

$$\Gamma[Q(\cdot)] + W[J(\cdot)] = \int_{t_0}^{T} dt \, J(t) Q(t) \,, \quad \frac{\delta W[J(\cdot)]}{\delta J(t)} = Q(t) \tag{5}$$

In this formula the second relation is used to express $J(\cdot)$ as a functional $J[Q(\cdot)]$ to define $\Gamma[Q(\cdot)]$. For a given value of $J(t)$ we have that $\delta W[J(\cdot)] / \delta J(t) = \langle q(t)\rangle^J$ is the mean value of the random variable $q(t)$ with the probability

$$\omega^J[q(\cdot)] = \frac{1}{Z[J(\cdot)]} \omega[q(\cdot)] \exp\left[\frac{1}{\eta} \int_{t_0}^{T} dt \, J(t) q(t)\right] , \quad \int \mathcal{D}q \, \omega^J[q(\cdot)] = 1 \tag{6}$$

and for $J(t) = 0$ we obtain the mean value $\langle q(t)\rangle$ of the original stochastic process with probability $\omega[q(\cdot)]$ for the trajectories. From (5) we have the dual relations

$$\frac{\delta W[J(\cdot)]}{\delta J(t)} = Q(t) \,, \quad \frac{\delta \Gamma[Q(\cdot)]}{\delta Q(t)} = J(t) \tag{7}$$

Since $W[J(\cdot) = 0] = 0$ and $\delta W[J(\cdot)] / \delta J(t)|_{J(t)=0} = \langle q(t)\rangle$ one has the fundamental properties

$$\Gamma[Q(t) = \langle q(t)\rangle] = 0 \,, \quad \left.\frac{\delta \Gamma[Q(\cdot)]}{\delta Q(t)}\right|_{Q(t) = \langle q(t)\rangle} = 0 \tag{8}$$

$$\Gamma[Q(\cdot)] = \sum_{n\geq 2} \frac{1}{n!} \int_{t_0}^{T} d\tau_1 \dots d\tau_n \left. \frac{\delta^n \Gamma[Q(\cdot)]}{\delta Q(\tau_1)\dots\delta Q(\tau_n)} \right|_0 \prod_{i=1}^{n} (Q(\tau_i)) - \langle q(\tau_i)\rangle) \tag{9}$$

Here $|_0$ stands for evaluation at $Q(t)) = \langle q(t)\rangle$. We put

$$\left. \frac{\delta^n \Gamma[Q(\cdot)]}{\delta Q(\tau_1)\dots\delta Q(\tau_n)} \right|_0 = \Gamma^{(n)}(\tau_1, \dots, \tau_n) \tag{10}$$

For a Markov process generated without an explicit time dependence with conditional probability density $P(q', t' \mid q'', t'') = \tilde{P}(q' \mid q''; t' - t'')$ we have (see[3] as a general reference for functional integration and our treatment of Markov processes) for the discretized version of $\omega[q(\cdot)]$

$$\omega[q(\cdot)] = P(q_{N+1}, T \mid q_N, t_N) \cdots P(q_1, t_1 \mid q_0, t_0) p_0(q_0) \tag{11}$$

where $p_0(q)$ is a probability density, $\int dq\, p_0(q) = 1$, representing the initial condition at time t_0. If one wants to consider the Markov process in the stationary state there are two alternatives: a) to replace $p_0(q)$ by the stationary probability $p^{\text{st}}(q) = \lim \tilde{P}(q \mid q'; \tau)$, $\tau \to \infty$; or b) to take the limit $t_0 \to -\infty$ with $p_0(q)$ arbitrary. This second method is the most useful since in general $p^{\text{st}}(q)$ is not known. Let $\omega^{\text{st}}[q(\cdot)] = P(q_{N+1}, T \mid q_N, t_N) \cdots P(q_1, t_1 \mid q_0, t_0) p^{\text{st}}(q_0)$ be the discretized version of the probability of trajectories in the stationary state, and $Z^{\text{st}}[J(\cdot)]$, $W^{\text{st}}[J(\cdot)]$, $\Gamma^{\text{st}}[Q(\cdot)]$ the corresponding quantities defined by (2), (4) and (5). One can show that the functional $\Gamma^{\text{st}}[Q(\cdot)]$ has the form[4]

$$\Gamma^{\text{st}}[Q(\cdot)] = \int_{t_0}^{T} dt \left[V(Q) + \dot{Q} B_1(Q) + \ddot{Q} B_2(Q) + \dot{Q}^2 B_3(Q) + \cdots \right] \tag{12}$$

where the terms in $(\cdots)$ contain higher derivatives of $Q(t)$. The function $V(q)$ appearing in (12) is the effective potential. When $Q(t) = q$ =constant formula (12) gives

$$\Gamma^{\text{st}}[Q(\cdot) = q] = (T - t_0) V(q) \tag{13}$$

In the stationary state $\langle q(t)\rangle = \bar{q}$ is time independent and from (8) and (12) we conclude

$$V(\bar{q}) = 0, \qquad \left. \frac{dV(q)}{dq} \right|_{q=\bar{q}} = 0 \tag{14}$$

For Markov processes we have explicit expressions of $Z[J(\cdot)]$ in terms of functional integrals for diffusion processes and for jump processes described by a master equation of a canonical form[5,6,7]. In this last case the parameter $\eta = \Omega^{-1}$, where Ω is the volume of the system, and if X is an extensive variable we consider $x = X/\Omega$ as continuous in the limit of big volume $\eta << 1$. We have

$$Z[J(\cdot)] = \int_{\gamma(0)} \mathcal{D}q \mathcal{D}^{\eta} p \exp\left[\frac{i}{\eta} \int_{t_0}^{T} dt \, [p\dot{q} - H(p, q; \eta) - iJq] \right] \cdot p_0(q(t_0)) \tag{15}$$

where

$$H(p,q;\eta) = H_0(p,q) + \eta H_1(p,q) + \eta^2 H_2(p,q) + \cdots \tag{16}$$

In (15) the function $H(p,q;\eta)$ is determined by the master (or Fokker-Planck) equation, $\gamma(0)$ stands for prepoint discretization which means that $H(p,q;\eta)$ is discretized as $H(p_j, q_{j-1};\eta)$, and $p_0(q(t_0))$ is the initial condition which we shall take deterministic as $\delta(q(t_0) - \alpha_0)$. The discretized version of (16) is (after integrating dq_0)

$$Z[J(\cdot)] = \int_{q_0=\alpha_0} \prod_{i=1}^{N+1} dq_i \prod_{j=1}^{N+1} \frac{dp_j}{2\pi\eta} \exp \frac{i\varepsilon}{\eta} \sum_{j=1}^{N+1} \left[p_j \frac{\Delta q_j}{\varepsilon} - H(p_j, q_{j-1};\eta) - iJ_j q_j \right] \tag{17}$$

with $\Delta q_j = q_j - q_{j-1}$. Let us consider de diffusion process defined by the stochastic differential equation

$$\dot{q} = A(q(t)) + \sqrt{\eta}\,\sigma(q(t))\xi(t) \tag{18}$$

where $\xi(t)$ is a gaussian white noise with zero mean and correlation $\langle \xi(t)\xi(t')\rangle = c\delta(t-t')$. The discretized version of (18) is

$$\Delta q_j = \varepsilon A(q_{j-1}) + \sqrt{\eta}\,\sigma(q_{j-1} + s\Delta q_j)\Delta w_j \tag{19}$$

where $\Delta w_j = w_j - w_{j-1}$ is the increment of the Brownian motion with correlation $\langle \Delta w_j \Delta w_k \rangle = c\varepsilon\delta_{jk}$ and $s = 0$ (s=1/2) correspond to the Ito interpretation (Stratonovich interpretation). Then (see(16))[3]

$$H_0(p,q) = pA(q) - \frac{i}{2}g(q)p^2\,, \quad H_1(p,q) = \frac{s}{2}pg'(q) \tag{20}$$

with $g(q) = c\sigma(q)^2$, $H_j = 0$, $j \geq 2$. In the case of the master equation for one-step processes one has[5,6]

$$H(p,q;\eta) = i\left[\left(e^{-ip} - 1\right)a(q) + \left(e^{ip} - 1\right)b(q)\right] \tag{21a}$$

$$a(q) = a_0(q) + \eta a_1(q) + \cdots\,, \quad b(q) = b_0(q) + \eta b_1(q) + \cdots \tag{21b}$$

We calculate now $W[J(\cdot)]$ from (15-17) using the method of steepest descent[8]. We do the change of variables

$$p(t) = v(t) + \sqrt{\eta}\,P(t)\,, \quad q(t) = u(t) + \sqrt{\eta}\,Q(t) \tag{22}$$

and choose $(v(t), u(t))$ as solution of

$$\dot{u}(t) = \frac{\partial H_0}{\partial v}\,, \quad \dot{v}(t) = -\frac{\partial H_0}{\partial u} - iJ\,, \quad u(t_0) = \alpha_0\,, \quad v(T) = 0 \tag{23}$$

Keeping terms only up to order η in $W[J(\cdot)]$ we obtain

$$Z[J(\cdot)] = \exp\left[\frac{i}{\eta}\int dt\,[v\dot{u} - H_0(v,u) - iJu] - \int dt H_1(v,u)\right]\cdot G[u(\cdot)] \qquad (24a)$$

$$\begin{aligned} G[u(\cdot)] = &\int_{\gamma(0)} \mathcal{D}Q\mathcal{D}P \\ &\times \exp i\int_{t_0}^{T} dt\left[P\dot{Q} - \frac{1}{2}\frac{\partial^2 H_0}{\partial p^2}P^2\Big|_0 - \frac{\partial^2 H_0}{\partial p\partial q}PQ\Big|_0 - \frac{1}{2}\frac{\partial^2 H_0}{\partial q^2}Q^2\Big|_0\right] \\ &\times \delta(Q(t_0)) \end{aligned} \qquad (24b)$$

Here $G[u(\cdot)]$ is independent of η and the terms $O(\eta^{-1/2})$ in the argument of the exponential in $Z[J(\cdot)]$ vanish due to (23). From the first equation in (23) one obtains $v(t)$ as function of $(u,\dot{u})$. The notation $F(p,q)|_0$ means that one replaces $p = v$, $q = u$, and since v is function of $(u,\dot{u})$ one finally has that the coefficients of the quadratic form in the integrand in (24b) are functions of $(u,\dot{u})$ and then the functional integral is a functional $G[u(\cdot)]$ of $u(\cdot)$. We see now that $W[J(\cdot)] = W_0[J(\cdot)] + \eta W_1[J(\cdot)] + O(\eta^2)$ with

$$W_0[J(\cdot)] = i\int_{t_0}^{T} dt\,[v\dot{u} - H_0(v,u)] + \int_{t_0}^{T} dt J(t)u(t) \qquad (25a)$$

$$W_1[J(\cdot)] = -i\int_{t_0}^{T} dt H_1(v,u) + \ln G[u(\cdot)] \qquad (25b)$$

and the dependence on $J(t)$ comes through $u(t)$ which is a functional of $J(t)$. From the definition of $\Gamma[Q(\cdot)]$ one has $\Gamma_0[Q(\cdot)] + \eta\Gamma_1[Q(\cdot)] + O(\eta^2)$ with

$$\Gamma_0[Q(\cdot)] = -W_0[J(\cdot)] + \int dt J(t)Q(t)\,, \qquad \frac{\delta W_0[J(\cdot)]}{\delta J(t)} = Q(t) \qquad (26)$$

But (25a) shows that $\delta W_0[J(\cdot)]/\delta J(t) = u(t)$ and then at this order $Q(t) = u(t)$. This implies then

$$\Gamma_0[Q(\cdot)] = -i\int_{t_0}^{T} dt\,[v\dot{u} - H_0(v,u)]\Big|_{u(\cdot)=Q(\cdot)} \qquad (27)$$

In the case of multiplicative noise one has from (20-23) $v = i[\dot{u} - A(u)]/g(u)$ and then

$$\Gamma_0[Q(\cdot)] = \int_{t_0}^{T} dt\frac{1}{2g(Q(t))}\left[\dot{Q} - A(Q(t))\right]^2 \qquad (28)$$

In the next order one can still use $Q(t) = u(t)$. This can be easily seen as follows. Consider a function $W(y) = W_0(y) + \eta W_1(y) + O(\eta^2)$ and its Legendre transform $\Gamma(x)$ defined by

$$\Gamma(x) + W(y) = xy\,, \quad W'(y) = x \tag{29}$$

Putting $W^{(1)} = W_0 + \eta W_1$, $\Gamma = \Gamma_0 + \eta\Gamma_1 + O(\eta^2)$, $\Gamma^{(1)} = \Gamma_0 + \eta\Gamma_1$, one has

$$\Gamma_0(x) = -W_0(\varphi_0(x)) + x\,\varphi_0(x) \tag{30}$$

where $y = \varphi_0(x)$ results from the inversion of $W'(y) = x$. At the next order

$$\Gamma^{(1)}(x) = -\left(W_0(y) + \eta W_1(y)\right) + xy\,, \quad y = \varphi_0(x) + \eta\varphi_1(x) \tag{31}$$

where the last formula results from the inversion of $W_0'(y) + \eta W_1'(y) = x$. Keeping terms up to $O(\eta)$ in (31) and using the identity $x = W_0'(\varphi_0(x))$ one obtains

$$\Gamma^{(1)}(x) = -W_0(\varphi_0(x)) + x\,\varphi_0(x) - \eta W_1(\varphi_0(x)) \tag{32}$$

and using (30) one has $\Gamma_1(x) = -W_1(\varphi_0(x))$. We have then from (25b)

$$\Gamma_1\left[Q(\cdot)\right] = i\int dt\, H_1(v, Q) - \ln G\left[Q(\cdot)\right] \tag{33}$$

with $v(t)$ a function of $(u, \dot{u}) = (Q, \dot{Q})$ as explained after equation (24). The problem is then to calculate the gaussian functional integral (24b) which can be done using the methods in[3]. Since we are interested now in the effective potential we consider the stationary state of the Markov process taking the limit $t_0 \to -\infty$ and evaluate $\Gamma^{\mathrm{st}}\left[Q(\cdot)\right]$ at $Q(\cdot) = q =$ constant. This means that we have to evaluate $G\left[u(\cdot) = q\right]$ in (24b). Putting

$$\left.\frac{\partial^2 H_0}{\partial p^2}\right|_0 = -iD\,, \quad \left.\frac{\partial^2 H_0}{\partial p \partial q}\right|_0 = B\,, \quad \left.\frac{\partial^2 H_0}{\partial q^2}\right|_0 = -iE \tag{34}$$

we have that these coefficients (D, B, E) will now be constants and functions of q. In order to give their values we write explicitly equations (23) putting $v = ir$ since $v(t)$ is purely imaginary. We have from (20) (multiplicative noise case i))

$$\dot{u} = A(u) + g(u)r \tag{35a}$$

$$\dot{r} = -rA'(u) - \frac{1}{2}g'(u)r^2 - J \tag{35b}$$

and from (21) (master equation case ii))

$$\dot{u} = e^r a_0(u) - e^{-r} b_0(u) \tag{36a}$$

$$\dot{r} = -\left[\left(e^r - 1\right) a_0'(u) + \left(e^{-r} - 1\right) b_0'(u)\right] - J \tag{36b}$$

$$H_j = i\left[\left(e^r - 1\right) a_j(u) + \left(e^{-r} - 1\right) b_j(u)\right]\,, \quad j = 0, 1, \ldots \tag{36c}$$

One has in case $i)$ $r = -A(u)/g(u)$ and in the case $ii)$ $e^r = \sqrt{b_0(u)/a_0(u)}$ since now we have to put $\dot{u} = 0$ because we are evaluating at $u(\cdot) = q =$ constant. We have then in case $i)$ $(r = -A(q)/g(q))$

$$D(q) = g(q) \tag{37a}$$
$$B(q) = A'(q) + g'(q)r \tag{37b}$$
$$E(q) = -rA'' - \frac{1}{2}g''r^2 \tag{37c}$$

and in case $ii)$ $\left(e^r = \sqrt{b_0(q)/a_0(q)}\right)$

$$D(q) = e^r a_0(q) + e^{-r} b_0(q) \tag{38a}$$
$$B(q) = e^r a_0'(q) - e^{-r} b_0'(q) \tag{38b}$$
$$E(q) = -\left(e^r - 1\right) a_0''(q) - \left(e^{-r} - 1\right) b_0''(q) \tag{38c}$$

The functional integral (24b) becomes now a function $\tilde{G}(q) = G^{\text{st}}\left[u(\cdot) = q\right]$ with discretized form $(q_0 = 0)$

$$\tilde{G}(q) = \int \prod_{i=1}^{N+1} dq_i \prod_{j=1}^{N+1} \frac{dp_j}{2\pi} \exp i\varepsilon \sum_{j=1}^{N+1} \left[p_j \frac{\Delta q_j}{\varepsilon} + \frac{i}{2} Dp_j^2 - Bp_j q_{j-1} + \frac{i}{2} Eq_{j-1}^2\right] \tag{39}$$

Doing the integrals over dp_j one has

$$\tilde{G}(q) = \int \prod_{i=1}^{N+1} \frac{dq_i}{\sqrt{2\pi\varepsilon D}} \exp\left[-\frac{\varepsilon}{2} \sum_{j=1}^{N+1} \left[\frac{1}{D}\left(\frac{\Delta q_j}{\varepsilon} - Bq_{j-1}\right)^2 + Eq_{j-1}^2\right]\right] \tag{40}$$

and in the continuous limit

$$\tilde{G}(q) = \int_{\gamma(0)} \mathcal{D}q \exp\left[-\int_{t_0}^{T} dt \left(\frac{1}{2D}(\dot{q} - Bq)^2 + \frac{1}{2}Eq^2\right)\right] \cdot \delta(q(t_0)) \tag{41}$$

The cross term $\dot{q}q$ cannot be integrated by parts in the $\gamma(0)$ discretization but one has

$$\int_{\gamma(0)} dt\, \dot{q}q = \sum_{j=1}^{N+1} q_{j-1} \Delta q_j = \sum_{j=1}^{N+1} \left(q_{j-1} + \frac{1}{2}\Delta q_j\right) \Delta q_j - \frac{1}{2} \sum_{j=1}^{N+1} \Delta q_j^2 \tag{42}$$

Under the functional integral one can replace[3] $\Delta q_j^2 \doteq \varepsilon D$ and then

$$\int_{\gamma(0)} dt\, \dot{q}q = \int_{\gamma(\frac{1}{2})} dt\, \dot{q}q - \frac{D}{2} \sum_{j=1}^{N+1} \varepsilon = \int_{\gamma(\frac{1}{2})} dt\, \dot{q}q - \frac{D}{2}(T - t_0) \tag{43}$$

where $\gamma(1/2)$ stands for midpoint discretization since $q_{j-1}+\frac{1}{2}\Delta q_j = \frac{1}{2}(q_{j-1}+q_j)$. Putting $\tau \equiv T - t_0$ (remember that at the end we must take the limit $\tau \to \infty$ for the stationary state) we obtain from (41) that

$$\tilde{G}(q) = \exp\left(-\frac{B}{2}\tau\right) \cdot \int dQ \tilde{P}(Q,T \mid 0,t_0) \tag{44a}$$

$$\begin{aligned}\tilde{P}(Q,T \mid 0,t_0) =& \frac{1}{\sqrt{2\pi\varepsilon D}} \int_{\gamma(\frac{1}{2})} \prod_{i=1}^{N} \frac{dq_i}{\sqrt{2\pi\varepsilon D}} \\ &\times \exp\int_{t_0}^{T} dt \left[-\frac{1}{2D}\left(\dot{q}^2 + B^2 q^2\right) + \frac{B}{D}\dot{q}q - \frac{1}{2}Eq^2\right] \\ &\times \delta(q(T)-Q)\,\delta(q(t_0))\end{aligned} \tag{44b}$$

and the δ-functions mean $q_0 = 0$, $q_{N+1} = Q$. Since now we are in the midpoint we can integrate by parts

$$\int_{t_0}^{T} dt\, q\dot{q} = \int_{t_0}^{T} dt\, \frac{d}{dt}\left(\frac{1}{2}q^2\right) = \frac{1}{2}Q^2$$

and then $\tilde{P}(Q,T \mid 0,t_0) = \exp\left(\frac{B}{2D}Q^2\right) P(Q,T \mid 0,t_0)$ with

$$P(Q,T \mid 0,t_0) = \int \mathcal{D}q \exp\left(-\int dt\, L(q,\dot{q})\right) \cdot \delta(q(T)-Q)\,\delta(q(t_0)) \tag{45a}$$

$$L(q,\dot{q}) = \frac{1}{2D}\left(\dot{q}^2 + \lambda^2 q^2\right), \quad \lambda^2 = B^2 + ED \tag{45b}$$

and we have omitted $\gamma(1/2)$ in (45a) since now the functional integral is independent of the discretization[3]. In (45a) we make the displacement $q(t) = \alpha(t) + Q(t)$ with $\alpha(t)$ solution of the Euler-Lagrange equations of $L(q,\dot{q})$

$$\ddot{\alpha} = \lambda^2 \alpha, \quad \alpha(t_0) = 0, \quad \alpha(T) = Q \tag{46}$$

Putting $\beta(t) = \partial L(\alpha,\dot{\alpha})/\partial\dot{\alpha} = \dot{\alpha}/D$ one has

$$\int dt L(\alpha,\dot{\alpha}) = \frac{1}{2}\beta(T)\alpha(T) = \frac{\lambda Q^2}{2D}\frac{e^{\lambda\tau} + e^{-\lambda\tau}}{e^{\lambda\tau} - e^{-\lambda\tau}} \tag{47}$$

and then $P(Q,T \mid 0,t_0) = \exp\left(-\frac{1}{2}\beta(T)Q\right) P(0,T \mid 0,t_0)$. The functional integral (45a) for $Q = 0$ gives $P(0,T \mid 0,t_0)$ and its value is (see Ref. [3], Chapter IX. Appendix 9.2)

$$P(0,T \mid 0,t_0) = \frac{1}{\sqrt{2\pi J(t_0)}} \tag{48}$$

with $J(t)$ satisfying the Euler-Lagrange equations of L, i.e. $\ddot{J}(t) = \lambda^2 J(t)$, with boundary conditions $J(T) = 0$, $\dot{J}(T) = -D$. Then

$$J(t_0) = \frac{D}{2\lambda}\left(e^{\lambda\tau} - e^{-\lambda\tau}\right) \tag{49}$$

Putting all this together

$$\tilde{G}(q) = \sqrt{\frac{\lambda}{\pi D}} \frac{e^{-\frac{B}{2}\tau}}{\sqrt{e^{\lambda\tau} - e^{-\lambda\tau}}} \int_{-\infty}^{\infty} dQ\, e^{-\frac{Q^2}{2D}\left(\lambda \frac{e^{\lambda\tau}+e^{-\lambda\tau}}{e^{\lambda\tau}-e^{-\lambda\tau}} - B\right)} \tag{50a}$$

and doing the gaussian integral over Q we obtain

$$\tilde{G}(q) = \sqrt{\frac{2\lambda}{D}} \frac{e^{-\frac{B}{2}\tau}}{\sqrt{(\lambda - B)e^{\lambda\tau} + (\lambda + B)e^{-\lambda\tau}}} \tag{50b}$$

We recall that $\lambda(q) = \sqrt{B^2 + ED}$ is a positive number for the values of q (see (37-38)) such that $(B^2 + ED) > 0$ and will be purely imaginary for the values of q such that $(B^2 + ED) < 0$. In this last case there is no limit of (50b) when $\tau \to \infty$ and we cannot determine the effective potential through this expansion. When $\lambda(q) > 0$ and $\lambda(q) \neq B(q)$ formula (50b) gives $(\tau \to \infty)$

$$\tilde{G}(q) = \sqrt{\frac{2\lambda}{D(\lambda - B)}}\, e^{-\frac{\tau}{2}(B+\lambda)} \left[1 + \mathrm{O}\left(e^{-2\lambda\tau}\right)\right] \tag{51a}$$

and for values $q = \bar{q}$ such that $\lambda(\bar{q}) = B(\bar{q})$ one gets

$$\tilde{G}(q) = \frac{1}{\sqrt{D}} \tag{51b}$$

From formula (13) defining $V(q) = V_0(q) + \eta V_1(q) + \mathrm{O}(\eta^2)$ and using (33) we see that $V_1(q) = V_{11}(q) + V_{12}(q)$ has a constribution $V_{11}(q)$ due to the term with H_1 in (33) and a term $V_{22}(q)$ given by

$$V_{12}(q) = \lim_{\tau \to \infty} \frac{1}{\tau} \ln \tilde{G}(q) = \begin{cases} \frac{1}{2}(B(q) + \lambda(q)), & B \neq \lambda \\ 0, & B(\bar{q}) = \lambda(\bar{q}) \end{cases} \tag{52}$$

We can now give the final result for $V(q)$ using (27), (33), putting there $u(\cdot) = q =$ constant (then $\dot{u} = 0$) and expressing v as a function $v(q)$ as explained after equations (36). One gets $(V = V^{(1)} + \mathrm{O}(\eta^2),\ V^{(1)} \equiv V_0 + \eta(V_{11} + V_{12}))$

$$V_0(q) = iH_0(v(q), q)\,, \quad V_{11}(q) = iH_1(v(q), q) \tag{53a}$$

$$V_{12}(q) = \begin{cases} \frac{1}{2}(B(q)+\lambda(q)), & B \neq \lambda \\ 0, & B(\bar{q}) = \lambda(\bar{q}) \end{cases} \tag{53b}$$

In case *i*) (multiplicative noise) one has

$$V_0(q) = \frac{1}{2}\frac{A(q)^2}{g(q)}, \quad V_{11}(q) = \frac{s}{2}\frac{g'(q)A(q)}{g(q)} \tag{54a}$$

$$V_{12}(q) = \frac{1}{2}\left(A' - \frac{g'}{g}A\right) + \frac{1}{2}\sqrt{\left(A' - \frac{g'}{g}A\right)^2 + \left(AA'' - \frac{1}{2}\frac{g''}{g}A^2\right)} \tag{54b}$$

and in case *ii*) (master equation)

$$V_0 = \left(\sqrt{a_0} - \sqrt{b_0}\right)^2, \quad V_{11} = \left(1 - \sqrt{\frac{b_0}{a_0}}\right)a_1 + \left(1 - \sqrt{\frac{a_0}{b_0}}\right)b_1 \tag{55a}$$

$$V_{12} = \frac{1}{2}\left(a_0'\sqrt{\frac{b_0}{a_0}} - b_0'\sqrt{\frac{a_0}{b_0}}\right) + \\ + \frac{1}{2}\sqrt{\left(\sqrt{\frac{b_0}{a_0}}a_0' - \sqrt{\frac{a_0}{b_0}}b_0'\right)^2 + 2\sqrt{a_0 b_0}\left[\left(1 - \sqrt{\frac{b_0}{a_0}}\right)a_0'' + \left(1 - \sqrt{\frac{a_0}{b_0}}\right)b_0''\right]} \tag{55b}$$

In both cases we have written V_{12} when $B \neq \lambda$. We prove now that points $\bar{q}$ such that $B(\bar{q}) = \lambda(\bar{q}) = \sqrt{B^2 + ED}$ are unstable fixed points of the deterministic system. In case *i*) the deterministic system is $\dot{q} = A(q)$, if $\bar{q}$ is an unstable fixed point $A(\bar{q}) = 0$, $A'(\bar{q}) > 0$, and from (37) $r(\bar{q}) = 0$, $E(\bar{q}) = 0$, $B(\bar{q}) = A'(\bar{q}) = \lambda(\bar{q})$. If $q = q_s$ is an stable fixed point $A(q_s) = 0$, $A'(q_s) < 0$, one has $B(q_s) = -\lambda(q_s) < 0$. In case *ii*) the deterministic system is $\dot{q} = a_0(q) - b_0(q)$ (see Ref. [5]), if $\bar{q}$ is an unstable fixed point $a_0(\bar{q}) = b_0(\bar{q})$, $a_0'(\bar{q}) - b_0'(\bar{q}) > 0$, and from (38) we see that $r(\bar{q}) = 0$, $E(\bar{q}) = 0$, $B(\bar{q}) = a_0'(\bar{q}) - b_0'(\bar{q}) = \lambda(\bar{q})$. If $q = q_s$ is an stable fixed point the only change is that $a_0'(\bar{q}) - b_0'(\bar{q}) < 0$ and then $B(q_s) = -\lambda(q_s) < 0$. From (53) we can conclude now that when $q = q'$, q' a fixed point of the deterministic system, one has $v(q') = 0$ and $V^{(1)}(q') = V_0(q') + \eta V_1(q') = 0$. However when $q' = \bar{q}$ is an unstable fixed point $V^{(1)}(q) \neq 0$ in an arbitrary neighborhood of $\bar{q}$ since $B = \lambda$ only at q' and the potential is discontinuous there. The value of the discontinuity can be calculated from (52) and is the positive quantity

$$\lim_{q \to \bar{q}} V^{(1)}(q) = \lim_{q \to \bar{q}} V_{12}(q) = \eta B(\bar{q}) = \begin{cases} \eta A'(\bar{q}) \\ \eta\left(a_0'(\bar{q}) - b_0'(\bar{q})\right) \end{cases} \tag{56}$$

since $V_0(\bar{q}) = V_{11}(\bar{q}) = 0$.

We shall discuss now some possible interpretations of the effective potential $V(q)$ from the point of view of probability theory. This subject is related to results on large deviations[9]. Let X be a random variable taking values in a finite set of real numbers $\{x_1, x_2, \ldots, x_n\}$, $x_j < x_{j+1}$, with probability measure $(\omega_1, \omega_2, \ldots, \omega_m)$ and mean value $\bar{x} = \sum \omega_i x_i$. If we make n independent observations of X we shall observe n_i times the value x_i, $i = 1, \ldots, m$, $\sum n_i = n$. Putting $p_i = \frac{n_i}{n}$, we can ask for the probability $P(\{p_i\})$ of observing a given measure $(p_1, \ldots, p_m)$. One has[10]

$$P_n(\{p_i\}) = n! \prod_{i=1}^{m} \frac{\omega_i^{n_i}}{n_i!} \approx n^n \prod_i \left(\frac{\omega_i}{n_i!}\right)^{n_i} \tag{57}$$

using $N! \approx N^N$ for N big enough. A simple calculation shows that $\ln P_n = -n\,K(p,\omega)$, where the relative entropy $K(p,\omega)$ is defined by

$$K(p,\omega) = \sum_{i=1}^{m} p_i \ln \frac{p_i}{\omega_i} \geq 0 \tag{58}$$

where the equality holds only for $p_i = \omega_i$. Then the probability P_n is given for large n by

$$P_n(\{p_i\}) = e^{-n\,K(p,\omega)}, \quad n \to \infty \tag{59}$$

Suppose now that we ask for the probability of observing a given mean value x in the n previous observations, i.e. if $X^{(j)}$ is the observed value of X in the jth measurement we impose $\frac{1}{n}\sum X^{(j)} = x$. This probability $Q_n(x)$ will be $\exp(-n\,I(x))$ where $I(x)$ is the minimun of $K(p,\omega)$ over all the measures $(p_1, \ldots, p_m)$ such that $\sum p_i x_i = x$. We find the minimizing measure $\{\tilde{p}_i\}$ using Lagrange multipliers λ and J to take into account $\sum p_i = 1$ and $\sum p_i x_i = x$. We put

$$\tilde{K} = \sum p_i \ln \frac{p_i}{\omega_i} + \lambda\left(1 - \sum p_i\right) + J\left(x - \sum p_i x_i\right) \tag{60}$$

and $\partial\tilde{K}/\partial p_k = 0$ gives $\left(Z^{-1} = e^{\lambda - 1}\right)$

$$\tilde{p}_k = \frac{1}{Z(J)}\,\omega_k e^{J x_k}, \quad Z(J) = \sum_{k=1}^{m} \omega_k e^{J x_k} \tag{61}$$

with J determined as a function $J(x)$ of x by $(W(J) = \ln Z(J))$

$$\frac{\partial W}{\partial J} = \sum_{k=1}^{m} \frac{1}{Z}\,\omega_k e^{J x_k} x_k = x \tag{62}$$

Then $I(x) \geq 0$ is

$$I(x) = K(\tilde{p}, \omega) = -W(J) + Jx\,, \quad \frac{\partial W}{\partial J} = x \tag{63}$$

The results

$$P_n(\{p_i\}) = e^{-n\,K(p,\omega)}\,, \quad Q_n(x) = e^{-n\,I(x)}\,, \quad n \to \infty \tag{64}$$

are large deviation results in the sense that they give in some limit (here the number of observations) the probability of a deviation from an "average". In the first case it is the probability of observing the experimental measure $\{p_i\}$ of the random variable X with probability $\{\omega_i\}$ and in the second the probability of observing the mean value x of X which has mean value $\bar{x}$. Formula (63) says that $I(x)$ is the Legendre transform of $W(J)$ and gives the connection with the effective potential. To see this we replace the random variable X by a stochastic process $\{q(t), t \in [t_0, T]\}$ taking values in a function space. The values are the trajectories in $[t_0, T]$, ω_j is replaced by the probability $\omega\,[q(\cdot)]$ of the trajectory $q(\cdot)$ and $Z(J)$ in (61) by $Z\,[J(\cdot)]$ given by (1) (the only new thing is the factor η whose origin we have already explained). Comparing (5) and (63) we conclude that the probability of observing the trajectory $q(\cdot)$ behaves as $\exp\left(-n\frac{\Gamma[q(\cdot)]}{\eta}\right)$ when the number of observations $n \to \infty$. If we consider the constant trajectory $q(\cdot) = q$ in the stationary state we have $\Gamma^{\rm st} = (T - t_0)V(q)$ and the probability will be $\exp\left(-n\frac{T-t_0}{\eta}V(q)\right)$. We remark at this point that a large deviation property in the limit $\eta \to 0$ can be proved with the functional $\Gamma_0\,[q(\cdot)]$ given by (28) in the role of $I(x)$ in (64)[11,12].

Another possible interpretation is obtained considering the random variables ($q(t)$ is a stochastic process in the stationary state)

$$X_T = \frac{1}{T - t_0}\int_{t_0}^{T} dt\, q(t) \tag{65}$$

which has a large deviation property[9] in the limit $\tau = T - t_0 \to \infty$ with a function $I(x)$ in the sense that

$$\lim_{\tau \to \infty} \text{prob}\,[x < X_T < x + dx] = e^{-\frac{1}{\eta}\tau I(x)}dx \tag{66}$$

the function $I(x)$ is the Legendre transform of $C(J)$ defined by (see (1) to (12))

$$\begin{aligned} C(J) &= \lim_{\tau \to \infty} \frac{\eta}{\tau} \ln Z^{\rm st}\,[J(\cdot) = J = \text{constant}] \\ &= \lim_{\tau \to \infty} \frac{1}{\tau} W^{\rm st}\,[J(\cdot) = J] \end{aligned} \tag{67}$$

We shall see now that $I(x)$ is just the effective potential $V(x)$. Let us evaluate $\Gamma^{\mathrm{st}}[q(\cdot)]$ (see (5)) at $q(\cdot) = q = \text{constant}$. One has

$$\Gamma^{\mathrm{st}}[q(\cdot) = q] = \tau V(q) = W^{\mathrm{st}}[J(\cdot)] + q\int_{t_0}^{T} dt\, J(t)\,, \quad \frac{\delta W^{\mathrm{st}}}{\delta J(t)} = q \tag{68}$$

But $\delta W^{\mathrm{st}}/J(t) = q = \text{constant}$ can only be realized by a constant $J(t) = J$ in the limit $\tau \to \infty$, and in this limit $W^{\mathrm{st}}[J(\cdot) = J] = \tau C(J)$. Then (68) gives

$$\tau V(q) = -\tau C(J) + \tau q J \tag{69}$$

which together with

$$\frac{\partial C}{\partial J} = \frac{1}{\tau}\frac{\partial W^{\mathrm{st}}[J(\cdot) = J]}{\partial J} = \frac{1}{\tau}\int_{t_0}^{T} dt\, \frac{\delta W^{\mathrm{st}}[J(\cdot)]}{\delta J(t)}\bigg|_{J(\cdot)=J} = q \tag{70}$$

shows that $V(q) = I(q)$.

References

[1] W.Horsthemke y R. Lefever: *Noise Induced Transitions: Theory and Applications in Physics, Chemistry and Biology* , (Springer-Verlag, Berlin 1984)

[2] R. Graham: in *Stochastic Processes in Nonequilibrium Systems*, L. Garrido et al., eds. (Springer-Verlag, Berlin 1978)

[3] F. Langouche, D. Roekaerts and E. Tirapegui, *Functional integration and Semiclassical expansions* (Reidel, Dordrecht 1982).

[4] S. Coleman and E. Weinberg, Phys. Rev. D7, 1888 (1972).

[5] E. Tirapegui: *Stochastic processes in macroscopic physics* lectures in *Proceedings of the Latin American School of Physics 1987*, ELAF'87, J. J. Giambiagi et al., eds. (World Scientific, Singapore 1988).

[6] H. Calisto and E. Tirapegui, J. Stat. Phys. (May 1993)

[7] H. Calisto, E. Cerda and E. Tirapegui, J. Stat. Phys. (May 1993)

[8] H. Calisto, E. Cerda and E. Tirapegui: J. Stat. Phys. 69, 1115 (1992)

[9] R. S. Ellis: *Entropy, Large Deviations and Statistical Mechanics*, (Springer-Verlag, Berlin 1985)

[10] F. Schlögl, Ann. Phys. 45, 155 (1967), Z. Phys. 243, 303 (1971) and Z. Phys. 244, 199 (1971)

[11] M. E. Vares: *Grandes desvios em processos Markovianos*, ed. Instituto de Matemáticas Pura e Aplicada (IMPA), (Río de Janeiro 1985)

[12] D. W. Strook: *An Introduction to the Theory of Large Deviations*, (Springer-Verlag, Berlin 1984)

PART II

INSTABILITIES IN NONEQUILIBRIUM SYSTEMS

SOME ASYMPTOTIC TIME BEHAVIOR FOR THE REAL GINZBURG LANDAU EQUATION

P.COLLET
Centre de Physique Théorique
UPRA 14 du CNRS
Ecole Polytechnique
F-91128 Palaiseau Cedex (France)

1. Introduction

It is a well known fact that the Ginzburg Landau evolution equation

$$\partial_t u = \partial_x^2 u + u - u|u|^2 \tag{1}$$

is a first order normal form for the absolute instability of a continuous spectrum when this instability develops at a non zero wave number [C.E.1]. In this equation, u is a complex amplitude, t is the time variable, and x is the one dimensional space variable. It is also known [C.E.1] that this equation defines a dynamical system in the space of regular and bounded functions in the following sense. If the initial function u and it's first two space derivatives are bounded on the whole line, there is a constant C (which depends on these initial data) such that for any positive time t and any x, we have

$$|u(x,t)| + |\partial_x u(x,t)| + |\partial_x^2 u(x,t)| \le C \ .$$

In other words, the parabolic equation (1) gives rise to a well defined dynamical system in the space of regular and bounded functions. One of the main difference with respect to the theory of dynamical systems on compact manifolds is that here the linear part of the equation is an operator with a continuous spectrum. As a consequence, most of the standard techniques are of no help in the study of this problem. For example, it is true that equation (1) is in some sense a gradient dynamic, since the right hand side is the functional derivative of the formal potential

$$\mathcal{F}(u) = -\int [|\partial u|^2 - |u|^2 + |u|^4/2]dx \ .$$

However this formal expression diverges if the function u is non trivial at infinity and there is in general no way to renormalize the functional $\mathcal{F}$ for interesting classes of solutions.

E. Tirapegui and W. Zeller (eds.), Instabilities and Nonequilibrium Structures IV, 177–180.

The simplest solutions of (1) are of course the stationary solutions. There is a two parameter family of such solutions which is given by

$$u_q(x) = \sqrt{1-q^2}e^{iq(x-x_0)} \qquad \text{for} \quad q \in [-1,1] .$$

These solutions are unstable for $|q|^2 > 1/3$ (this is the well known Eckhaus instability), and they are marginally stable otherwise. This marginal instability is due to the invariance by translation of the equation.

Out of these stationary solutions we can construct other interesting classes of solutions. A well known case are the front solutions which are of the form (at least to lowest order)

$$u_q(x)A(x-ct) ,$$

and which interpolate in space and time between different stationary solutions, for example u_q at $-\infty$ and 0 at $+\infty$. The envelope moves at speed c.

We will consider below solutions which describe the competition between stationary solutions with different wave numbers at $-\infty$ and $+\infty$.

2. Competition between stationary solutions

A first natural problem about the stationary solutions u_q with $|q|^2 < 1/3$ is in which sense they are stable. Recall that since the spectrum of the linearized operator around these solutions is only marginal, we cannot conclude anything yet about their stability or instability. A first result can be formulated in terms of the phase and amplitude of the solution. If the solution $u(x,t)$ is different from zero at all times, we can define it's modulus R and phase ϕ by

$$R(x,t) = |u(x,t)| \quad \text{and} \quad e^{i\phi(x,t)} = u(x,t)/R(x,t) .$$

Note that the stationary solution u_q corresponds to $R = \sqrt{1-q^2}$ and $\phi(x,t) = q(x-x_0)$. It is easy to rewrite the evolution equation (1) in terms of the quantities R and ϕ, and one obtains a pair of coupled evolution equations. A particularly interesting quantity for the discussion below is the local wave number $\eta(x,t)$ which is defined by $\eta(x,t) = \partial_x\phi(x,t)$. We can now formulate a stability result.

Theorem 1. *For $|q| < 1/\sqrt{3}$, small and local variations of the amplitude and wave number of u_q relax diffusively to zero.*

We refer to [C.E.E] for the precise definition of the topology of the function space of the perturbations which is used in this result. The terminology local refers in particular to the fact that these perturbations vanish at infinity.

The main point of the proof is to show that the amplitude R of the solution does not vanishes. If this is the case, this amplitude is in some sense slaved to the local wavenumber. In order to illustrate this let us consider the case $q = 0$. Define r by $R = 1 + r$, then neglecting higher order terms one gets the system of equations

$$\begin{aligned}\partial_t r =& \partial_x^2 r - 2r - \eta^2 \\ \partial_t \eta =& \partial_x^2 \eta + 2\partial_x \left(\frac{\eta\partial_x r}{1+r}\right) .\end{aligned}$$

If we replace r by $-\eta^2/2$ in the second equation, we get a non linear diffusion equation for η, namely

$$\partial_t \eta = \partial_x \left[\left(\frac{1 - 5\eta^2/2}{1 - \eta^2/2} \right) \partial_x \eta \right] = \partial_x [D \partial_x \eta] \,, \tag{2}$$

where

$$D(x) = \frac{1 - 5\eta^2/2}{1 - \eta^2/2} \,.$$

Assume now that the initial condition satisfies $|\eta(x,0)| < \sqrt{2/5}$, then at $t = 0$ we have $D > 0$. By the Maximum Principle, these two inequalities will remain true for all times. In particular, R does not vanish. Moreover, by a theorem of Nash [F.S.], if the initial data $\eta(x,0)$ is bounded and decays to zero at infinity, the solution $\eta(x,t)$ will vanish diffusively with time.

We can formulate similarly a more general and non local problem. Let q and q' be two small real numbers. We can now consider an initial condition with modulus near 1 and which interpolates between u_q at $-\infty$ and $u_{q'}$ at $+\infty$. If we let this initial condition evolve with time what happens? The first result concerns the absence of zeroes of the solution (also called phase shifts).

Theorem 2. *If q and q' are small enough and if at time zero, the function*

$$z = R^2 - (1 - \eta^2)$$

is small enough in a suitable function space, then the function R is bounded below away from zero uniformly in time.

We refer to [C.E.2] for the details on the function spaces involved. We simply mention that the function z vanishes at $+\infty$ and $-\infty$. Note also that this function describes the local discrepancy between the amplitude R and the amplitude associated to the local wave number η.

The question of convergence is more subtle and before stating the result we will consider several related but simpler questions. For simplicity we will only discuss diffusion equations for the local wave number η. Consider first the usual diffusion equation

$$\partial_t \eta = \partial_x^2 \eta \tag{3}$$

with an initial condition $\eta(x,0)$ bounded and with two limits η_+ and η_- at $+\infty$ and $-\infty$ respectively. Using the heat kernel, it is easy to prove that for a fixed position x, the solution $\eta(x,t)$ converges to $(\eta_+ + \eta_-)/2$ when t tends to $+\infty$. Note however that for any finite time the limits at $+\infty$ and $-\infty$ of the solution are fixed to η_+ and η_- respectively. A slightly more involved problem is to consider the equation

$$\partial_t \eta = \partial_x (D(x) \partial_x \eta)$$

where D is a strictly positive bounded and regular function with two positive limits D_+ and D_- at $+\infty$ and $-\infty$ respectively, and the initial condition for η is as before. Then one can prove [C.E.1.] that for any fixed x, the solution converges to the number

$$\frac{\sqrt{D_+}\eta_+ + \sqrt{D_-}\eta_-}{\sqrt{D_+} + \sqrt{D_-}} \,.$$

These two examples have the property that locally the solution converges to a well defined limit. One can construct however an initial condition for the diffusion equation which is bounded and regular but such that the solution (although of course uniformly bounded) does not converge at any fixed location. These initial conditions have a derivative whose modulus is not integrable over the whole line. The rate of variation of these solutions is smaller and smaller with time, typically $1/\sqrt{t}$, however this last function is not integrable and this is why there is no convergence. We refer to [C.E.1] for more details.

The three above examples illustrate some of the asymptotic time behavior for diffusion equations in infinite volume. In the case of the Ginzburg Landau equation the convergence problem was solved recently by J.Bricmont, A.Kupianen and G.Lin using a powerful renormalisation method (see [B.K.L.] and [B.K.]). For simplicity, we will describe their technique mostly using the simplified equation (2) with boundary conditions η_+ and η_- at $+\infty$ and $-\infty$ respectively. The first remark is that this equation has a so called scaling solution, namely a solution which is only a function of the quantity $x/\sqrt{t}$. If we denote by s this function, we have $\eta(x,t) = s(x/\sqrt{t})$, and by plugging this ansatz into equation (2) we obtain an equation for $s(y)$ which can be written

$$(1 - 5s^2/2)s'' - 4ss'^2 + y(1 - s^2/2)s = 0 \,.$$

This second order equation has a two parameter family of solutions and one determines these parameters by the conditions at $+\infty$ and $-\infty$. Note that by time translation invariance of the equation, if $s(x/\sqrt{t})$ is a solution, so is the function $s(x/\sqrt{(t+\tau)})$ for any positive number τ. Therefore if we start with the initial condition $\eta(x,0) = s(x/\sqrt{\tau})$, for any fixed x the solution converges to $s(0)$. The idea of [B.K.L.] is to consider the stability of this solution. They prove that if the initial condition is a small perturbation of the function $s(x/\sqrt{\tau})$, then the perturbed solution relaxes diffusively towards the unperturbed one. This implies immediately the convergence. Note that since the perturbed and unperturbed initial conditions should have the same limit at $+\infty$ and $-\infty$ respectively, the perturbation tends to zero at infinity, and this is essentially why it diffuses away.

The case of the Ginzburg Landau equation is treated similarly. The main difference is that one obtains convergence for the amplitude and local wave number. Since the phase is a primitive of the local wave number, one can get for this phase a term which depends only on time and diverges like $\sqrt{t}$.

References

[B.K.] J.Bricmont, A.Kupiainen. Renormalisation group and Ginzburg-Landau equation. Commun. Math. Phys. (to appear)

[B.K.L.] J.Bricmont, A.Kupiainen and G.Lin. Renormalisation group and asymptotic of solutions of nonlinear parabolic equations. Preprint 1992.

[C.E.1] P.Collet, J.-P.Eckmann. Space-Time Behavior in Problems of Hydrodynamic Type: A Case Study. Preprint, University of Geneva 1992.

[C.E.2] P.Collet, J.-P.Eckmann. Commun. Math. Phys. **145**, 345 (1992).

[C.E.E.] P.Collet, J.-P.Eckmann and H.Epstein.Helvetica Physica Acta **65**, 56 (1992).

[F.S.] E.B.Fabes, D.W.Stroock. Arch. Rat. Mech. and Analysis **96**, 327 (1986).

ANOMALOUS TRANSPORT IN HETEROGENEOUS MATERIALS

E. GUYON, J.P. BOUCHAUD
Ecole Normale Supérieure
24 rue Lhomond
75231 Paris CEDEX 05

1. INTRODUCTION

Mixing is an essential process in practically all fields of natural sciences and in industrial operations. But it is also a basic component of the field of non linear science where it is at the origin of experimental tools for diagnosis of reversibility . The notions of diffusion and mixing [1] are closely associated with the random walk of a Brownian particle which is such that the square of the distance R to the origin taken at a time t = 0 , averaged over many independent walks varies linearly with time

$$\overline{R^2} = D_m t$$

where D_m is the molecular diffusion coefficient.

This is a robust definition as it applies if the mean free paths, l, are distributed randomly, provided the statistical law P(l) is limited enough (we will come back to this point later). The expression "*black hole of statistical physics*" has been coined by Bouchaud & Georges to express the fact that a broad range of distributions of mean free paths all lead to a gaussian spreading . Applications of brownian statistical laws are too numerous to even be listed here. Two extreme cases deal with continuous geometries or regular lattices where the randomness comes from thermal noise and with geometrically disordered systems with deterministic flows (hydrodynamic dispersion).

2. MIXING IN FLOWS

2.1. MIXING AND TURBULENCE

The historical development of turbulence has been closely associated with that of turbulent diffusion starting from the work of G.I. Taylor in 1915 [2], and the classical book of Monin & Yaglom [3], is largely devoted to it. Schematically, one can see two origins to turbulent mixing : Particles are being passively convected by eddies of various extent and life time. This causes adjacent particles to separate in time. Richardson's law deals with the rate of separation of two particles distant of l, which is given by

$$d\, l^2 / dt = D(l)$$

E. Tirapegui and W. Zeller (eds.), Instabilities and Nonequilibrium Structures IV, 181–192.

D (l) would be equal to $2D_m$ for ordinary brownian diffusion. In homogeneous 3D turbulence, D (l) α $l^{4/3}$ which is consistent with Kolmogorov law, as D (l) α u (l).$l^{1/3}$, where the velocity of an eddy of size l scales as u (l) ~ $l^{1/3}$. The increase of diffusion coefficient with scale - we talk in this case of *hyperdiffusion* - comes from the fact that only eddies of the order of magnitude or smaller than l are effective in separating particles distant of l . Larger eddies carry convectively the two particles together. As l increases, the range of eddy sizes increases as the rate of separation.

Turbulent pair diffusion has been probed extensively using *forced Rayleigh scattering* [4], a powerful tool to study diffusion in the bulk of a transparent material, as a function of scales l and orientations (a).

A second effect comes from the stretching effect of elongational flows. A particle of fluid will become progressively elongated and distorted. One can define a lower length scale λ of the stretching such that when the *Peclet number* $Pe = U\lambda / D_m$ becomes of the order of unity molecular diffusion takes over convective effects (b). An important and recent development is the study of *Lagrangian chaos* [5], which deals with the study of low Reynolds number flows in the limit of very large Peclet numbers (this can be easily obtained by using viscous fluids or by placing large passive but still brownian particles in the flow). It has been shown that, despite the fact that the flow is deterministic and reversible in time (a basic property of the low Reynolds number limit), the motion of particles becomes irreversible if hyperbolic points (such as that around a stagnation point in a flow) are present in the flow. Actually the problem mimics exactly the solution of chaotic problems, and the classical terminology applying to phase trajectories in chaos (*baker's transform, folding...*) takes a direct explicit meaning in mixing. Here again, student experimental projects can be easily devised [6] . Recent developments on the subject have included such effects as noise (or small oscillations of the flow pattern around a mean position) and inertia when particles present in the fluid have a different density from the fluid . Such parameters modify the mixing characteristics and, in particular, can add to the effect of molecular diffusion in the process of exchange between flow lines . Although the process is of fundamental nature, it has indeed very practical implications in engineering processes as well as environmental science .

2.2. CORRESPONDANCE WITH RANDOM WALKS

In the following, we will not consider the problem of turbulent flows but will focus our attention to the random walk approach and, more specifically, to situations where a gaussian behaviour is not attained. An example of *hyperdiffusive* behaviour has been described in the case of turbulent flows. A corresponding situation in random walks is that of the *self avoiding walk* which is used to mimic the effect of swelling in the study of the conformation of a polymeric chain in a good solvent. Typically, one gets a growth in "time" (i.e. number of monomers) as [7]

$$R \;\alpha\; t^{\nu}$$

where the Flory exponent $\nu = 3/5$ - is such that - $1/2 < \nu < 1$. This implies that the monomers get away from the origin faster than in a diffusive process (where $\nu = 1/2$).

On the other hand, slower than gaussian - or *hypodiffusive* - behaviours can be observed. An example was presented in the present workshop and is discussed by Pina & al. in these proceedings. It dealt with the diffusion of a passive scalar across the rolls in a Taylor Couette instability such that the particle remains trapped within a roll. A corresponding situation for a brownian particle would be that of a random walk on a restricted geometry such as the fractal infinite cluster formed at the percolation threshold - the so called *ant in a labyrinth* problem. In this case one gets the relation for the average distance [8] :

$$R^{2+\theta} \;\alpha\; t$$

where $\theta = 0.87$ is a combination of exponents of percolation which can be obtained by studying the crossover, away from threshold, between the anomalous walk on a fractal at small distances smaller than the correlation length of percolation and the ordinary walk on restricted lattices at larger scales.

3. GENERAL FORMULATION

Thus, in general one can expect a general relation

$$R^2 = D_m t^x \qquad \text{or} \qquad R^{d_w} \sim t$$

where d_w can be seen as the fractal dimension of the walk such that
for diffusive process $x = 1$, $d_w = 2$;
for hyperdiffusive ones, $x > 1$ and $d_w < 2$;
and for hypodiffusive ones, $x < 1$, $dw > 2$.

The maximum number of sites within a radius R which can be visited

$$S(t) \sim R^{d_F}$$

where the fractal dimension d_F identifies with the Euclidian one, d, for ordinary lattices, or

$$S(t) \sim t^{d_F/d_w}$$

For ordinary brownian walk in 3d, $d_F / d_w = 3/2$. The number of sites, S (t), thus grows faster than linear with time : only an infinitesimal fraction of the accessible can be visited by the t-long walk. On the other hand, in a percolation problem it is found that $d_F / d_w \approx 2/3$ independently of the dimension of space on which the percolation lattice is imbedded : the exploration of the lattice is *compact* and the probability for a site to be visited is unity . This can have direct applications for reactions on fractal lattices as studied experimentally by Kopelman from triplet annihilation on disordered lattices [9]. There is a time dependent reaction rate or - in other terms - a "fractal like" reaction kinetics.

4. WHAT GOES WRONG WITH THE BROWNIAN SOLUTION ?

A detailed general answer to this problems is given in the review article by Bouchaud & Georges [10] and we will only indicate a few results in connection with experimental facts dealing with hydrodynamic examples. We will distinguish problems where the anomalous behaviour comes directly from the distribution function of the walk itself and those where the disorder in the geometry enters explicitly.

4.1. ANOMALOUS WALKS

The effects can be observed as being due to the existence of long time correlations as in the example of the S.A.W. problem, unlike the ordinary Markovian process where a complete loss of memory of the previous step is obtained at each collision. It can also come as a result of large enough distribution of the waiting times τ between jumps or of the step lengths l. The arguments leading to the diffusion behaviour are similar in the two last cases on which we focus now.

4.1.1. *Large distribution of steps*

We suppose that the distribution decays for large values of the steps as a power law

$$p(l) \sim l^{-(1+\mu)}$$

For a walk of N steps, one can expect that the maximum length of a step met will be such that

$$N \int_{lc}^{\infty} P(l)\, dl = 1$$

or, using the scaling form of p(l), $l_c \; \alpha \; N^{1/\mu}$.

This expresses that the maximum value l_c (N) has been obtained roughly once out of the N individual step lengths.

For $0 < \mu < 1$, the average distance for a walk of N step is given

$$X_N \sim N \int^{lc} l \; p(l)\, dl = N l_c^{1-\mu} \sim N^{1/\mu}$$

Similarly, for $1 < \mu < 2$ the dispersion is given by :

$$(X_N - X_N)^2 \sim N \int_o^{lc} (l - l)^2 \, p(l)\, dl \sim N^{2/\mu}$$

In both cases, the integral is governed by the upper bound l_c (N) and the average distance, or the *dispersion*, varies more rapidly with N that in the brownian case which is recovered for $\mu > 2$.

The *Levy flight* , illustrated in figure 1, corresponds to a very anomalous distribution of length steps; it has been observed recently by an elegant experiment using a solution of micellar (cylindrical, wormlike) objects [1,11] : A photochromic dye is included in the solution, and serves as a tracer for the micelles solution. The micelles undergo diffusion with a diffusivity which is a rapidly decreasing function of size. But micelles will also deform, break and recombine as a function of time. Then the dye will belong to different objects during the observation time and will undergo a motion with widely varying length steps : long jumps correspond to short micelles. A forced Rayleigh scattering technique has been used to probe the diffusivity of the dye and lead to a decay rate of the "printed" grid of wavelength l as l^{μ} with $\mu < 2$. This is the signature of a "faster than brownian" diffusion.

4.1.2. *Large distribution of waiting traps*

There exists an ample literature initiated by experiments on the anomalous transport of photo electrons in disordered semiconductors [12]. Continuous time random walk describes the effect of the distribution of waiting times $\psi(\tau_n)$ of the electrons in such traps. One can express the large τ limit by a power law distribution

$$\psi(\tau) \propto \tau^{-(1+\mu)}$$

Following the same reasoning as above one finds that, for slowly converging distribution such that $0 < \mu < 1$, the average square of the path length is found to vary as

$$X^2 \sim l^2 t^{\mu}$$

which corresponds to an hypodiffusive law.

Dispersion with recirculating zones is an example of such a behaviour, as already mentioned in the introduction with the example of a Taylor Couette flow . Quantitative data have also been obtained in the case of Rayleigh Benard convection by Gollub & Solomon [13]. Cardoso and Tabeling [14] have used an electrically conducting liquid film along which an electric current flows and which is submitted to a magnetic field with alternate + and - poles. The Laplace force induces a set of parallel counterrotating rolls of radius l . The diffusion of a passive dye injected along the axis of a roll is followed in time. The results giving the concentration as a function of distance from the origin clearly show a spreading varying as $t^{1/3}$, i.e slower than the gaussian result in $t^{1/2}$.

Qualitatively, one can understand the problem in terms of three time scales: a short one $\tau \sim l/v$ where v is the order of magnitude of the convective velocity ; a long one $T = l^2/D_m = \tau$ Pe above which ordinary diffusion takes place . For intermediate time scales (typically of order $\tau Pe^{1/2}$) a particle remains trapped for a long time in a roll due to the slow motion near the walls.The random walk approach can be used at the scale of the rolls. An equivalent step of length l takes place each time the separation line between rolls is small enough for a particle in one roll to make a diffusive step to the next roll. The ψ (τ) α $\tau^{-(1+\mu)}$, with μ = 2/3. The quantitative analysis of the problem is quite involved and implies a detailed knowledge of the structure of the boundary layer flow near the walls and of the corner flow near the stagnation point ; we will refer to the analysis of [15] for the solution of theproblem (see also [10]).

4.2. ANOMALOUS TRANSPORT IN RANDOM MATERIALS

We only consider strongly disordered material structures, such as the pore structure of a partially clogged *porous medium* (in the case where two immiscible fluids are present simultaneously in the medium, one can think of an experiment with transport particles soluble in only one phase, the other one playing then the role of random impervious obstacles).

Another example is the case of *gels* . They form random structures for which the solid phase has a fractal structure near the *gel point*, where a continuous reticulated form first appears. On the other hand their pore space is quite open anddoes not show any singular behavior at the gel point. In this particular case, the anomalous transport occurs in the solid phase. It is more easily studied in terms of the frequency response to an imposed vibration. Extensive experiments have been carried on *aerogels* (a solid form obtained from gels by eliminating the solvent in supercritical conditions) using the *Brillouin scattering* at large frequency [15]. We have also performed a table top experiment by vibrating a metal grid at low acoustic frequency [16]. The grid is initially degradated at random. When the percolation threshold is approached from above (that is when the lattice has been sufficiently weakened randomly) one observes a crossover between extended resonant modes at low frequencies and non propagating ones at larger frequencies at a cross over frequency going towards zero at threshold. This last regime corresponds to localized vibrations on the fractal structure (whose extent becomes infinite at percolation threshold) called *fractons*.

The electrical analogue is that of a composite with conducting and insulating elements mixed together at random [17]. The conductivity is proportional to the diffusion coefficient as stated by the linear Einstein relation between diffusion and mobility [18,10]. At a high enough frequency, such that typically the travelling time of the E.M.wave over a correlation length is smaller than the inverse wave frequency, only localized fracton modes exist. The anomalous response of a fractal lattice has been probed by a simple teaching lab. experiment using a conducting percolation grid deposited on an aluminized foil and looking for its frequency response [19]. The experiment in turn was used as a model to understand the relaxation of pressure waves in a porous medium.

In all the various above experiments the transport is *hypodiffusive* because, as the separation increases, more and more trapping occurs in large loops or in the dead ends.

Some direct experiments have been made in porous media using forced Rayleigh scattering to study this anomalous regime. However, in practical cases where the pore geometry is well connected (away from the limit of the strongly sintered of a ceramics material or of the clogging of a filter) the fractal scales are too small to be seen directly and mostly act to renormalize the ordinary transport coefficients.

An important development of these studies in the case of a porous medium is the study of *hydrodynamic dispersion*. In this problem, there is a continous average flow **U** across the medium ; the spreading of a marker injected in the fluid - say a concentration step in a direction parallel to **U** - is due to the combination of the spreading due to the repeated effect of separation of the flow lines in branches of the porous medium, of trapping in dead flow regions and of molecular diffusion. The problem of anomalous spreading (i.e. deviation from a gaussian dispersivity around a mean line corresponding to the average flow velocity U) has been a subject of long dispute. It seems that, in most case , it arises from insufficient mixing due to a finite size effect [20]. For example if there is trapping with a maximum trapping time τ_m (for example in molecular diffusion in long dead ends l such that $\tau_m \sim l^2 / D_m$), to recover a brownian response would require a length L , such that $L/U >> \tau_m$ or $Pe << L/l$.

There is a possibility that genuine anomalous dispersion effect can take place. One particular example imagined by Matheron & de Marsily [21] deals with dispersion in a two dimensional stratified porous medium made with parallel layers of randomly variable permeabilities K(z) . If a pressure gradient is applied parallel to the layer the fluid flow will take place at different local velocities V(z) . In the same time, a marker will have a transverse diffusive motion from layer to layer with a diffusivity which we assume to have a constant value D . In order to simplify the description, one chooses a referential moving with the average longitudinal velocity so that the spreading occurs symetrically across the origin. Thus, in a time t , a marker will have explored a transverse distance z such that $\overline{z^2} \sim D\,t$. Or, in other terms, the probability to return to the origin after a time t will be given by $P_0(t) = 1/\sqrt{D_\perp t}$. Thus $\sqrt{D_\perp t}$ layers will have been visited $t/\sqrt{D_\perp t}$ times. The velocity auto-correlation function which is a measure of the longitudinal dispersion will decay as a function of time as

$<V(0)V(t)> = V^2 / (D\,t)^{1/2}$

Finally one gets the average dispersion around the mean velocity as [21,10]

$$\Delta x^2 \sim D_{//}\ t$$

with

$$D_{//}(t) = \int_0^t <V(0)\ V(t)>dt = V^2\,(t / D_\perp)^{1/2}$$

and thus

$$\Delta x^2 \sim V^2\ D^{-1/2}\ t^{3/2}$$

The exponent 3/2 instead of 1 in ordinary diffusion is a signature of an hyperdiffusive process. It comes from the fact that the frequent return to the origin creates a long-ranged correlation between the random longitudinal velocities. It would not occur if, say a transverse flow component would be added as, then, the probability of return to the origin, and hence the velocity auto-correlation, decays exponentially with time. This is what takes place if the layers of the medium are oblique to the applied pressure gradient.

The action of a flow on the transport of particles met in important processes such as *bulk filtration* and *electrophoresis* leads to a rich (and yet poorly understood from the point of statistical physics) variety of behaviours . The applied driving force may have an effect directly on the particles : gravity for non buoyant objects; electric field on charged particles . It may also come from the effect of convective transport by the flow. In the latter case, one can consider that the force experienced by the particle varies with its position in the medium. We will not consider this problem in detail but will sketch some features of it using point like particles moving in a disordered lattice.

Two main classes of trapping phenomena can take place . One is trapping in a *dead ends* such that the external field will push the particle into it (fig.2a). In practice, the particle could free itself by over passing the adverse potential barrier with a probability given by the Maxwell Boltzmann factor. De Gennes has suggested that this problem could be similar to that of the anomalous transport in disordered semiconductor with a distribution of traps [22].

Another factor is the existence of back bends (fig. 2b) where particles get pushed at the bottom point and stay there forever. However, if interactions exist between the particles injected they will push each other and will circulate freely in back bends. Similarly a trap, once filled, can be omitted from the flow paths. A last factor is the presence of the particles themselves. There is a reaction to the flow field as a particle gets transported in the flow. Experiments clearly show that the trapping in a porous medium takes place very differently if particles are injected one after the other one or several at the same time. This implies that a correlation takes place due to the modified flow field in addition to the direct correlations between particles. Experimentally, the main effect seems to be that the particles penetrate further in a porous medium if they are injected collectively than in the other case, which is somewhat counterintuitive !

In the different problems of biased transport of particles under the effect of an external field, the response is highly non monotonic : for low fields, the drift velocity of particles increases linearly with the field but, as the field pushes more efficiently particles in the different kinds of traps the drift starts decreasing as the field increases until the drift velocity is zero. Above this point the motion is characterized by a sublinear relation between distance and time. An example of such treatment is the numerical solution by Roux & Guyon of a uniformly biased diffusion on a percolation square lattice between the two sides of a strip of width L [23].

In zero field the time to cross the sample increases faster than L^2 as in hypodiffusive processes

$$\overline{t} \sim L^{2+\theta}$$

the result already given for the "ant in a labyrinth problem".

In finite field and for a probability p slightly larger than threshold, the average time varies non linearly with field as given in the figure 3 in agreement with the above discussion with the three regimes of diffusion (D), facilitated transport (FT) and trapping (T). One can imagine a rich phase diagram for the problem as a function of the two variables L/ξ where ξ is the correlation length of the percolation lattice, and the Peclet number Pe .

By analogy with the problem of electrophoresis of polymers in gels, Bouchaud & Georges [24] have considered the possibility to limit the effect of trapping by using a periodic sequence of fields of opposite direction. There is in fact a large class of problems opened to numerical as well as to analog studies. In the later case, one has to take into account precisely the local hydrodynamic effects (such as stagnation points or recirculating zones) which are very non universal and may completely govern the properties of transport, trapping and collection of particles.

(a) An interference grid of wavelength 1 is impressed in the fluid by shining an intense short pulse of light which causes a temperature or a composition modulation. In the later case, this is due to light induced chemical change reversible (if photochromic molecules are present) or irreversible (photobleaching). The fading of the grid which takes place with an exponential rate ($e^{-t/\tau(l)}$) where is followed from the variation diffraction intensity of a second continuous laser of small intensity incident of the grid. The experiment has been used in graduate experimental programs.

(b) In liquid flows the Prandtl number, $Pr = \nu / K$, ratio of kinematic viscosity to heat diffusivity, is often larger than unity. The inertial regime has a bound at small lenght scales where viscous diffusion limits the range of inertial scales (the so called *Kolmogorov limit*). This is different from the regime where the Peclet number is of the order of unity which occurs at even smaller length scales. The intermediate range - so called *viscoconvective* - was the one investigated in experiments.

REFERENCES

[1] *"disorder and mixing"* edited by E. Guyon, J.P. Nadal & Y. Pomeau, Kluwer (1988)

[2] G.I. Taylor, *Collected Works*, Cambridge University Press, 1963.

[3] A.S. Monin, A.M. Yaglom, *Statistical Fluid Mechanics: "Mechanics of Turbulence"* (MIT Press, Cambridge, 1971).

[4] A. Darco, J.C. Charmet, M. Cloître, P. Jenffer, E. Guyon, *Eur. Jl. Phys.* 2 (1981) 193.

[5] J.M. Ottino et al., *Nature* 333, (1988) 419.

[6] E. Villermaux & J.P. Hulin . *Eur. Jl Phys.* 11 (1990) 179

[7] P.G. de Gennes, *"Scaling Concepts in Polymer Physics"*, Cornell University Press (1985).

[8] See e.g. S. Havlin, D. ben Avram, *Adv. Phys.* 36 (1987) 695.

[9] R. Kopelman, J. Prasad, S.J. Parns, in *"Molecular Dynamics in Restricted Geometrics"*, J. Klafter, J.M. Drake eds J. Wiley (1989) p. 145.

[10] J.P. Bouchaud, A. Georges, *Phys. Rep.* 195 (1991) 127.

[11] J.P. Bouchaud, A. Ott, D. Langevin, W. Urbach, *J. Phys.* II (France) 1 (1991) 1465.

[12] H. Pfister, H. Scher, *Adv. Phys.* 27 (1978) 747.

[13] H. Gollub, T. Solomon , *Phys. Fluids* 31 (1989) 1372.

[14] O. Cardoso, P. Tabeling, *Europhys. Letters* 7 (1988) 225.

[15] Proceedings of the International Symposium on aerogels (Montpellier 1989) *Revue de Physique appliquée* 24, 4 (1989).

[16] E. Guyon, J. Wu, A. Palevski, S. Roux, I. Rudnick, *CRAS* 305 II (1987) 323.

[17] See e.g. *ETOPIM* 2, M. Gadenne, J.C. Lafait eds, North Holland (1988).

[18] B. Derrida, in [1

[19] P. Rigord, J.P. Hulin, *Europhys. Lett.* 6 (1988) 389.

[20] See *"Hydrodynamics of dispersed media"*, A.M. Cazabat, E. Guyon, J.P. Hulin, Edts, North Holland (1989).

[21] G. Matheron, G. de Marsily, *Water Resour*, Res. 16 (1980) 901.

[22] P.G. de Gennes, *CRAS* 284 B (1979) 329.

[23] S. Roux, E. Guyon, *Europhys. Lett.* 4 (1987) 175.

[24] J.P. Bouchaud, A. Georges, *J. Phys.* A 23 (1990) 4003 .

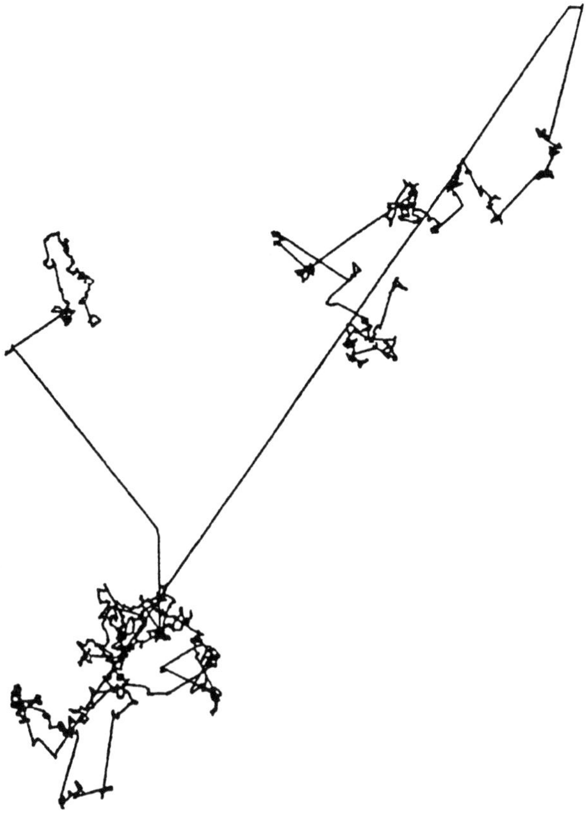

FIGURE 1.- Levy flight.

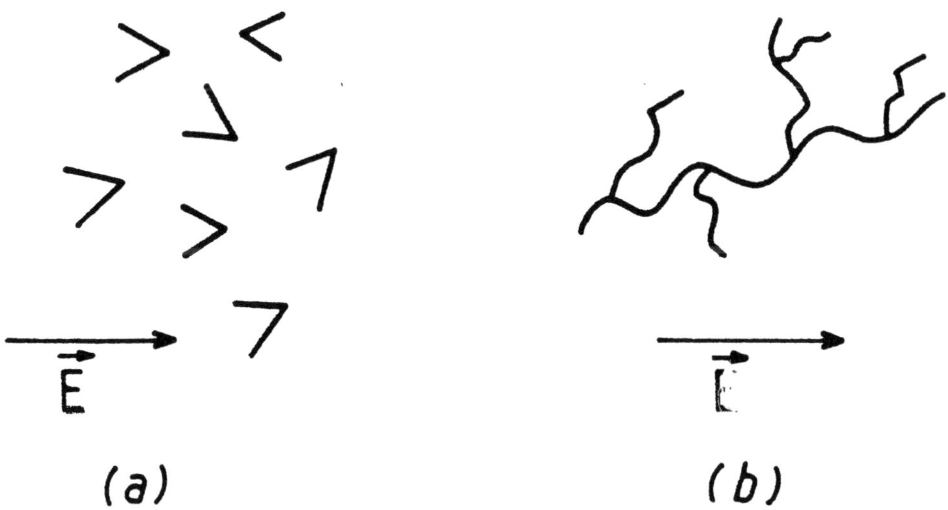

FIGURE 2.-Trapping phenomena.

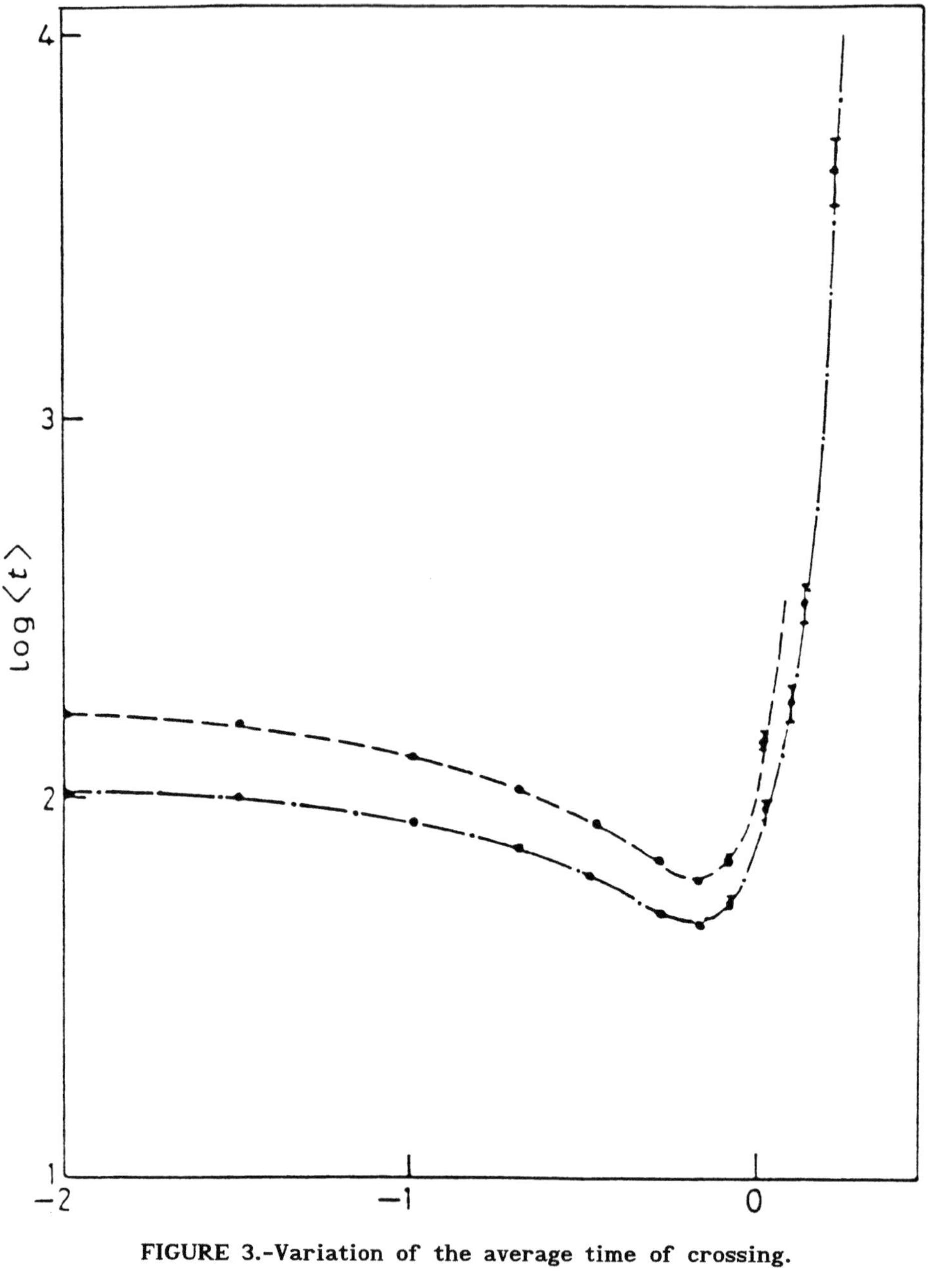

FIGURE 3.-Variation of the average time of crossing.

Subcritical traveling pulses and pattern formation in a film dragging experiment

F. Melo.
Laboratoire de Physique de l'Ecole Normale Supérieure de Lyon,
69364 Lyon cedex 07

ABSTRACT. In the lower portion of a horizontal rotating cylinder, the front created by dragging a viscous fluid can display the subcritical nucleation of traveling pulses. The pulse is a robust state of well defined shape and velocity. The number of pulses traveling along the front increases with angular velocity; when the angular velocity is small the front presents a chaotic behavior of colliding pulses. At a critical value the front displays a stationary periodic pattern of regular cells. This periodic state consists of a collection of interacting pulses, resulting from a new form of pattern formation.

I. Introduction

In the study of pattern formation, we usually encounter experimental situations for which a spatially homogeneous state of the system loses stability at a certain value of the control parameter. Beyond the threshold the state of the system becomes spatially periodic [1]. In many cases, for a larger value of the control parameter, the periodic state also undergoes secondary instabilities or spatio-temporal chaotic transitions. Well-known examples are convection [2], or Couette flow between two horizontal coaxial cylinders with a partially filled gap [3]. Similar behavior have been also observed in a film dragging experiment [4] and in a directional crystal growth experiment [5]. In all of these systems, complexity increases with control parameter.

In this paper we report the observation of a transition from a homogeneous state to one characterized by the nucleation of solitary traveling pulses. We also report a different and surprising new form of pattern formation, in which the periodic state observed experimentally is a dense packing of interacting pulses. In contrast with usual systems, the transition is from a disordered to an ordered state.

Our experimental set-up is described in part II. The dynamics of colliding pulses and the transition to the regular pattern is presented in part III, followed by a short conclusion.

II. Experimental Set-up

Our experimental set-up consists of a horizontal rotating cylinder of inner radius 5cm and lengh 60cm, driven by a variable-speed motor at a rotation frequency in the range 0.1-3.0Hz, with velocity fluctuations less than 3%. Vibrations produced by the motor are damped by the coupling mechanism with the cylinder, reducing the fluctuations from horizontal to less than $5*10^{-4}$rad. The cylinder is made of glass reducing the variation of its inner radius to less than 1%.The fluid used is silicon oil (47V20), with kinematic viscosity ν=0.2cm^{2}s^{-1}, surface tension γ=21erg/cm^2 at 25°C and density ρ=0.97gr/cm^3. This oil insures a perfect wetting of the glass substrate.

E. Tirapegui and W. Zeller (eds.), Instabilities and Nonequilibrium Structures IV, 193–197.

III. Results

SUBCRITICAL TRAVELING PULSES

The principal experimental parameters of the system are the quantity of liquid Ω in the cylinder and its angular velocity ω. We consider only the case: $\Omega << \Omega_T$, where Ω_T is the total capacity of the cylinder. For a given Ω and small enough angular velocity a portion of the fluid is dragged by rotation, forming a homogeneous viscous layer on the inner wall of the cylinder. The portion of liquid that remains near the bottom of the cylinder forms a bump. We can consider this bump to be a stationary linear front, parallel to the principal axis of the cylinder, propagating on the wetted substrate.

Above a critical value of $\omega=\omega_{c1}$ the linear front become unstable to the nucleation of traveling pulses. These are first generated via spatial oscillations of the front near the lateral boundaries of the cylinder. Below the instability threshold, these propagating waves are damped, i.e. the linear front is stable to finite amplitude perturbations. Above threshold the amplitude of the perturbations introduced by boundaries is larger than a critical value and the local wave is amplified to give rise to pulses of well-defined size and shape (cf. fig 1a). This shows that the nucleation of pulses is subcritical. The propagation velocity of the pulses is well-defined, about ten times smaller than the driving tangential velocity. The pulses are asymmetric: from left to right on fig. 1a, we observe first a gradual modulation of the front, followed by a steep rise, a cusp, and finally a smooth decrease. For $\omega > \omega_{c2} > \omega_{c1}$, the edge of the pulse nearer to the cusp emits a propagating wave. This wave moves in the direction opposite to that of the pulse; they are apparently generated at the cusp of the pulse, and quickly damped.

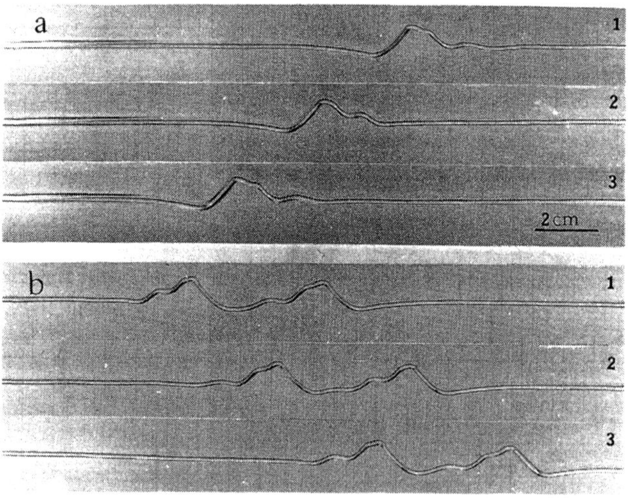

FIG. 1 (a) Time evolution of a single pulse. Its shape and its velocity of propagation are well defined. (b) Two traveling pulses moving meet in the same direction. They can not propagate if their distance of separation is smaller than a critical value. The experimental parameters are Ω=140cm^3 and ω=11.5s^{-1}. The time interval between two successive photographs is 30s.

When the number of pulses is small, one can observe two traveling pulses, generated by opposite boundaries and traveling in opposite directions, colliding in the central region of the cylinder. The dynamics of the collision is complex but one can distinguish two situations depending on the value of ω:

-For $\omega < \omega_{c2}$, where the cusps do not emit waves, the pulses vanish upon colliding. More detailed observation shows that the pulse velocity starts to decrease when the distance between them is on the order of their size.

-For $\omega > \omega_{c2}$, the amplitude of the waves emitted by the cusps is sufficient to trigger two new pulses after the collision (cf. fig. 2). Careful observation shows that the new pulses are generated by the splitting of the cusps of the initial pulses. As shown in figure 2.4, the cusps begin to split when the distance between them is on the order of the size of one pulse. Because the cusps split before they actually meet, the new pulses generated appear shifted with respect to what their position would have been had the initial pulses merely passed through one another.

If a single pulse emits a wave of sufficiently high amplitude, this wave can generate a new pulse propagating in the opposite direction. It is possible that this type of emission is correlated with some perturbation produced by the neighboring pulses. A study of this correlation is in progress. Our observations show that the cusps dominate the dynamics of pulses and their interactions.

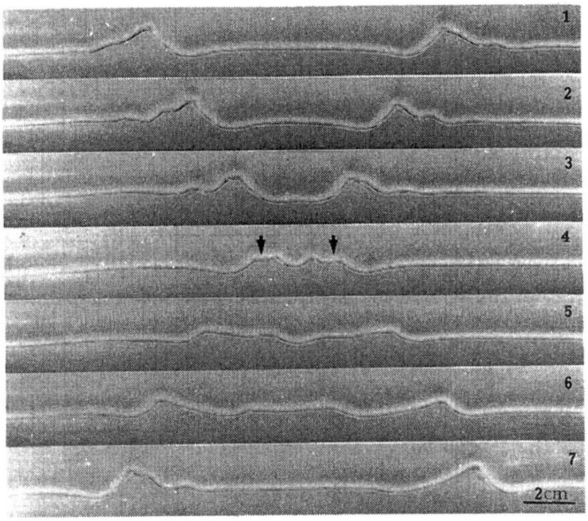

FIG. 2 Time evolution of two colliding pulses. When the pulses emit large-amplitude waves, two new pulses are generated, which separate and whose phases appear shifted with respect to the initial pulses. The arrow on fig. 2.4 shows the cusps spliting and forming the new pulses. The parameters are Ω=140cm^3 and ω=11.5s^{-1}. The time interval between two successive photographs is 30s.

FROM TRAVELING PULSES TO STATIONARY PATTERN

As ω is increased (cf. fig. 3a-d), the number of pulses created by lateral boundaries and by the amplification of the wave emitted from a cusp increases quickly. Fig. 3b shows many triangular cells bounded by two cusps; each of these was originally generated by a pair of colliding pulses. The size of these cells is not uniform but it is on the order of the size of one pulse. For low values of ω these cells are unstable to splitting of the cusps (see discussion of fig. 4).

For larger ω, most of the front is occupied by triangular cells which are still unstable to splitting of the cusps but present a more uniform size (cf. fig. 3c). Some regions of the front form a periodic pattern. Nevertheless, some cells are still unstable and can destroy the periodic domain which may reappear after a transient.

Finally, for a large enough $\omega \geq \omega_{c3}$, the triangular cells become stable with respect to splitting of the cusps (cf. fig. 3d). The front displays a periodic stationary pattern of uniform well selected wavelength.

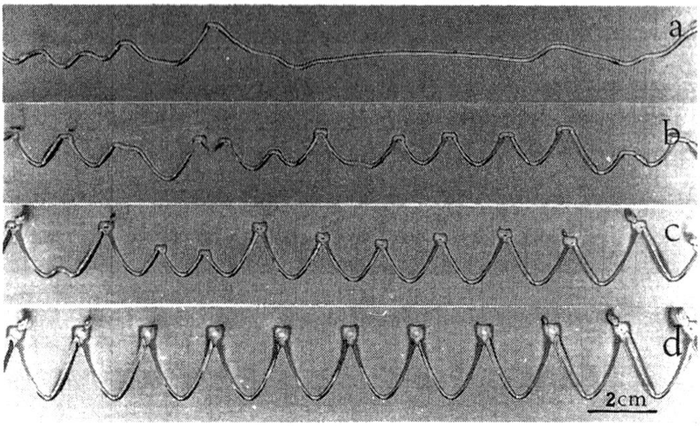

FIG. 3. Photographs of a portion of the front for several increasing values of ω, for Ω=140cm^3 a) ω= 11.93s^{-1}; b) ω=12.25 c) ω=12.56 d) ω=13.07 : stable pattern of stationary cells

The instability mechanism -- and hence also the mechanism of wavelength selection -- is the spliting of the higher cusp of the larger cells, as illustrated in fig. 4. Very near ω_{c3}, most of the front is occupied by "normal" cells of uniform size but some larger cells may also be present (cf. fig. 4a). In order to describe the mechanism, we focus on a single larger cell. When the higher cusp, which is at the left side of the larger cell (middle of fig. 4a), splits (cf. fig. 4b), it forms two new cusps which quickly decrease in height and separate to define a new small cell (fig. 4c). The neighboring cells are slightly compressed by this process. After a short time, the two new cusps destabilize (fig. 4d): the left cusp rises, while the right cusp decreases and vanishes. This cusp-splitting process yields a larger cell with the higher cusp now at the right side (cf.fig. 4e). The process repeats in one of two ways. If the cusp-splitting is followed by its time-reversed mirror-image, then the higher cusp regains its initial position on the left of the larger cell, giving a local oscillation. More commonly, the cusp-splitting continues in the same direction, giving a net lateral motion of the abnormal large cell. The lateral velocity of a cusp is smaller than the velocity of a free pulse, and is associated with a collective phenomena rather than a local one.

Cusps may appear or disappear; they always appear after the splitting of the higher cusp of a larger cell. The two cusps created by splitting first separate, compressing the neighboring cells to obtain enough space to create a new smaller cell. In contrast, a disappearing cusp removes a smaller cell, creating a larger one. This is the fundamental mechanism of wavelength adjustment.

In this regime, it is important to note that traveling pulses can still exist, as long as a flat front is available into which they can propagate. We demonstrate this by quickly increasing ω from $\omega<\omega_{c1}$, where the linear front exists, to $\omega=\omega_{c3}$, where a periodic pattern exists. One first observes two pulses, generated by the boundaries, propagating towards one another. The velocity of the free pulses, measured in this way, is then non-zero at $\omega=\omega_{c3}$. Behind the pulses are converging patterns of moving triangular cells. After the collision between the pulses, a stationary periodic pattern of stable triangular cells occupies the entire domain. This behavior shows that, in the periodic pattern regime, the propagation of pulses is inhibited only by their interaction. This interaction between converging pulses generates triangular cells which together form a stable periodic pattern.

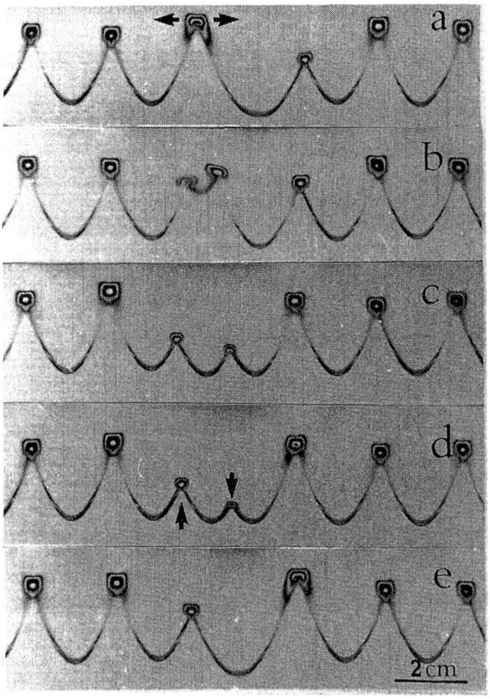

FIG. 4. Photographs of a portion of the front showing a larger cell, for Ω=140cm^3 and ω=13.071s^{-1}. a) The higher cusp bounds the left side of the larger cell The arrows indicate that it is about to split. b) The higher cusp splitting. c) A new smaller cell is created between two larger ones. d) One of the cusps decreases and disapears to generate a larger cell while the other one rises. The arrows indicate the direction of motion of the cusps. e) The higher cusp is now at the right of the larger cell.

Concluding Remarks

We have reported a new form of spatially periodic pattern formation, in which the periodic state consists of a collection of interacting propagative pulses. The wavelength of the pattern is given by the range of this interaction which is on the order of the size of the pulses. In contrast with usual systems, the transition is from a disorded state of colliding pulses to an ordered pattern as the control parameter is increased.

Acknowledgments

I thank S. Douady, S. Fauve and L. Tuckerman for helpful discussions.

References

[1] See J. S. Langer, Rev. Mod. Phys. 52, 1 (1980)
[2] R. W. Walden, P. Kolodner, A. Passner and C. Surko. Phys. Rev. Letters 55 (1985) 496-499.
[3] I. Mutabazi, J. J. Hegseth, C. D. Andereck and J. E. Wesfreid. Phys. Rev. A38 (1988) 4752-4760.
[4] M. Rabaud, S. Michalland and Y. Couder. Phys. Rev. Lett. 64 (1990) 184.
[5] A. J. Simon, J. Bechhoefer and A. Libchaber. Phys. Rev. Lett. 61 (1988) 2574-2577.

EMPIRICAL DETERMINATION OF THE ONSET OF CONVECTION FOR A HARD DISK SYSTEM

Dino Risso* and Patricio Cordero†
Departamento de Física
Facultad de Ciencias Físicas y Matemáticas
Universidad de Chile, Casilla 487, Santiago 3, Chile.

ABSTRACT. In this paper we present molecular-dynamics results on Rayleigh-Bénard convection in two dimensions for a system of hard particles contained in a rectangular box. The system shows significant compressibility effects. Important density and temperature variations are observed when different values for the gravitational acceleration and temperature gradient are considered. We have studied the influence of these effects on the onset of convection. The onset takes place for Rayleigh numbers significantly different to the values predicted by the Oberbeck-Boussinesq approximation. We give an empirical curve describing the onset of convection in terms of our control parameters. The results are compared with those obtained from the Navier Stokes equations when considering constant transport coefficients. There is a qualitative agreement with the theory when the work of compression is considered.

1. INTRODUCTION

Thermal convection in a fluid heated from below is one of the simplest cases of hydrodynamic instability [1, 2]. The computational experiments of Rayleigh-Bénard convection by means of event driven molecular-dynamics (MD) using hard disks [3, 4, 5, 6] has opened a new way to study dissipative structures that emerge

*Partially supported by grant *FONDECYT* 90-0005 and Project Fundación Andes C-10003.
†Partially supported by grant *FONDECYT* 90-1240.

E. Tirapegui and W. Zeller (eds.), Instabilities and Nonequilibrium Structures IV, 199–211.

when the system is driven off equilibrium. For the system under study the compressibility effects are important for the determination of the onset of convection curve in terms of the control parameters. In fact we have taken this as an interesting problem to be studied in more detail since as it has been stressed: "*..classifying the behavior observed in MD simulation has yet to be established*" [6].

The study of Rayleigh-Bénard convection when the compressibility effects are important is a research area where relatively few authors have come into. The analytic problems for the linear system [7, 8, 9] and the nonlinear one [10, 11] as well as the numerical simulations of the Navier-Stokes (NS) equations [12] have centered their attention in fluids that obey the ideal gas equation of state but with constant transport coefficients. In these studies, contrary to the Oberbeck-Boussinesq (OB) approximation for incompressible fluids, a strongly stratified fluid is considered with temperature differences ΔT which can be comparable with the characteristic temperature of the system.

The study of the Rayleigh-Bénard convection using molecular-dynamics techniques is relatively recent. Studying a system of a few thousand hard disks a group in Brussels was able to achieve a Rayleigh-Bénard molecular-dynamics simulation [3, 13]. Sometime later a similar result but for other values of the number density, number of disks and boundary conditions was obtained in [4] (see also [14]).

In [5] it was seen that it is possible to get good agreement between the molecular-dynamics experiments and the results obtained with the NS equations. Later on an interesting result for time dependent convection for a high Rayleigh number (7.8×10^4) with a large number of particles (57600 hard disks) was obtained in [6].

In this paper we report on our recent results on Rayleigh-Bénard (or thermal) convection in a two dimensional system of hard disks in a rectangular box. Contrary to what has been done in [3, 4, 6, 14] where the gravitational acceleration g has been chosen so as to minimize the density variation with height, in the present work we have studied the convective phenomenon for values of g such that the non-Oberbeck-Boussinesq (NOB) effects due to variations of the density with height become notorious.

The simulations were carried out using an efficient code based on a *event scheduling algorithm* inspired in [15]. The code – written in C – was run in an AViiON 200 workstation and a DEC 5400. In the workstation we get about 4 million collisions per hour of CPU time.

In our simulations the disks collide elastically among themselves. When they hit a wall the tangential component of the velocity is preserved. Preserving the tangential component of the velocity at wall collisions corresponds to setting stress free boundary conditions in all four walls. The normal component of the velocity is chosen from a random distribution to simulate contact with heath baths at temperatures T_b at the top and T_t at the bottom. The collisions at the lateral walls are elastic hence no heat can be exchanged through them.

This paper is organized as follows. In section 2 hard particle molecular-dynamics is described. In section 3 the experimental setting and the results are presented.

2. THE COMPUTER SIMULATIONS

The hard particle molecular-dynamics simulations are *event-driven* that is, the particles move free of interparticle interactions except for the moments when they become in contact with each other or with obstacles. The movement between collisions obeys Newtons's equation with whatever external force (e.g. gravity) may exist. The time intervals by which the simulation proceeds are the increments between events. Our program is based on a new highly efficient $O(\log N)$ algorithm in which only the times of the most probable collisions are compared when searching for the next collision [16].

Experimental setting. We are interested in simulating two-dimensional convection in a rectangular box. The experiment is defined giving the value of eight parameters. Besides the temperatures at the bottom and top, T_b and T_t, these are: the width L_X and the height L_Z of the box. Inside the box N hard disks of mass m and diameter D collide elastically with each other and move under the effect of gravity g.

There is no loss of generality fixing the units of temperature such that the Boltzmann constant is unity ($k_B = 1$) and fixing the units of length, mass and time such that: $D = 1$, $m = 1$, $T_b = 1$. With the free parameters it is useful to define the *number density* $n = \frac{N}{L_X L_Z}$ and the *aspect ratio* $\lambda = \frac{L_X}{L_Z}$.

Control parameters. We fixed part of the parameters as follows: the number of disks $N = 1521$, the aspect ratio $\lambda = 1$ and the density number $n = 0.25$.

This leaves as control parameters the gravitational acceleration g and the top temperature T_t.

Further we rewrite the Rayleigh number Ra in terms of the control parameters,

$$Ra = \frac{g\,\alpha\, n\, \Delta T\, L_Z^3}{\nu k} = \left(\frac{\Delta T}{T}\right)^2 \frac{N\,Fr}{\lambda} J(n) \tag{2.1}$$

$J(n)$ is a function of the number density and involves the viscosity ν, the thermal conductivity k and the thermal expansion coefficient α (see the Appendix of [17]). T is a mean temperature $\frac{1}{2}(T_t + T_b)$ and $\Delta T = (T_b - T_t)$ is the imposed temperature gradient. The Froude number Fr — which measures the ratio between the potential energy gained by a particle raised from the bottom to the top of the box, over the kinetic energy release — is defined by

$$Fr = \frac{g\,L_Z}{\Delta T}. \tag{2.2}$$

To distinguish our different experiments then, it suffices to give the values of Ra and Fr.

We fixed the gravitational acceleration g and the top temperature T_t through (2.1) and (2.2) choosing different values for the Rayleigh and Froude numbers. The Rayleigh and Froude numbers considered in our simulations ranged from 100 to 2600 and from 0.5 to 1.7 respectively.

The reduced temperature gradient $\Delta T/T$ in our experiments is not small (typically at the onset of convection its value being about 1.5). From (2.1) fixing Ra, n, λ and Fr - only when N is large enough one can have a small reduced temperature gradient. In spite of having a non negligible $\Delta T/T$ it is possible to see (Appendix of [17]) that taking a Froude number of about 1.56 corresponds to a region where the density does not vary sensibly with height and hence we can say that the OB approximation holds. The set of Froude numbers considered in our experiments goes into regions where the OB approximation does not apply.

Measurements. The hydrodynamic time scale can be taken as the characteristic time of the thermal diffusion and in our units it is given by $t_{\rm diff} = \frac{L_Z^2}{\kappa} = \frac{N}{\kappa n}$. Here κ is the thermal diffusivity. In our simulations this $t_{\rm diff}$ implies about one million total collisions (about 1300 collisions per particle).

To simulate the system a time large enough compared with the diffusion time in every simulation the system was left to evolve during at least a total of 20 million collisions (20 thermal diffusion times).

In the following we summarize and comment the results on the series of computational experiments that we have been making in Rayleigh-Bénard convection with hard disks.

3. THE RESULTS

The onset of convection for a given number density n and geometry depends only on two independent adimensional parameters: the Rayleigh and Froude numbers. This is different to the situation of a incompressible fluid when the Navier-Stokes equations are reduced by means of the OB approximation to a simple form. In that approximation the Rayleigh number alone determines whether there is convection or not.

Roughly speaking in our simulations the Rayleigh number controls the buoyancy force while the dependence with height of the density and temperature is driven by the Froude number.

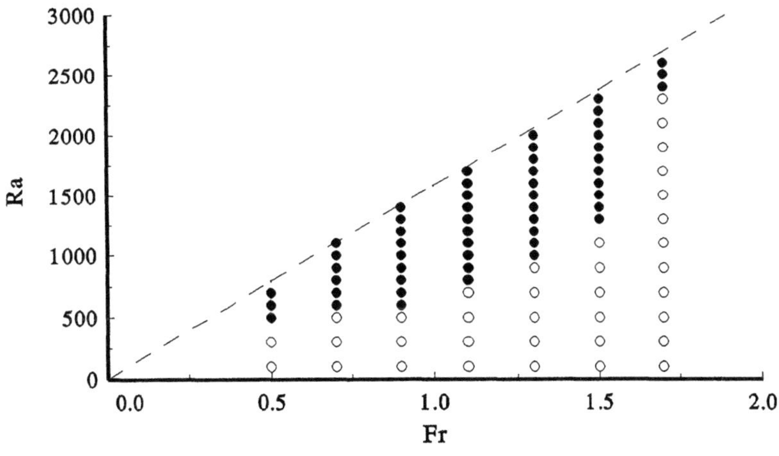

Figure 1. Summary of experimental observations. Solid circles are points where developed convection is observed. Open circles indicate that convection has not developed. The dotted line gives the maximum Rayleigh number in our setting.

To be able to measure the intensity of the convection in a simple way we use – as in the study of equilibrium phase transitions – an *order parameter*. This order parameter is chosen to be the *circulation of the velocity field* or rather the average in concentric paths of the circulation that we call C. This choice is possible near the onset of convection when the aspect ratio guarantees that a single convective roll develops.

In our experiments the dependence of Ra on Fr was studied classifying as non-convective or convective the regimes observed according to whether the circulation C did or did not change sign. To decide whether there was a stable convective pattern we observed the circulation values before making any time average. This is summarized in figure 1 in the Rayleigh-Froude plane. Also we show the curve corresponding to the largest possible Rayleigh number (that can be reached in our experimental setting due to the finite size of the system).

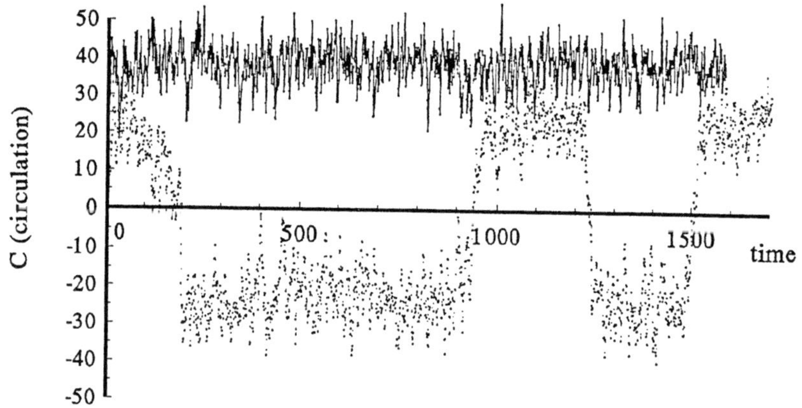

Figure 2. Time evolution of the circulation C. The solid line correspond to $Ra = 1200$ and the dotted line to $Ra = 870$. The Froude number is 1.0 in both cases. Each time unit corresponds to 200,000 collisions.

A typical plot of the evolution in time of the circulation C can be seen in figure 2. In this figure we show the evolution of C in experiments with two different values of the Rayleigh number sharing the same Froude number ($Fr = 1.0$) and the same initial conditions. The results correspond to two particularly large simulations (600 millions collisions for $Ra = 870$ and $Ra = 1200$). Of these experiments one ($Ra = 1200$) has a clear evolution towards a stable nonzero circulation. It is an undisputed convective situation. On the other extreme the situation with ($Ra = 200$ not shown in the figure) oscillates about zero circulation and it is clearly a nonconvective situation. The case in between with $Ra = 870$ corresponds, as we shall see, to a point slightly above the transition (or critical) value Ra_C.

The probability distribution $P(C)$ for having a circulation value between C and $C + dC$ has a double-peaked structure. In the limit of a continuous systems the width of these peaks tends to zero but for the discreet nature and finite size of our system it has two implications: the circulation will always fluctuate, and it can also switch sign. Therefore in one long run the circulation histogram will show the doubled peaked structure of the probability distribution (see figure 3). This structure can be seen even below the Ra_C. In a quite different context there is a clarifying discussion in [18], §6.B.2. See also [1], §VII.B.2. As we mention below the presence of the transition is further validated by the existence of a peak in the fluctuations of C about $Ra_C \approx 700$.

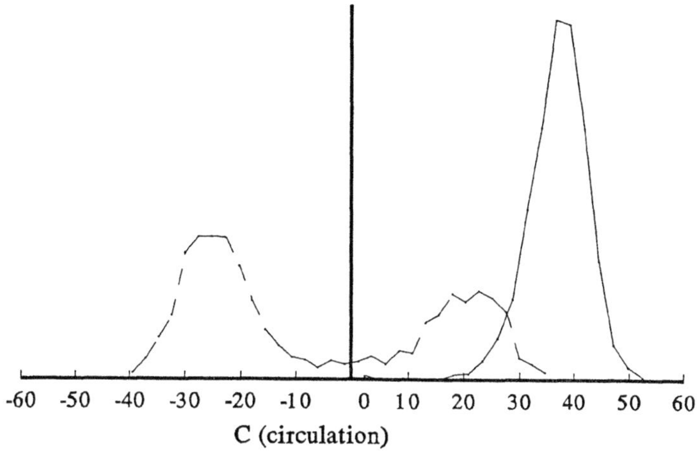

Figure 3. Distribution of the circulation C. The solid line corresponds to $Ra = 1200$ and the dotted line to $Ra = 870$. The Froude number is 1.0 in both cases.

It is interesting to point out that in the OB approximation the critical Rayleigh number, for the aspect ratio 1.0 system with stress free boundary conditions [2, 5], is $Ra_C \sim 780$ while in the simulations shown here convection is observed to develop for Rayleigh numbers below 700 when the Froude number is smaller than 1.0 and above 700 for Froude numbers larger than 1.0. The fact that Ra_C depends on other parameters is well known, see [7, 8, 19] or more recently [10]. Let us see what is the value for Ra_C in our cases with $Fr = 1.0$.

Experimental measures [20, 21, 22, 23, 24] and theoretical analysis [1, 2, 25,26] show that the square of the velocity increases linearly with Ra near the onset of convection, $v^2 = (\frac{Ra - Ra_C}{Ra_C})\, v_0^2$ therefore in figure 4 we have plotted C^2 and the time fluctuations (r.m.s) of C against Rayleigh number for the case of a Froude number $Fr = 1.0$. Each point in the figure corresponds to an average over a set of three experiments each one corresponding to different random initial conditions. The experiments lasted for about 20 million collisions and the figure shows the time average over the last 10 million collisions.

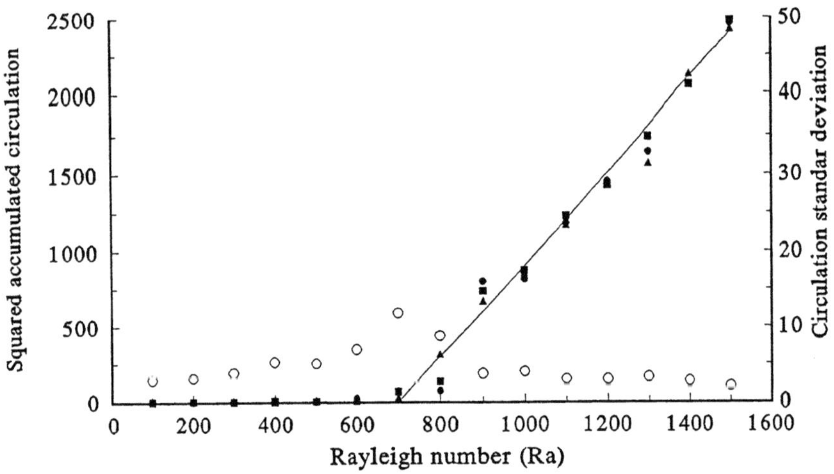

Figure 4. Squared circulation C^2 and the fluctuations of C in time versus Rayleigh number in arbitrary units. Solid symbols corresponds to C^2 for three different initial conditions. Open circles indicate the time fluctuations of the circulation (r.m.s.). The straigth line is a linear regression of the experimental values of C^2.

Assuming that C^2 grows linearly with Ra after the onset of convection a linear regression (the straight line in figure 4) is used to determine Ra_C from the experimental points. This gives a critical Rayleigh number of $Ra_C = 712 \pm 56$.

As we have mentioned before, one should expect that the fluctuations of the order parameter have a peak at the transition as in fact it is observed in figure 4. Physically this means that near the onset of convection there is plenty of circulation, only that it is constantly switching sign, hence giving a negligible average but strong fluctuations.

Let us consider now a compressibility effect in marginal stability. Next and following [7] we study experimentally if it is possible to redefine Ra in such a way that it plays a role similar to that in the incompressible OB approximation, namely, the only parameter that determines whether there is convection.

In [7] the authors study the equations that govern the thermal convection of a compressible fluid with constant transport coefficients and check the circumstances under which the OB approximation can be applied. These conditions are: a) that the characteristic variation lengths be much larger that the height of the container and b) the density and pressure variations induced by the convective motion do not significantly exceed the static variations of these quantities. When these conditions are met the NS equations take the same form as in the case of an incompressible fluid in the OB approximation provided the temperature gradient ΔT is replaced by its excess over the adiabatic gradient $(\Delta T - \alpha_0 c_p/g)$ (here $\alpha_0 = \alpha/T$) and the constant volume specific heat c_v replaced by the constant pressure specific heat c_p. That is the *Ra* appearing in the NS equations — as given in (2.1) — is replaced by

$$Ra^* = Ra\,(c_p - \alpha_0 Fr) \tag{3.1}$$

(here $\alpha_0 = \alpha/T$).

In our experiments however the ratio between the height of the vertical dimension and the observed scale variations of temperature (D_T), pressure (D_p) and density (D_n) — near criticality — are of order unity: $L_Z/D_T \sim 1.2$, $L_Z/D_p \sim 1.2$ and $L_Z/D_n \sim 0.5$.

Even though our experiments do not satisfy the assumption that the static variations are small and also assuming that — near criticality — convective variations in density and pressure are small compared with the static variations, let us see that our results tend to be consistent with $Ra_C^{-1} \propto Fr$ as predicted by (3.1).

In figure 5 we summarize our measurements of the order parameter C. We have plotted C^2 against *Ra* and *Fr*. It may be noticed that in the region where convection is well developed, C^2 varies linearly with respect to both variables [27]

$$C^2 \propto (Fr - Fr_C)(Ra - Ra_C). \tag{3.2}$$

It is also seen that when *Fr* approaches a value about 2.0 — namely the temperature excess over the adiabatic temperature gradient approaches zero — the intensity of the circulation changes rapidly being almost zero about $Fr \sim 1.7$. On

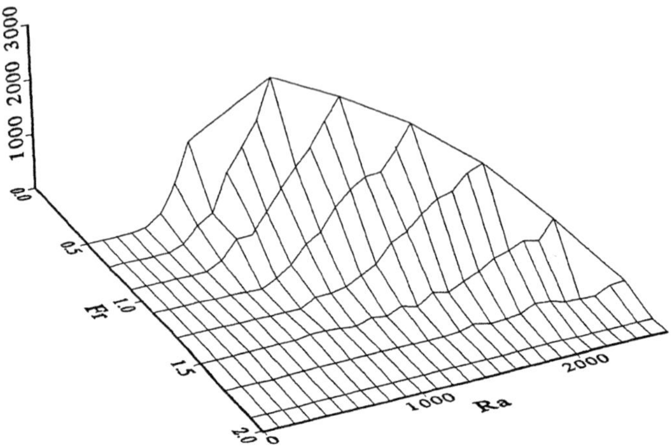

Figure 5. Squared circulation C^2 (in arbitrary units) in the Rayleigh-Froude plane. The dashed line corresponds to the extrapolated critical curve.

the other hand, when the Froude number goes down to about 0.5 — now the temperature excess is large — the critical Rayleigh number becomes small, reaching values as low as 400.

Starting from the values of C^2 given in figure 5 we extrapolate the critical curve associated to the onset of convection proceeding as follows. We use a linear regression in the region about $Fr \sim 1.7$ with fixed values for Ra to obtain a critical value for Fr. In the region $0.5 \leq Fr \leq 1.3$ a linear regression is used with fixed values for Fr to determine a critical value for Ra. The critical curve obtained in this way can be seen in figure 6. In that figure we have plotted the inverse of the experimental Ra_C against Fr (solid circles) — but correcting these to take into account finite size effects (see below) — showing good agreement with the predicted linear behavior.

Finite size effects are important. In our experiments when $Fr \sim 1.1$ the density variations are small contrary to the theoretical prediction ($Fr \approx 1.56$) obtained using a static analysis of the NS equations (see the Appendix of [17]). This can be partially understood from the balance of kinetic energy flow. In fact, having a finite number of particles the temperature gradient that the system selects is smaller than the externally imposed gradient. This implies that the condition $Fr \approx 1.56$ — to have a constant density profile — is modified to a Fr value smaller than 1.56. This also means that the Rayleigh and Froude number definitions (2.1) and (2.2) must be corrected to take into account the effective temperature gradient ΔT and the effective mean temperature T. Hence, in figure 6 we have plotted the experimental Ra and Fr correcting these to take into account the mean temperature and temperature gradient chosen by the system. It gives effective Rayleigh and Froude numbers Ra^{eff} and Fr^{eff} respectively.

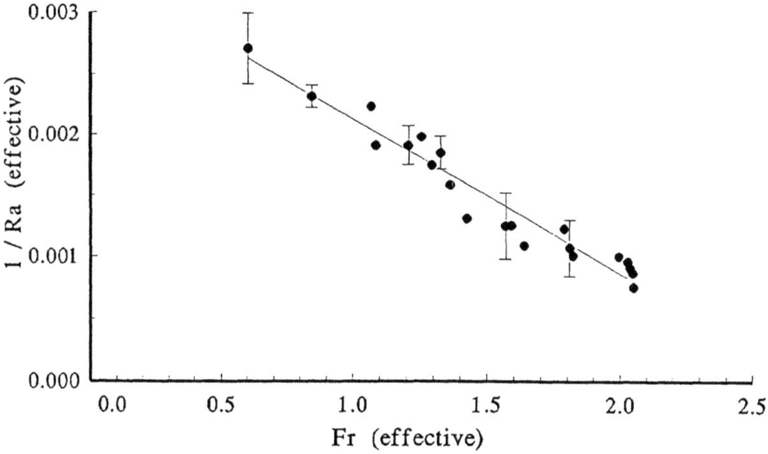

Figure 6. Inverse of the critical effective Rayleigh number versus the effective Froude number (see text).

It is worth remarking that the Ra_C^{eff} that is obtained from the simulations coincides with the theoretical Ra_C in the OB approximation (780) for Fr^{eff} slightly above 1.6 but in this case the observed density profile changes appreciable with height. On the other side almost homogeneous density profiles are observed for $Fr^{\text{eff}} \sim 1.1$ and one could say that the OB approximation applies. However here $Ra_C^{\text{eff}} \sim 550$. If we correct this value to take into account the work of compression — using (3.1) and the state equation for hard disks of Henderson [28] (in which case $\alpha_0 \approx 0.667$ and $c_p \approx 2.03$) — we obtain $Ra_C^{*\,\text{eff}} \sim 650$ which is well below the theoretical 780. The theoretical predictions for other Froude numbers can be seen in figure 6 (dashed line) where we have plotted the inverse of Ra_C as predicted by (3.1) using $Ra^* = 780$ and the state equation of Henderson. It is possible to see that for low Froude numbers there is no good quantitative agreement between the experimental and theoretical values. This is probably due to the strong height dependence of the transport coefficients and that Henderson's state equation cannot be taken to be valid far away from equilibrium. We must also take into account that multiple height scale corrections [8, 9] — which in the OB approximation are usually of order $(\Delta T/T)^2$ — may be of importance.

Making a linear regression of the experimental data adjusting values for Ra_C^0 and Fr^0 in

$$\frac{1}{Ra_C} = \frac{1}{Ra^0}\left(1 - \frac{Fr}{Fr^0}\right) \tag{3.3}$$

we obtain $Ra^0 = 368 \pm 14$ and $Fr^0 = 2.02 \pm 0.12$. Here Fr^0 is the extrapolated critical value for the adiabatic convective instability [29] and Ra^0 is the extrapolated Ra_C in the limit of very small Fr. That is, the region in which the temperature gradient exceeds the adiabatic temperature gradient — and therefore convection may occur — is $Fr < Fr^0$. If we consider the corrections derived from considering the temperature gradient chosen by the system, these parameters modify to $Fr^0 = 2.68 \pm 0.18$ and $Ra_C^0 = 295 \pm 10$.

In short, we have studied the influence of the compressibility effects on the onset of convection in MD simulations for a system of hard disks. In particular we have found an empirical curve describing the onset of convection in the Rayleigh-Froude plane. These results have been compared with those obtained from the NS equations. There is a qualitative agreement with the theory when compression work is considered but there is no easy way to get a good quantitative agreement. One should take into account that the temperature and the density variations are significant across the system and consequently the transport coefficients and other fluid properties cannot be taken to be constant. Another question that we have left open is the size effect that the velocity field introduces to the critical stability curve in the *Ra*-*Fr* plane. As it has been observed in MD simulations of RB convection [5] the velocity field — when convection is well developed — presents important effects due to the size of the system. These boundary effects may have an influence in the eigenfunctions (coming from the linear stability analysis) and consequently in the determination of the critical Rayleigh number. An important open question is to understand the relative importance of all these factors. Work along this line are in progress.

REFERENCES

[1] C.Normand and Y.Pomeau, Rev. Mod. Phys. **49** 3 (1977) 581

[2] S.Chandrasekhar, *Hydrodynamic and Hydromagnetic stability*

Clarendon Press, Oxford, 1961.
[3] M.Mareschal, E.Kestemont, J. Stat. Phys., **48** (1987) 1187
[4] D.C.Rapaport, Phys. Rev. Lett., **60** (1988) 2480
[5] A.Puhl, M.Malek-Mansour and M.Mareschal, Phys. Rev. A, **40** (1989) 1999.
[6] D.C.Rapaport, Phys.Rev.Lett **43** (1991) 7046.
[7] E.Spiegel, G.Veronis Astrophysics J. **131** (1960) 442.
[8] E.Spiegel, Astrophysics J. **141** (1965) 1068.
[9] D.Gough, D.Moore, E.Spiegel, N.Weiss, Astrophysics J. **206** (1976) 536.
[10] S.Gauthier, T.Desmarais, G.Ioos, Europhysics Lett. **10** (1989) 543.
[11] S.Gauthier, A.Gammes, G.Ioos, Europhysics Lett. **13** (1990) 117.
[12] N.Hurlburt, J.Toomre and M.Massaguer, Astrophysical J. **282**, (1984) 557.
[13] M.Mareschal, E.Kestemont, Nature, **427** (1986) 329
[14] M.Mareschal and E.Kestemont, Phys Rev Lett **62** (1989) 691;
[15] D.C.Rapaport, J.Comp.Phys **34** (1980) 184.
[16] M.Marin, D.Risso and P.Cordero, submitted to J.Comp.Phys. (1992).
[17] D. Risso and P. Cordero in *Proceedings of the VIth Medyfinol Conference* H.S. Wio, E.S. Hernández and A.N. Proto eds. Dept. Pub. Centro Atómico Bariloche, 1992
[18] O.L.Mouritsen, *Computer Studies of Phase Transitions and Critical Phenomena*, Springer, 1984.
[19] F.H.Busse, J.Math.Phys. **46** (1967) 149.
[20] P.Berge and M.Dubois, J. Fluid Mech., **32** (1974) 1041.
[21] P.Berge *Fluctuations, Instabilities, and Phase Transitions* T Riste ed, Plenum NY 1975
[22] M.Dubois and P.Berge, J. Fluid Mech., **85** (1978) 641.
[23] J.Wesfreid, Y.Pomeau, M.Dubois, C.Normand and P.Berge, Le Journal des Physique-Letters, **39** (1978) 725
[24] M.P.Arroyo and J.M.Savirón, J.Fluid Mech. **235** (1992) 325.
[25] L.P.Gorkov, Sov. Phys. JEPT **6** (1957) 311
[26] W.V.T Malkus and G.Veronis, J.Fluid Mech. **4** (1958) 225
[27] P.Coullet in *Non Linear Phenomena in Physics* F.Claro ed. Springer-Verlag, 1984.
[28] L.Landau and E.M.Lifschitz, *Fluid Mechanics* Pergamon, 1959.
[29] D.Henderson, Mol. Phys. **30** (1975) 971

1D AND 2D NONLINEAR EVOLUTION EQUATIONS FOR BÉNARD-MARANGONI CONVECTION

A. N. GARAZO *
Instituto de Física de Líquidos y Sistemas Biológicos (IFLYSIB)
Universidad Nacional de La Plata
Casilla de Correo 565, 1900 La Plata, Argentina

M. G. VELARDE
Facultad de Ciencias
Universidad Nacional de Educación a Distancia (UNED)
Apartado 60141, Madrid 28080, España

ABSTRACT.The 1D Korteweg-de Vries and the corresponding quasi 2D Kadomtsev-Petviashvili theory for solitary waves in shallow liquid layers is generalized to account for viscous and heat dissipation, and energy supply provided by interfacial tension forces. A connection is stablished with closely related experimental work where the essential features of soliton-like collisions is observed.

1. Introduction

In Marangoni-Bénard convection it is a well known experimental fact that steady cellular convective motions arise by heating an horizontal liquid layer, usually from below. However another interfacial instability may appear as an oscillatory mode [1,2] tunnig adequately the control parameters, in this case the thermal gradient and the depth of the fluid layer (the last one adjusts a ratio between buoyancy stresses -Rayleigh-Bénard problem- and the tractions due to non uniformity of the surface tension -Marangoni problem-).
The linear stability of this coupled problem was extensively analized, but with especial emphasis in the steady state convection. However more recent studies include the overstable case [3,4,5], and although any conclusions deserve experimental confirmation, we can use a lot of established qualitative information, especially the possibility to excite *long interfacial waves* heating from *above a thin* layer of viscous and incompressible fluid.
Our analysis is aimed to looking for the *nonlinear* evolution of these surface excitations. In a recent letter one of us [6] has shown that the Kadomtsev-Petviashvili (K-P) equation [7-11] governs solitary waves in Bénard convection (this analysis was made without any consideration of the Marangoni effect). This equation (strictly speaking a hierarchy of re-

E. Tirapegui and W. Zeller (eds.), Instabilities and Nonequilibrium Structures IV, 213–223.

lated nonlinear equations) displays a set of solutions being *solitary waves* the most striking of them. Then for a 2-D geometry there exists a subspace (defined by a suitable change of variables) where we would expect the appearence of regular wavetrains or "pulses". The possibility to observe these waves without damping is closely related with a mechanism to supply energy to the convecting fluid. Remembering that this is esentially a surface perturbation, a reasonable assumption is that the Marangoni effect is actually the driving force in *shallow* liquid layers [12]. Thus the present work, which consider the competition between driving forces (Marangoni-Bénard convection), attempts to make the early two dimensional picture [6] more complete, thus performing predictions relevant to experiment. In previous papers [13-15], following two different approaches we have shown that the Korteweg-de Vries (KdV) equation, modified by dissipation, is able to account for the evolution of the nonlinear excitation in 1-D. The present work is the natural extension when the second transversal direction in the surface plane is *weakly* taken into account.
The turbulence associated to many experiments of the required type, e.g. those done by Linde [16] and coworkers, makes difficult compare and check quantitatively our findings. A recent analysis of his old films and pictures [17] at least qualitatively supports a possible collision between soliton-like pulses but we expect more conclusive evidence in the future.

2. Problem Formulation

We consider a horizontal liquid layer, of thickness d subject to a transverse vertical temperature gradient.
Let x, y and z denote the two horizontal and one vertical coordinates respectively. With (u, v, w) the corresponding three velocity components, ρ, p, T and σ the density, pressure, temperature and air-liquid interfacial tension, the evolution of the liquid layer is governed by the following set of Boussinesquian-Newtonian thermohydrodynamic equations: [18-21]

2.1. Balance Laws (in the bulk)

2.1.1. *Equations of state*

$$u_x+v_y+w_z=0, \tag{1}$$

$$\rho_0\ (u_t+uu_x+vu_y+wu_z)=-p_x+\mu(u_{xx}+u_{yy}+u_{zz}), \tag{2a}$$

$$\rho_0\ (v_t+uv_x+vv_y+wv_z)=-p_y+\mu(v_{xx}+v_{yy}+v_{zz}), \tag{2b}$$

$$\rho_0\ (w_t+uw_x+vw_y+ww_z)=-p_z+\mu\,(w_{xx}+w_{yy}+w_{zz})-g\ \rho(z)\,, \tag{2c}$$

$$T_t+uT_x+vT_y+wT_z=\kappa\,(T_{xx}+T_{yy}+T_{zz}) \tag{3}$$

$$\rho=\rho_0\ [1-\alpha\,(T-T_0)]\,, \tag{4a}$$

$$\sigma=\sigma_0+\gamma\,(T-T_0)\,,\qquad \gamma=\left(\frac{\partial\sigma}{\partial T}\right), \tag{4b}$$

where g denotes the gravitational acceleration acting along the z-direction downwards, ρ_0 and σ_0 are some reference values, e.g., at the motionless quiescent state at the temperature $T_0=T_{z=d}$, and subscripts t, x, y and z refer to time and space derivatives. Note that generally α is positive (except water around 4º C) and γ though generally negative may according to the liquid become positive [22-25]; μ is the dynamic viscosity and κ the thermometric conductivity or thermal diffusivity of the liquid.
The above given evolution equations must satisfy the following boundary conditions (**b.c.** in what follows):[18,19]

2.1.2. *Boundary conditions.* At $z=d+\eta\,(x,y,t)$,with $\eta\,(x,y,t)$ the time and space dependent interfacial deformation,

$$w=\eta_t+v\ \eta_y+u\ \eta_x\ , \tag{5}$$

$$(p-\rho_0 g\eta)\,\eta_x\ -2\ \mu u_x\eta_x+\mu\,(u_z+w_x)-\sigma_x-\eta_x\sigma_z-(\bar\kappa+\bar\varepsilon)\ [u_{xx}+v_{yx}] +\bar\varepsilon\,[v_{xy}-u_{yy}]=0\,, \tag{6}$$

$$(p-\rho_0 g\eta)\,\eta_y\ +\ \mu\ [\eta_x(v_x+u_y)+(v_z+w_y)-\sigma_y-\eta_y\sigma_z-(\bar\kappa+\bar\varepsilon)\ [u_{xy}+v_{yy}] +\bar\varepsilon\,[-v_{xx}+u_{yx}]=0\,, \tag{7}$$

$$p\ -\ \rho_0 g\eta-\frac{\rho_0 g\alpha\beta\eta^2}{2}+\ \mu\ [u_z\eta_x-2\ w_z]+\eta_x\,(\sigma_x+\eta_x\sigma_z)+\sigma\eta_{xx}=0\,, \tag{8}$$

$$\frac{k}{N}\ [-\eta_x T_x-\eta_y+T_z]=-k\beta-h\,(T-T_0)\qquad \beta d=T_b-T_0\qquad T_b=T(0)\,,\quad T_0=T(d) \tag{9}$$

at $z=0$,

$$w=0\ , \tag{10a}$$

$$u_z=v_z=0 \qquad (\textit{stress-free}), \tag{10b}$$

$$T=T_b \qquad (\textit{"ideal" conductor})\ , \tag{11a}$$

or

$$T_z=\textit{constant} \quad (\textit{"ideal" insulator}) \tag{11b}$$

The alternative to Eq.(10b) is the more realistic case of a rigid boundary, $u=v=0$. However we take the stress-free case for simplicity. We have introduced the two coefficientes $\bar{\kappa}$ and $\bar{\epsilon}$, the surface shear and dilational viscosities, respectively. $N^2=1+\eta_x^2+\eta_y^2$ is the norm of the surface metric. The quantity h/k accounts for the air-liquid heat transfer characteristics, with $k=\kappa/(\rho_0 c)$ the thermal conductivity and c the specific heat. Later on we shall use the temperature gradient $\beta=(T_b-T_0)/d$, taken positive when the heating is from the liquid side as in the traditional Bénard set-up. As before T_0 is a reference value taken at the undeformed liquid-air interface in the quiescent state and $T_b=T_{z=0}$. Note that for simplicity we have considered the air above the liquid layer weightless and dynamically passive and at the pressure p_{air} coincident with that of the liquid surface at $z=d$.
The motionless quiescent steady state denoted with subscript s is a solution of the above posed problem. We have

$$T_s(z) = T_b - \beta z\ , \tag{12}$$

$$\rho_s(z) = \rho_0\,[1+\alpha\beta(z-d)]\ , \tag{13}$$

$$u_s = 0\ , \tag{14}$$

$$v_s = 0\ , \tag{15}$$

$$w_s = 0\ , \tag{16}$$

$$p_s(z) = p_{air} - \rho_0 g \left[(z-d) + \frac{\alpha\beta(z-d)^2}{2} \right] \tag{17}$$

In the remaining of this work we shall consider the conditions for the above motionless steady state to be convectively unstable to time-dependent disturbances of the solitary type described by the dissipation modified K-P equation (up to the sign of the second order derivative in the transverse y -direction this is an equation derived in plasma theory for the study of disturbances upon the Korteweg-de Vries (KdV) equation) [7-11].

The nondimensionalisation of the problem and an *a priori* estimate of its pertinent scales is useful now. As new *units* we choose d for length, βd for temperature, κ/d for velocity, d^2/κ for time and $\mu\kappa/d^2$ for pressure, respectively. For later convenience we also introduce the following dimensionless groups: $P=\mu/\kappa\rho_0$ (Prandtlnumber), $Bi=hd/\kappa$ (Biot number), $K=\mu\kappa/\sigma_0 d$ (Capillary number), $B=\rho_0 g d^2/\sigma_0$ (Bond/Eötvös number), $V_1=\bar{\kappa}/\mu d$ and $V_2=\bar{\epsilon}/\mu d$ (Surface viscosity numbers), $R=\alpha g\beta d^4/\kappa\nu$ (Rayleigh number), $M=-\gamma\beta d^2/\kappa\nu$ (Marangoni number or surface velocity Reynolds number) with $\nu=\mu/\rho_0$, the kinematic viscosity. Note that $\Gamma=M/R$ is a ratio of surface forces to buoyancy forces in the layer, thus it appears as a quantity not depending on the thermal gradient but merely connected with the liquid layer's depth and its material properties. Incidentally, Γ is proportional to both α^{-1} and $(-\gamma)$, two quantities that although generally positive may under appropiate circumstances either vanish or change sign.

Following the above stated b.c.(11a), we analyze the case when the solid bottom plate is an ideal heat conductor, e.g. a copper plate. For the air-liquid interface a reasonable assumption is that the heat transfer is rather weak. Introducing ϵ, a smallness parameter we can set $Bi=\epsilon^2$. This estimate is consistent with a search of long wavelength excitations because these convective modes generally dissipate the least. For the remaining variables we shall use: [26-29]

$$\xi=\epsilon(x-Ct),\ \upsilon=\epsilon^2 y,\ \tau=\epsilon^3 t,$$

$$u=\epsilon^2(u_0+\epsilon u_1+\ldots),\ v=\epsilon^3(v_0+\epsilon v_1+\ldots),\ w=\epsilon^3(w_0+\epsilon w_1+\ldots),$$

$$p-p_s=\epsilon^2(p_0+\epsilon p_1+\ldots),\ T-T_s=\epsilon^3(\theta_0+\epsilon\theta_1+\ldots),\ \eta=\epsilon^2(\eta_0+\epsilon\eta_1+\ldots),$$

$$C=C_0+\epsilon^2 C_2+\ldots,\ R=R_0+\epsilon^2 R_2+\ldots,\ M=M_0+\epsilon^2 M_2+\ldots,$$

$$V_1=\epsilon^2(V_{10}+\epsilon^2 V_{12}+\ldots)\quad \text{and}\quad V_2=\epsilon^2(V_{20}+\epsilon^2 V_{22}+\ldots)$$

Using these scalings, inserting the above given expansions in the evolution equations and b.c., and noticing that ϵ becomes an *ordering* parameter, from the original nonlinear problem we get a hierarchy of linear inhomogeneous equations that have

to be sequentially solved subjected to a solvability condition (Fredholm alternative) here reduced to integrating the x-component of Navier-Stokes equation over the liquid layer.
To the lowest order of approximation, i.e., at the ϵ^0 order we get the basic linear heat conduction steady state. Then at the following order of approximation in ϵ, the solvability condition provides $C_0 = \sqrt{BP/K - M_0 P}$ which shows the correction in the celerity of the wave due to the Marangoni effect appearing at the lowest-order coefficient in the expansion of the phase velocity C in powers of ϵ.
To the order ϵ^2 from the solvability condition we get the neutral stability locus

$$-\frac{M_0}{12} + \frac{R_0}{30} = 1 \,, \tag{18}$$

that for vanishing buoyancy, $R_0 = 0$, reduces to $M_0 = -12$. For a standard liquid, i.e., for a liquid with negative $\gamma = (\partial\sigma/\partial T)$, a negative Marangoni number demands that the heating be done from the *air side*. Thus we recover the experimental structure for which an earlier prediction was reported concerning on the one hand the onset of capillary-gravity waves in a deep, actually an infinitely deep liquid layer [30], and on the other hand KdV excitations in the one dimensional analysis of a shallow liquid layer subjected to Marangoni or Bénard stresses [13-15,31,32].
The value $R_0 = 30$ at vanishing Marangoni effect is just the same number that appears either in the one-dimensional KdV description [31,32] or in the two-dimensional K-P theory [6]. This critical Rayleigh number defines the onset of an excitation in a Rayleigh-Bénard liquid layer heated from below and open to the ambient air with, however, no Marangoni stresses involved. Although for comparison with experiment the model seems quite an unrealistic one yet it shows the ubiquity of either the KdV or the K-P equations when the shallow layer approximation is used.
Note that if γ is negative, when M_0 is negative we are heating from below and assuming α positive this demands that R_0 be negative as expected. If, however, M_0 is positive the heating is from below and R_0 is also positive, then Eq.(18) allows for a region of values where instability is possible.
When γ is positive, like in some water-alcohol mixtures and other liquids [22-25], if the heating is from above M_0 is positive. Then Eq.(18) demands R_0 to be positive which is not allowed. However, if the heating is from below M_0 is negative and R_0 is now positive. Then Eq.(18) is satisfied thus allowing for convective instability in the form of a K-P excitation.
The above given results can be reinterpreted using the above

mentioned parameter $\Gamma=M/R$. The neutral stability locus is defined by $M_0=60\Gamma/(2-5\Gamma)$. For standard liquids Γ is positive. Then in the region $0<\Gamma<0.4$ we have M_0 positive which demands heating the layer from below. Thus Marangoni stresses compete for instability with the stabilizing Rayleigh buoyancy forces (R_0, **positive**). In this narrow region bouyancy is the dominating agent in the neighborhood of $\Gamma=0$. When Γ tends to the value 0.4, the influence of buoyancy is reduced and accordingly M_0 rises.
A qualitative change appears at $\Gamma=0.4$, or more generally when Γ exceeds such value. Indeed when both M_0 and R_0 are negative, the Marangoni stresses tend to dominate the evolution of the liquid layer and as the heating is from above the effect of buoyancy looses relevance. When Γ increases, M_0 asymptotically approaches the limiting negative value earlier indicated.
The region where Γ is negative should not be excluded from our analysis. A liquid with non-positive α is for example water at and below $4°C$. Another interesting and thermodynamically more usual case is that of a liquid solution with γ positive. In this case, assuming that α is positive, if we heat from below R_0 is positive and M_0 is negative. Then there is a wide region where instability is possible due to the fact that buoyancy and Marangoni stresses are competing driving forces. If, however, the heating is from above, Eq.(18) shows that instability is not predicted.
In dimensional form the celerity of the traveling waves is

$$C_0^2=gd+\left[\frac{60\gamma\mu\kappa}{(2\rho_0^2 g\alpha d^4+5\rho_0 d^2\gamma)}\right], \tag{19}$$

that for a *shallow* layer can be approximated by

$$(C_0^2)_{SW}\approx gd\left(1+\frac{12\mu\kappa}{\rho_0 g d^3}\right) \tag{20}$$

It shows a correction of order d^{-3} which differs from the *ideal, dissipation-free* KdV or K-P theory [8,27] where the correction scales with d^{-1}. For a liquid layer of silicone oil with $d=10^{-2}cm$ our predicted correction due to the dissipation is about 30%.
The next term in the hierarchy of equations is at the ϵ^3 order. The solvability condition in this case is equivalent to the following differential equation to be satisfied by η_0,

$$(\eta_{0\tau}+\delta\eta_{0\xi}+\alpha_1\eta_0\eta_{0\xi}+\alpha_2\eta_{0\xi\xi\xi})_\xi+\alpha_3\eta_{0\upsilon\upsilon}=0, \tag{21}$$

which is just the *K-P equation* [7]. The coefficients δ, α_1, α_2 and α_3 are:

$$\delta = - C_2 - \frac{6 + \frac{M_2}{2} - \frac{R_0}{30}}{\left[12 + \frac{B}{K} - \frac{2}{5} R_0\right]^{1/2}} \tag{22}$$

$$\alpha_1 = \frac{P^{1/2}\ (15B+120K+KR_0)}{10K\left[12 + \frac{B}{K} - \frac{2}{5} R_0\right]^{1/2}} \quad , \tag{23}$$

$$\alpha_2 = \alpha_{21} + \alpha_{22} \quad , \tag{24}$$

where

$$\alpha_{21} = \frac{P^{1/2}\ (-7875+2625B+88200K-2880KR_0-2KR_0^2)}{15750K\left[12 + \frac{B}{K} - \frac{2}{5} R_0\right]^{1/2}} \quad ,$$

and

$$\alpha_{22} = \frac{P^{3/2}\ (12600B+151200K+5BR_0-4980KR_0-2KR_0^2)}{15750K\left[12 + \frac{B}{K} - \frac{2}{5} R_0\right]^{1/2}} \quad ,$$

$$\alpha_3 = \frac{P^{1/2}\ (5B+60K-2KR_0)}{10K\left[12 + \frac{B}{K} - \frac{2}{5} R_0\right]^{1/2}} \tag{25}$$

As we are interested in the evolution equation obeyed by the surface deformation when dissipation competes and/or cooperates with nonlinearity and dispersion, we proceed to the ϵ^4 order. Then introducing the hat variable $\hat{\eta} = \eta_0 + \epsilon\eta_1 + \ldots$ we have

$$\hat{\eta}_\tau + \delta\hat{\eta}_\xi + \alpha_1\hat{\eta}\hat{\eta}_\xi + \alpha_2\hat{\eta}_{\xi\xi\xi} + \epsilon\left[\alpha_4\hat{\eta}_{\xi\xi} + \alpha_5\hat{\eta}_{\xi\xi\xi\xi\xi} + \alpha_6(\hat{\eta}\hat{\eta}_\xi)_\xi\right] + \alpha_3\hat{\eta}_{\upsilon\upsilon} = 0 , \tag{26}$$

where δ ,α_1 ,α_2 and α_3 have already given and α_4 ,α_5 and α_6 are

$$\alpha_4 = -\frac{P}{90}\ [360+15M_2-R_0-6R_2+45(V_1+V_2)] \quad , \tag{27}$$

$$\alpha_5 = \alpha_{51} + \alpha_{52} + \alpha_{53} \quad , \tag{28}$$

with

$$\alpha_{51}=\frac{(653400K+3979800KP+504900BP^2+6058800KP^2)}{1559250K},$$

$$\alpha_{52}=\frac{220BP^2R_0-199320KP^2R_0-22KR_0^2-112KPR_0^2}{1559250K},$$

$$\alpha_{53}=\frac{(21120KR_0-111375KPR_0-88KP^2R_0^2)}{1559250K},$$

and finally

$$\alpha_6=\frac{2}{15}P(R_0+30) \tag{29}$$

Eq.(26) is our *dissipation modified K-P equation* or two-dimensional dissipation modified KdV equation. Note that α_6 could be positive or negative according, in particular, to the value taken by the Rayleigh number. Thus Eq.(26) is the extension to the Marangoni-driven Bénard problem of the earlier result obtained by Kurcbart et al.[6].

In order to obtain an expression of Eq.(26) valid in the original reference frame, we return to the variables x,y,z and t. At the same time we are turning $\hat{\eta}$ into the surface deformation η measured in the fixed frame; so Eq.(26) turns into:

$$\eta_t+\overline{C_e}\,\eta_x+\alpha_1\eta\,\eta_x+\alpha_2\eta_{xxx}+\overline{\alpha_4}\,\eta_{xx}+\alpha_5\eta_{xxxx}+\alpha_6\,(\eta\,\eta_x)_x+\alpha_3\eta_{yy}=0, \tag{30}$$

where $\overline{\alpha_4}=\epsilon^2\alpha_4$ and $\overline{C_e}=C_0+\epsilon^2C_2+\epsilon^2\delta$, the velocity that one would observe in the laboratory frame of reference, are fixed by the experimental setup and measurable quantities. A Galilean transformation can be performed to eliminate η_x from Eq.(30) and put it in the standard form.

A numerical treatment of this equation is the following step to determine the nature and stability of the solutions.

* Member of the Carrera del Investigador Científico from CONICET, Argentina.

Acknowledgments

The first author benefited from a visiting postdoctoral fellowship from the European Economic Community where the first part of this work was carried out. The authors acknowledge the kind hospitality offered by Prof. A. Liñán at the Aeronautical Engineering Superior School, Polytechnic University of Madrid. This research has been sponsored by CICYT Grant PB 86-651 (Spain).

References

1. For an introduction to surface tension-driven convection, see Velarde,M.G. and Normand,C.,Sci.Am. **243** (1980) 78.
2. The relevance of Marangoni effect in several fields is ex tensively described in *Physicochemical Hydrodynamics: Interfacial Phenomena*, Vol. 174 of NATO Advanced Study Institute,Series B:Physics, ed. by Velarde, M.G. (Plenum, New York, 1988).
3. Benguria,R.D. and Depassier,M.C., Phys.Fluids A **1** (1989) 1123.
4. Gouesbet,G., Maquet,J., Rozé,C. and Darrigo,R., Phys.Fluids A **2** (1990) 903.
5. Perez García,C. and Carneiro,G.,Phys.Fluids A **3** (1991) 292.
6. Kurcbart,S.M., Manna,M.A., Pereira,J.G. and Garazo A.N., Phys.Lett. A **148** (1990) 53.
7. Kadomtsev,B.B. and Petviashvili,V.I., Sov.Phys.Dokl. **15** (1970) 539.
8. Drazin,Ph. and Johnson,R.S.,*Solitons:An Introduction* (Cambridge Univ. Press, Cambridge, 1989).
9. Novikov,S., Manakov,S.V., Pitaevskii,L.P. and Zakha rov,V.E., *Theory of solitons* (Consultants Bureau, New York, 1984).
10. Hammack,J.,Scheffner,N. and Segur,H., J.Fuid Mech. **209** (1989) 567.
11. Segur,H. and Finkel,A.,Studies Appl. Maths. **73** (1985) 183.
12. Normand,C.,Pomeau,Y. and Velarde,M.G.,Rev. Mod. Phys. **49** (1977) 581.
13. Chu,X.-L. and Velarde,M.G.,Phys.Rev.A **43** (1991) 1094.
14. Velarde,M.G., Chu,X.-L. and Garazo,A.N.,Phys.Scr. **T35** (1991) 71.
15. Garazo,A.N., and Velarde,M.G., Phys.Fluids A **3** (1991) 2295.
16. Linde,H.,Schwartz,P. and Wilke,H. *Dynamics and Instability of Fluid Interfaces,* (ed. by Sorensen, T.S., Springer-Verlag, Berlin,1979, and references therein).
17. Weidman,P.,Linde,H. and Velarde,M.G. (unpublished).
18. Aris,R.,*Vectors, Tensors and the Basic Equations of Fluid Mechanics* (Prentice-Hall, Englewood Cliffs, NJ, 1962).
19. Levich,B., *Physicochemical Hydrodynamics* (Prentice-Hall, Englewood Cliffs, NJ, 1962).
20. Pérez Cordón,R. and Velarde,M.G., J.Phys.(París) **36** (1975) 591.

21. Velarde,M.G. and Pérez Cordón,R., J.Phys.(París) **37** (1976) 177.
22. Desré,P.J. and Joud,J.C., Acta Astronaut. **11** (1984) 143.
23. Legros,J.C.,Limbourg,M.C. and Petré,G., Acta Astronaut. **11** (1984) 143.
24. Motomura,K.,Iwanaga,S.-I.,Hayami,Y.,Uryu,S. and Matuura,R., J.Colloid Interface Sci. **80** (1981) 32.
25. Vochten,R. and Petré,G., J.Colloid Interface Sci. **42** (1973) 320.
26. Jeffrey,A. and Kawahara,T.,*Asymptotic Methods in Nonlinear Wave Theory* (Pitman, London, 1982).
27. Whitham,G.B., *Linear and Nonlinear Waves* (J.Wiley, New York, 1974).
28. Ioos,G. and Joseph,D.D., *Elementary Stability and Bifurcation Theory*, 2nd. Edition, chapter VIII (Springer-Verlag, Berlin, 1990).
29. Joseph,D.D.,in:*Hydrodynamic Instabilities and the Transition to Turbulence*, 2nd. Edition (eds. Swinney,H.L. and Gollub,J.P.) sec.3.7,pp.27-76 (Springer-Verlag, Berlin,1985).
30. Chu,X.-L. and Velarde,M.G.,Physicochem.Hydrodyn., **10** (1988) 727.
31. Alfaro,C.M., and Depassier,M.C. Phys. Rev. Lett. **62** (1990) 2597.
32. Aspe,H. and Depassier,M.C., Phys.Rev. A **41** (1990) 3125.

A GENERALIZED SWIFT-HOHENBERG MODEL FOR SEVERAL CONVECTIVE PROBLEMS

C. PEREZ-GARCIA ‡, J. MILLÁN-RODRIGUEZ, H. HERRERO
Departamento de Física y Matemática Aplicada,
Universidad de Navarra, 31080 Pamplona, Navarra, Spain.
‡ Departament de Física Fonamental, Universitat de Barcelona,
Avgda. Diagonal 647, 08028 Barcelona, Catalonia, Spain.

M. BESTEHORN.
Institut für Theorestische Physik und Synergetik der Universität Stuttgart,
Pfaffenwaldring 57/4, 7000 Stuttgart 80, Germany.

ABSTRACT. Some generalizations of the Swift-Hohenberg equation are proposed in order to study the convective patterns that appear under different conditions. By means of this generalized equation two situations are considered. The first one concerns changes of symmetries in stationary patterns. In the second one a convective pattern under rotation is studied. This effect must be included by a non variational term and the corresponding solutions are not stationary. A detailed analysis of the corresponding amplitude equations is also made in order to compare the results obtained with the generalized Swift-Hohenberg equation. Analytical and numerical methods are used to determine the stability of the corresponding solutions. Simulations of these models can be compared with patterns obtained in recent experiments and suggest some situations in which some new dynamical effects could be obtained.

1. Introduction

Among all the aspects of the dynamics of complex systems, one of the most interesting is the temporal and spatial formation of structures in systems far from equilibrium. This kind of structures, present in many phenomena in nature, has been largely studied in thermal convection [1]. This is due to the fact that experiments are made in closed container where one can achieve very good thermal regulation and a good visualization of the patterns. We consider two different situations related to this problem in the nonlinear regime: convection in a layer where a pattern of squares can be found (with nearly insulating boundaries in the top and bottom [2,3]) and convection with a rotation around a vertical axis with constant angular velocity Ω [4].

We consider the system of balance equations for a horizontal layer of fluid under a vertical temperature gradient rotating around the vertical axis. As usual these equations are written in dimensionless form and under the Boussinesq approximation. The five non-dimensional equations for the perturbations of the five variables around a quiescent conductive state (velocity $\mathbf{u} \equiv (u, v, w)$, pressure p and temperature θ) read as [5]:

$$Pr^{-1}\ (\partial_t + \mathbf{u}\cdot\nabla)\mathbf{u} = -\nabla p + R\theta\ \hat{\mathbf{z}} + Ta\ (\mathbf{u}\times\hat{\mathbf{z}}) + \Delta\mathbf{u} \tag{1}$$
$$(\partial_t + \mathbf{u}\cdot\nabla)\theta = w + \Delta\theta \tag{2}$$
$$\nabla\cdot\mathbf{u} = 0 \tag{3}$$

E. Tirapegui and W. Zeller (eds.), Instabilities and Nonequilibrium Structures IV, 225–234.

where $\hat{\mathbf{z}}$ denotes unit vector in the vertical direction. The term Ta $(\mathbf{u} \times \hat{\mathbf{z}})$ accounts for the Coriolis acceleration. The dynamics of the problem is ruled by three dimensionless parameters $R = \alpha g \Delta T d^3/\kappa\nu$, $Pr = \nu/\kappa$ and $Ta = 2\Omega d^2/\nu$, called the Rayleigh, Prandtl and Taylor numbers respectively.

Some boundary conditions (bc) for the corresponding variables must be added in order to find complete solutions of the problem. These bc usually change the eigenfunctions of the linear problem and have an important role in the planform that can appear in convective patterns. For simplicity we don't discuss the different bc in detail. (The interested reader can find a derivation of the main bc in the book of Platten and Legros [6]).

The solution of these equations with the bc is usually a very difficult task. Therefore, some simplified models are proposed that reproduce the main features observed in experiments: the formation and evolution of patterns with different planforms. The main theoretical tool to study these situations is the so called amplitude method, widely used in convective problems [7]. Althought this method includes spatial variations, it is useful only when these variations are much smaller than the regularity of the pattern.

It is necessary to develop an alternative method that overcomes this difficulty. It is based in the so-called Swift-Hohenberg equation (SHE) [8]. This equation includes the fast spatial variations in the pattern and no preferred orientation is assumed. Its coeffcients can be obtained from the basic equations. However, this model is only approximated because one must assume some hypothesis on the couplings among the nonlinear modes [9]. The main advantage is that the most important characteristics of convective patterns can be obtained in this framework above (but very near) the critical convective treshold: the periodicity and the dynamics of the pattern (evolution of defects, large scale flows, some secondary instabilities), the existence of a band of unstables modes, some wavelength selection mechanism for very large aspect ratio (the ratio between a characteristic horizontal lenght and the liquid depth) convective cell, etc.

In the present paper we take advantage of the two methods in order to understand the dynamics of convective patterns in the two mentioned situations. The second section is devoted to discuss a generalization of the SHE that allows to cover these situations. From this generalization we restrict the analysis to the main modes. This gives the amplitude equations in a simple way. From these equations one can straightforward deduce the general stability properties of the different solutions in the case of a spatially homogeneous pattern. Althought this introduces an important restriction, we use the corresponding diagram to have an idea of the dynamics that can take place in the system. By identifying the parameters in both equations one can use the stability diagram of the amplitude equations to simulate the SHE with the appropriated values. We use this procedure in the two cases considered separately in section 3 (competition between different symmetries) and section 4 (convection in a rotating layer).

2. Generalized Swift-Hohenberg equation

The original model proposed by Swift and Hohenberg [8] can be generalized to more situations. Following a framework recently proposed [10] one can derive a generalized SHE for the convection-rotation problem from the hydrodynamical equations (1)-(3). For the sake of simplicity we restrict the analysis to the case $Pr = \infty$. This assumption allows to eliminate some cumbersome calculations.

This equation involves only one order parameter Ψ which is directly related with the temperature field or the velocity field in the pattern. This scalar field Ψ includes the fast spatial oscillation of the patterns. At the lowest order this GSHE for Ψ has the form:

$$\dot{\Psi} = [\epsilon - (1+\Delta)^2]\Psi + a\Psi^2 - b\Psi^3 + c\Psi\Delta\Psi^2 + d\nabla\Psi\cdot\nabla\Psi^2 - e\Psi\Delta^2\Psi^2 + f[\nabla\Psi\times\nabla\Delta\Psi^2]_z \quad (4)$$

The parameters that can be externally regulated are the supercritical heating $\epsilon = (R - R_c)/R_c$ and f which is proportional to the rotation speed. (The subscript in the last term $[]_z$ indicates the vertical component). The other terms are derived in a consistent way, taking into account the possible contribution of the orientations in the pattern. the quadratic term for example, accounts for some contributions that break the vertical symmetry (non-Bousinesq effects, Bénard-Marangoni instability). Under suitable conditions this term can favour a pattern of hexagons. The term with the coefficient d can give rise to the stabilization of a pattern of squares. (Althought other combinations of terms can be consistently included in the GSHE only the most important are retained in eq.(4)). The parameters a, b, c, d, e y f can be derived from the basic equations and strongly depend on the specific vertical bc of the problem considered.

This equation is supplemented with bc for the scalar field Ψ that correspond to the usual ones found in experiments. (Notice that these are lateral bc, because Ψ is defined in the plane). Usually we take the following bc are taken on the sidewall: $\Psi = \partial_{\mathbf{n}}\Psi = 0$. (The vector $\mathbf{n}$ is perpendicular to them).

With these bc Eq.(4) can be simulated numerically following some standard methods. (Combined semiimplicit and pseudospectral methods). More details can be obtained in ref.[11]. These simulations are made in a mesh of 128 $\times$ 128 points. The aspect ratio is indicated in each case. The pattern corresponding to a simulation is represented by taking some iso-Ψ contour lines between the maximum and the minimum values of Ψ.

3. Stationary convective patterns with different symmetries

Some physical conditions can lead to the transition between different symmetries in convective patterns. Recently, some attention has been payed to the competition between hexagons and straight rolls in convection under non-Boussinesq conditions [12]. The transition between patterns of hexagons and squares has been predicted by using some amplitude equation formalism [13,14] but has not been yet observed in convection. (In principle this could be done by combining poorly conducting walls with non-Boussinesq effects, but this problem has not been yet theoretically analysed). However, this transition has been theoretically predicted in ferrofluids [15] and observed experimentally under suitable conditions [16].

The SHE for this problem is obtained from the complete GSHE (4) under zero rotation and taking the most relevant terms in order to favour the presence of rolls, hexagons, squares and fronts between two of these symmetries, depending on the parameter values. Therefore, the simplified model takes the form:

$$\dot{\Psi} = [\epsilon - (1+\Delta)^2]\Psi + a\Psi^2 - b\Psi^3 - e\Psi\Delta^2\Psi^2 \quad (5)$$

As stressed in section 2, instead to attempting directly numerical simulation of this equation we study the amplitude equations for this problem without spatial variation, in order to determine the stability range for the different solutions.

3.1. AMPLITUDE EQUATIONS

In the amplitude equations model one must assume the planform that can appear in the pattern, as well as a well defined orientation. For the problem of the transition between different symmetries we take a development for Ψ that accounts for these facts:

$$\Psi(\bar{r},t) = (\sum_{j=1}^{N} A_j(t) exp(i\bar{k}_j \cdot \bar{x}) + c.c.) g(z) \tag{6}$$

where $\bar{x}$ denotes the horizontal coordinates and A_j are the slowly varying amplitudes, $N = 6$, $\bar{k}_1 + \bar{k}_2 + \bar{k}_3 = 0$, $\bar{k}_4 + \bar{k}_5 + \bar{k}_6 = 0$, $\bar{k}_i \cdot \bar{k}_{i+3} = 0$, $i = 1, 2, 3$ and $g(z)$ stands for the solution of the linear problem. After introducing this development into eq.(5) one obtains a set of equations:

$$\dot{A}_1 = \epsilon A_1 + \alpha_0 A_2 A_3 - A_1[\beta_0 A_1^2 + \beta_1(A_2^2 + A_3^2) + \beta_2 A_4^2 + \beta_3(A_5^2 + A_6^2)] \tag{7}$$

$$\dot{A}_4 = \epsilon A_4 + \alpha_0 A_5 A_6 - A_4[\beta_0 A_4^2 + \beta_1(A_5^2 + A_6^2) + \beta_2 A_1^2 + \beta_3(A_2^2 + A_3^2)] \tag{8}$$

along with other four equations are obtained by cyclic permutations of the indices (1, 2, 3) and (4, 5, 6) in eqs.(7)-(8). Notice that these equations can be also derived by using simple symmetry arguments [13,14].

It is straightforward to stablish the relationship between the coefficients in (5) and in the amplitude equations (7)-(8):

$$\alpha_0 = 2a,\ \beta_0 = 3b + 16e,\ \beta_1 = 6b + 20e,\ \beta_2 = 6b + 16e,\ \beta_3 = 6b + 28e$$

and then, the range of stability for the different structures with the coefficients of the GSHE (5).

It is well known that the possible solutions of these equations are hexagons, squares and rolls. (Some mixed states are also possible [17] but they are always unstable.) The stability of the steady state solutions of these three possible types of structures can be found in recent works [13,14,15].

We performed some numerical simulations by restricting the mesh of points to a circular geometry. (The typical time step was 0.5). The aspect ratio is $\Gamma \simeq 25$ and we fixed the supercritical heating to be $\epsilon = 0.1$. In Fig.1 some of these simulations are shown. Starting from an initial pattern of squares, but taking the parameter values in the region where hexagons are stable ($a = 0.1$, $b = -0.28, e = 0.24$) the pattern evolves to a hexagonal lattice after 2000 time steps. If we instead change a to be in the region where the squares are stable ($a = 0.01$) one sees the evolution from a perfect hexagonal pattern to a pattern of squares.

3.2. DOMAIN BOUNDARIES

But the main aim of this section is to show that by means of a GSHE, fronts between hexagons and patterns with other symmetries can be stable. As stressed by Pomeau [18] these fronts can be stable (i.e., with zero velocity) for a finite range of the control parameter ϵ due to a pinning of a domain boundary with the small-scale structure of the pattern. Malomed et al.[19] demostrated the existence of this parameter range for domain boundaries

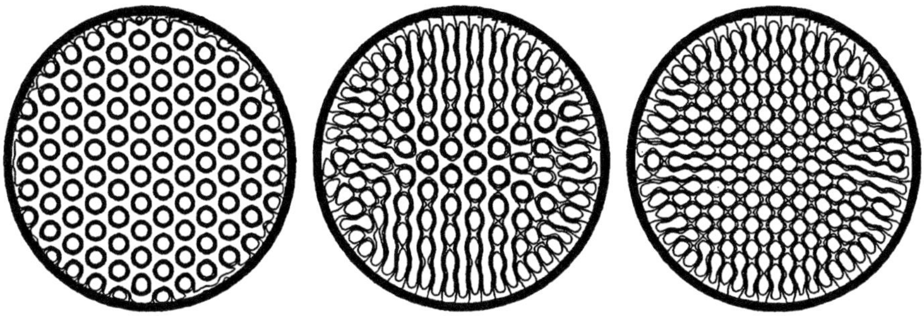

Figure 1: Squares are the stable configuration for $a = 0.01$ $b = -0.28$ and $e = 0.24$.

of the hexagon-roll type and the fact that the pinning is possible provided that the front is nearly perpendicular to any of the wave vectors $\bar{k}_l$ that form a hexagonal pattern. This pinning requires no change of the underlying wave number. It is obvious that the pinning effect cannot be calculated in the amplitude equation framework and, therefore, a GSHE could be useful to show (even by numerical simulations) the existence of this interesting effects. As stressed by Malomed *et al.*[19] the analytical calculations of the corresponding range is quite cumbersome, even for the SHE.

Therefore we can have the approximate values in that range by using some general potential methods based on the Lyapunov functionals. One can extend the treatment and include spatial dependency of the amplitudes in eqs.(6)(7). To do that we introduce the amplitude equations in a rescaled form [19]:

$$\dot{a}_1 = \delta a_1 + a_2 a_3 - a_1[a_1^2 + g_1(a_2^2 + a_3^2) + g_2 a_4^2 + g_3(a_5^2 + a_6^2)] \quad (9)$$

$$\dot{a}_4 = \delta a_4 + a_5 a_6 - a_4[a_4^2 + g_1(a_5^2 + a_6^2) + g_2 a_1^2 + g_3(a_2^2 + a_3^2)] \quad (10)$$

along with the four equations obtained from eqs.(9)(10) by cyclic permutations of the indexes (1, 2, 3) and (4, 5, 6). The scales have been changed in the usual way:

$$A_l = \frac{\alpha_0}{\beta_0} a_l, \quad \mathrm{l} = 1 - 6 \qquad \delta = \frac{\epsilon \beta_0}{\alpha_0^2}, \qquad g_i = \frac{\beta_i}{\beta_0}, \quad \mathrm{i} = 1, 2, 3$$

Now we restrict the analysis for the particular case of a hexagons-squares front. These two structures can coexist for $\epsilon > 0$. The hexagonal pattern can be stable for $\epsilon \leq s$ ($s = \frac{g_1+2}{(g_1-1)^2}$) and the squares for $\epsilon > r$ ($r = \frac{1+g_2}{(1-g_1+g_2-g_3)^2}$). Then the stability range of the hexagons and squares overlaps when $r < s$.

At $\epsilon \in (r, s)$ a front in eqs.(8)(9) matches the boundary conditions:
$a_1 = a_2 = a_3 = a_h \neq 0,\ a_4 = a_5 = a_6 = 0$, at $x = -\infty$
$a_1 = a_4 = a_s \neq 0, a_2 = a_3 = a_5 = a_6 = 0$, at $x = \infty$
where the spatially homogeneous stationary solutions are $a_h = \frac{1+\sqrt{1+4\delta(1+2g_1)}}{2(1+2g_1)}$ for the hexagons and $a_s = \sqrt{\frac{\delta}{1+g_2}}$ for the squares.

Figure 2: Starting from random initial conditions, $b = -0.28$, $e = 0.26$ and $a = 0.004$ one observe a front that is nearly perpendicular to one of the wavevector of the hexagons.

The stationary version of eqs.(8)(9) can be considered as the equation of motion of a mechanical system in a potential:

$$V = \sum_{i=1}^{N}(\frac{1}{2}\delta a_i^2 - \frac{1}{4}a_i^4) - \sum_{i,j,i\neq j}\frac{g_{ij}}{2}a_i^2a_j^2 + a_1a_2a_3 + a_4a_5a_6 \quad (11)$$

The motionless front corresponds to a separatrix trajectory of the mechanical system which connects two local extrema of the effective potential V for the hexagonal and the square solutions. This separatrix cannot exist unless the values of the potential V at the two points are equal. This is a condition to have a stable front, between the two patterns, but we insist that there is a full stability range that cannot be calculated in this framework. Therefore we rescale again the values of coefficients in eqs.(9,10) to values for the GSHE. Then numerical simulations are made for different values near these rescaled ones. We show only one of these fronts in Fig.2, but it remain stable by slight changes in the parameter values. Starting from a random initial pattern a stable square-hexagon domain boundary appears (Fig.2) for $b = -0.28, e = 0.26$ and $a = 0.004$.

A further step in this analysis must require the calculation of coefficients in the GSHE for convection in a non-Boussinesquian fluid or in ferrofluids under a magnetic field [15]. (In the last case this kind of fronts has been observed experimentally [16]).

4. Convection in rotating fluids

We shall discuss now the thermal convection problem under rotation. As a guide we consider the theoretical studies of Küppers-Lortz [20] that predicted that a pattern of rolls is unstable against another set of rolls oriented at 58° from the original ones (Küppers-Lortz instability [20]) and the recent experiments by Zhong *et al.*[21] for intermediate *Pr* and Bodenschatz *et al.* [22] for small *Pr*. From these experiments one may infer that the dynamics of defects present in the pattern is crucial to understand the dynamics of the pattern. For the sake of simplicity we take the GSHE (4) retainig only the most relevant terms for the interplay

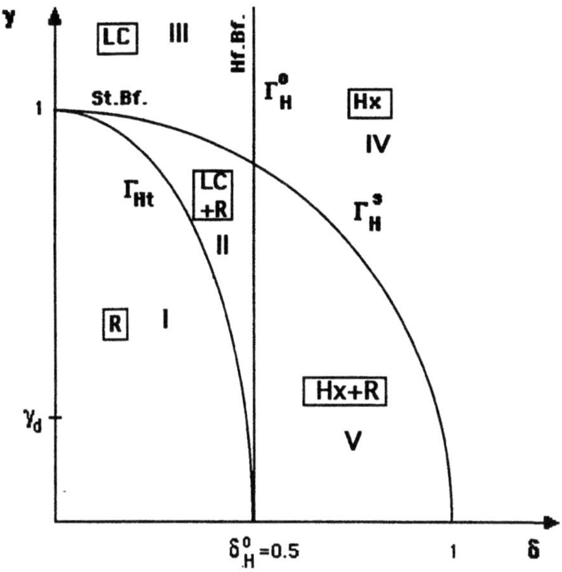

Figure 3: Stability diagram for the amplitude equation ($\epsilon = 0.1$).

between symmetries and rotation:

$$\dot{\Psi} = [\epsilon - (1+\Delta)^2]\Psi + a\Psi^2 - \Psi^3 - e[\nabla\Psi \times \nabla\Delta\Psi^2]_z \tag{12}$$

At this point it is interesting to stress that this model is non-variational and, therefore, the dynamics is not purely relaxational, and it can lead to periodic or even more complex behaviour in space and time. Here the cuadratic term Ψ^2 gives account of the vertical symmetry breaking effect [23] that lead to a hexagonal pattern in absence of rotation. Due to the rotation, the reflection symmetry is also broken. We carry out the analysis of the amplitude equations as in previous sections.

4.1. AMPLITUDE EQUATIONS

We consider the model eq.(12) with the development eq.(6), but restricted only to N= 1, 2, 3. (We only investigate the case of hexagons and rolls). Then the following amplitude equations are obtained

$$\dot{A}_i = A_i + \delta A_{i+1}A_{i+2} - A_i(A_i^2 + (2+\gamma)A_{i+1}^2 + (2-\gamma)A_{i+2}^2) \quad i = 1,2,3 \quad mod(3) \tag{13}$$

where the relations between coefficients in eqs. (12) and (13) are:

$$\delta = \frac{2a}{\sqrt{3\epsilon}} \qquad \gamma = \frac{2e}{\sqrt{3}} \tag{14}$$

The stability diagram for the possible patterns [24,25] are represented in Fig.3. The different regions in this diagram correspond to the following patterns: I) Rolls that become unstable

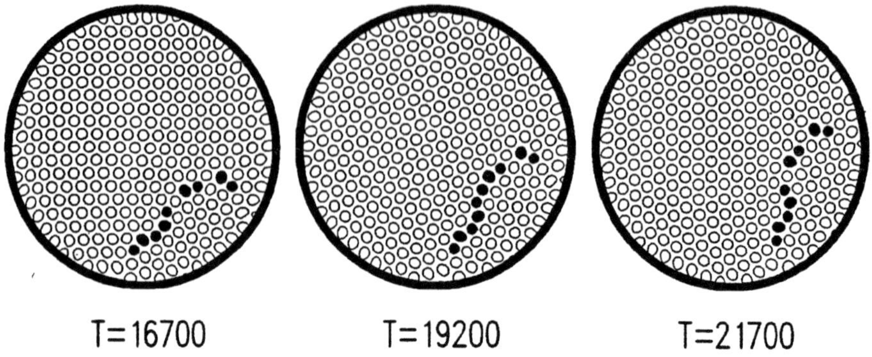

Figure 4: Simulation of the GSHE for $\delta = .8$ and $\gamma = 1$. The whole pattern rotates in the rotating reference frame.

through a stationary bifurcation ($A_i \neq 0$, $A_{i+1} = A_{i-1} = 0$). II) A limit cycle from a Hopf bifurcation whose frequency decreases when δ decreases and with three varying amplitudes. This can coexist with a roll solution. III) In this region only the limit cycle is stable. Its frequency decreases when γ increases. For $\delta = 0$ the Küppers-Lortz instability is recovered. IV) Hexagons ($A_i = A_{i+1} = A_{i-1} \neq 0$) are stable in this region. They become unstable through a Hopf bifurcation on Γ^o_H. V) Hysteretic regions of limit cycle-rolls and hexagons-rolls.

It is interesting to note that a heteroclinic line due to the collision of the limit cycle with three asymmetric unstable solutions ($A_i \neq A_{i+1} \neq A_{i-1} \neq 0$) appears in the stationary bifurcation line Γ_{Ht}.

4.2. NUMERICAL SIMULATIONS

Now, having in mind the diagram in Fig.3 we analyse the different dynamical behaviors by numerical simulations of the GSHE (12). All those simulations begin from random patterns. We gather the most relevant results in different subsections.

4.2.1. *Küppers-Lortz-like instability.* For $\delta = 0.2$ and $\gamma = 0.9$ we are near the heteroclinic line connecting the three asymmetric solutions. The dynamics is then very similar to that in the K-L instability, but in this last case the heteroclinic line connects three rolls forming an angle of 60° [22]. It is possible to reproduce how defects growing from the walls replace the old pattern of rolls by a new one [25]. There is not a global replacement of the orientation of rolls but a continuous change from the wall to the center of the pattern. In this process one or two grain boundaries can travel across the pattern and impose a new main orientation. This feature is very similar to that observed in experiments [21].

4.2.2. *Rigid rotation of the pattern.* For whatever values for δ and γ for which hexagons are stable (region IV), a rigid rotation of the pattern (proportional to Ta) can be obtained [10]. This is clearly due to the presence of sidewalls in the simulation and to the non-variational character of the last term in eq.(13). (We must point out that the sign of rotation can be

Figure 5: Simulation of the GSHE for periodic bc in the limit cycle region III ($\delta = 0.4$ and $\gamma = 1.1$).

changed by changing the sign of parameter γ in eq.(13)). The rotating pattern can be easily seen considering the line of defects (hepta-pentagons) that separates the two patches of different hexagonal orientations. We must stress that this effect has not been yet observed experimentally, but that rotation velocity (in the rotating frame of reference of the fluid) can be explicitly calculated for a given initial irregular pattern [25].

4.2.3. *Hexagonal oscillation.* In region III, the hexagons become unstable through a Hopf bifurcation and a limit cycle on which the three amplitudes are changing continuously appears. This situation is simulated in Fig.5 where the cylindrical sidewallls have been replaced by periodic bc, in order to show that this effect is not due to the sidewalls. During the cycle one can clearly see that a main set of rolls is replaced by another sihfted 60°. This effect have been not yet observed experimentally.

Acknowledgments

One of us (C.P-G) would like to thank Prof. E. Tirapegui (Santiago, Chile) and Prof. W. Zeller (Valparaiso, Chile) for their kind invitation to the 4th International Workshop on Instability and Nonequilibrium Structures where this work was presented. We also thank the suggestions and comments of Dr. M. Fantz and Dr. R. Friedrich (Stuttgart, Germany) and Prof. G. Ahlers (Santa Barbara, USA). This work was supported by the DGICYT (Spanish Government) under the grant PB90-0362 and by a Spain-Germany Integrated Action (26, 1991; 73, 1992)

References

[1] J. E. Wesfreid and S. Zaleski eds., *Cellular Structures in Instabilities* (Springer, Berlin, 1984).
[2] F. H. Busse and N. Riahi, J. Fluid Mech. **96**, 243 (1980).
[3] D. R. Jenkins and M. R. E. Proctor, J. Fluid Mech. **139**, 461 (1984).
[4] S. Chandrasekhar, *Hydrodynamic and Hydromagnetic Stability*, (Clarendon, Oxford, 1961).

[5] G. Veronis, J. Fluid Mech. **5**, 401 (1959).
[6] J. K. Platten and J. C. Legros, *Convection in Liquids*, (Springer, Berlin, 1977).
[7] P. Manneville, *Dissipative Structures and Weak Turbulence*, (Academic Press, New York, 1990).
[8] J. Swift and P. C. Hohenberg, Phys. Rev. **A15**, 319 (1977).
[9] H. S. Greenside, W. M. Coughran, Jr., and N. L. Schryer, Phys. Rev. Lett. **49**, 726 (1982).
[10] M. Fantz, R. Friedrich, M. Bestehorn and H. Haken, preprint (1992).
[11] M. Bestehorn and C. Pérez-García, to appear in Physica D.
[12] E. Pampaloni, C. Pérez-García, L. Albavetti and S. Ciliberto. J. Fluid Mech. **234**, 393 (1992).
[13] B. Malomed and M.I. Tribelsky, Zh. Eksp. Teor. Fiz. **92**, 539 (1987).
[14] H. Herrero, C. Pérez-García and M. Bestehorn, (to be published).
[15] A. Gailitis, J. Fluid Mech. **82**, 401 (1977).
[16] J.E. Wesfreid, private communication.
[17] L.A. Segel and J.T. Stuart, J. Fluid Mech. **13**, 289 (1962).
[18] Y. Pomeau, Physica **D23**, 3 (1986).
[19] B. Malomed, A. Nepomnyaschy and M.I. Tribelsky, Phys. Rev. **A42**, 7244 (1990).
[20] G. Küppers and D. Lortz, J. Fluid Mech. **35**, 609 (1969).
G. Küppers, Phys. Lett. **32A**, 7 (1970).
[21] F. Zhong, R. Ecke and V. Steinberg, Physica D **51**, 596 (1991).
[22] E. Bodenschatz, D. S. Cannell, J. de Bruyn, R. Ecke, Y. Hu, K. Lerman and G. Ahlers, to appear in Physica D.
[23] H. Haken, *Synergetics: An Introduction, 2nd ed.*, (Springer, Berlin, 1978); *Advanced Synergetics*, (Springer, Berlin, 1983).
[24] A. M. Soward, Physica **14D**, 227 (1985).
[25] J. Millán, C. Pérez-García, M. Bestehorn, M. Fantz and R. Friedrich, Phys. Rev. A (to appear).

TWO DIMENSIONAL PATTERNS IN A MODEL FOR THE ELECTROHYDRODYNAMIC INSTABILITY OF NEMATIC LIQUID CRYSTALS

P. DE BROUWER and D. WALGRAEF†
Service de Chimie-Physique,
Université Libre de Bruxelles,
Bd du Triomphe, CP 231,
B-1050, Brussels, Belgium.

ABSTRACT. The study of the different patterns which appear beyond the electrohydrodynamic instability of nematic liquid crystals is performed in the framework of a dynamical model of the Proctor-Sivashinsky type. This model describes the experimentally observed transitions between rolls, zig-zag and bimodal structures.

1. Introduction

In the recent years, there has been much interest for pattern forming instabilities in nematic liquid crystals, both experimentally [1-2] and theoretically [3-6]. It is now well known that a thin layer of nematic liquid crystal, submitted to an oscillating electrical field of increasing amplitude, presents a sequence of transitions between different types of structures ending in complex spatio-temporal regimes [1-5]. Usually, the following structures are successively observed : straight rolls, undulated rolls, oblique or zig-zag rolls, bimodal structures and finally states of high spatio-temporal complexity.

The theoretical description of these structures and of the electrohydrodynamic convection in general is extremely difficult, due to the complexity of the underlying nemato-electrohydrodynamics, and the time dependence of the external forcing. It has however been possible to perform the linear stability analysis of the homogeneous steady state and to compute the threshold for pattern formation, the

† Director of Research, National Fund for Scientific Research (Belgium).

E. Tirapegui and W. Zeller (eds.), Instabilities and Nonequilibrium Structures IV, 235–245.

neutral surface and the linear growth rate of the unstable modes [5], but a systematic derivation of the nonlinear terms of the correponding amplitude equations from the underlying dynamics is not available yet. Amplitude equations, based on symmetry arguments, have nevertheless been derived for each type of structure and they satisfactorily describe various aspects of electrically driven liquid crystals. Despite their interest, these equations are not able to decribe the transitions between the different structures, nor their relative stability. Hence it would be interesting to rely on a reduced dynamical model which encompasses the various instability thresholds and plays here the role of the Swift-Hohenberg equation for Rayleigh-Bénard convection [7].

It is why we propose to describe the behavior of nematic liquid crystals near the electrohydrodynamic instability with a reduced dynamical model for an order parameterlike variable which takes into account a minimal set of basic elements, namely the intrinsic anisotropy of the system, which is known to affect the selection and stability of spatial patterns [8], and the gradient dependence of the nonlinear couplings. This model reproduces the sequence of observed patterns but also suggests that bimodal structures should be stable in a closed domain of the parameter space. The fact that the boundary of this domain corresponds to a phase instability is consistent with the observed defect mediated disordered states that appear for increasing field intensities.

2. The model

Following the linear stability analysis and the near threshold description performed by Bodenschatz et al. for electrohydrodynamic convection in nematics [5], the linear growth rate of the order parameterlike variable may be written, near the instability threshold, as

$$-\left\{\epsilon[\partial_x^2 + s\partial_y^2] + [1 + \partial_x^2 + \partial_y^2]^2 + 2\eta\partial_x^2\partial_y^2 + \tau\partial_y^4\right\} \tag{1}$$

or

$$\left\{\epsilon[q_x^2 + sq_y^2] - [1 - q_x^2 - q_y^2]^2 - 2\eta q_x^2 q_y^2 - \tau q_y^4\right\} \tag{2}$$

in Fourier space. ϵ is the reduced distance to threshold $(V^2 - V_c^2)/V_c^2$ where V is the amplitude of the applied voltage and V_c its threshold value. s, η and τ are positive materials parameters reflecting the anisotropy of the system, and the critical wavenumber has been scaled to one. The x axis being the easy axis, s is larger than one, and, in the following, we will consider $s = 2$, $\eta = 0.25$ and $\tau = 1.5$, values that are consistent with the analysis of Bodenschatz et al. [5]

The corresponding marginal stability surface is thus given by (fig.1)

$$\epsilon = \frac{[1 - q_x^2 - q_y^2]^2 + 2\eta q_x^2 q_y^2 + \tau q_y^4}{q_x^2 + sq_y^2} \quad , \tag{3}$$

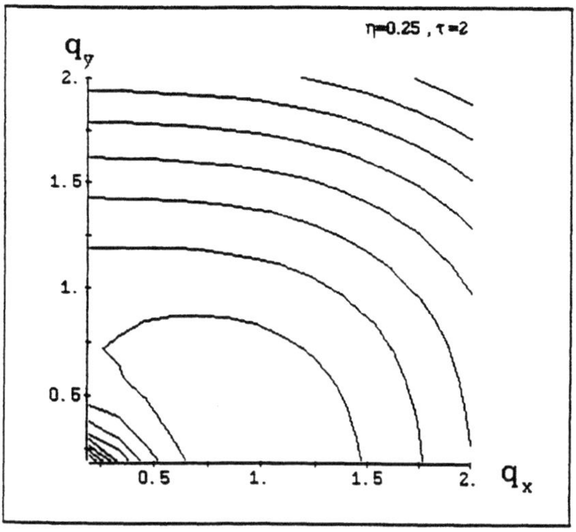

Contour plot of the marginal stability surface for the anisotropic PS model

Marginal stability curve of the anisotropic PS model

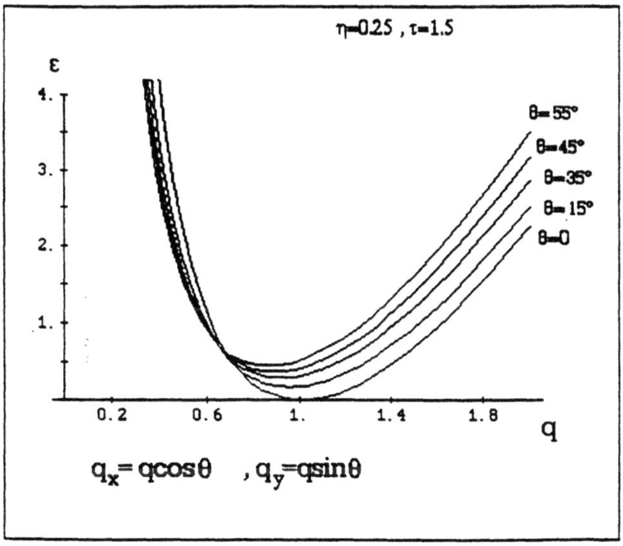

Figure 1 : Contour plot of the marginal stability surface and marginal stability curve for different wavevector orientations in the case of the linear growth rate given in (1).

and the fastest growing mode corresponds to

$$q_{0x}^2 = 1 + \frac{\epsilon}{2} \quad , \quad q_{0y}^2 = 0 \tag{4}$$

for $0 < \epsilon < \frac{2\eta}{s-1-\eta} = 0.667$, and

$$q_{0x}^2 = 1 + \frac{\epsilon}{2} - (1+\eta)\frac{1}{2}\frac{\epsilon(s-1-\eta) - 2\eta}{1+\tau-(1+\eta)^2} \quad , \quad q_{0y}^2 = \frac{1}{2}\frac{\epsilon(s-1-\eta)-2\eta}{1+\tau-(1+\eta)^2} \tag{5}$$

for $\epsilon > \frac{2\eta}{s-1-\eta} = 0.667$

Hence, the linear part of the dynamics provides a first pattern selection mechanism towards normal ($\epsilon < \frac{2\eta}{s-1-\eta} = 0.667$) or oblique rolls ($\epsilon > \frac{2\eta}{s-1-\eta} = 0.667$).

Of course, the eventual stabilisation of these structures depends on the nonlinear terms of the dynamics. The simplest nonlinearities compatible with the amplitude equations derived in [5] correspond to the cubic scalar nonlinearity of the Swift-Hohenberg equation, σ^3, where σ is the order parameterlike variable. However, this term is not able to stabilize bimodal structures [9] and one has to introduce gradient dependent nonlinearities. On performing a systematic expansion in the gradients, the first nonlinear terms able to stabilize bimodal structure are of the form

$$g_{\mu\nu}\partial_\mu[(\partial_\mu\sigma)(\partial_\nu\sigma)^2] \tag{6}$$

Let us note that in the case of the elliptically sheared nematics, where similar pattern sequences are observed, the introduction of a potential version of these terms, $g\vec{\nabla}.(\vec{\nabla}\sigma|\vec{\nabla}\sigma|^2)$, led to a quantitative description of the pattern formation phenomena [10]. Hence, since this term seems to capture the basics of the nonlinear part of the dynamics, we will also use it here and test its relevance to the description of the experimental observations. The proposed kinetic equation for the order parameterlike variable σ is thus, in scaled units :

$$\partial_t\sigma = -\left\{\epsilon[\partial_x^2 + s\partial_y^2] + [1 + \partial_x^2 + \partial_y^2]^2 + 2\eta\partial_x^2\partial_y^2 + \tau\partial_y^4\right\}\sigma + \vec{\nabla}.(\vec{\nabla}\sigma|\vec{\nabla}\sigma|^2) \tag{7}$$

3. Pattern Selection and Stability

3.1. NORMAL ROLLS

Since the first structure to bifurcate from the electrohydrodynamic instability correponds to normal rolls (i.e. with wavevector parallel to the easy axis 0x), let us derive their amplitude equation and study their stability. The normal rolls are defined by :

$$\sigma(x,y,t) = A(x,y,t)e^{iqx} + \bar{A}(x,y,t)e^{-iqx} \tag{8}$$

where $A(x, y, t)$ is the complex amplitude of the structure and may be slowly varying in space and time. On introducing this expression in the kinetic equation (7) and neglecting rapidly varying contributions and short scale effects, one obtains the following amplitude equation :

$$\begin{aligned}\partial_t A = &[\epsilon q^2 - (q^2-1)^2]A - 2iq[\epsilon - 2(q^2-1)]\partial_x A + [2(3q^2-1) - \epsilon]\partial^2_{xx} A \\ &+ [2(q^2-1) + 2\eta q^2 - \epsilon s]\partial^2_{yy} A - 3q^4 A|A|^2 + [12iq^3\partial_x A + 4q^2\partial^2_{xx} A \\ &+ 2q^2\nabla^2 A]|A|^2 - [q^2\nabla^2 \bar{A} + 2q^2\partial^2_{xx}\bar{A}]A^2 \end{aligned} \tag{9}$$

This equation admits uniform steady states defined by $A_0 = R_0 exp i\phi_0$, with $R_0^2 = [\epsilon q^2 - (q^2-1)^2]/3q^4$ and ϕ_0 arbitrary. As usual, since these patterns appear via a continuous symmetry breaking, one may expect a diffusive behavior of their phase perturbations. Effectively, on using the standard derivation procedure, one obtains the following phase dynamics :

$$\partial_t \phi = D_x \partial^2_{xx}\phi + D_y \partial^2_{yy}\phi \tag{10}$$

where

$$D_x = \frac{\epsilon(q^4+3) - (q^2-1)^2(3q^2+4+5q^{-2})}{\epsilon q^2 - (q^2-1)^2} \tag{11}$$

and

$$D_y = (1-s)\epsilon - \frac{1}{q}^2 + (1+2\eta)q^2 \tag{12}$$

Hence, there is an Eckhaus instability at

$$\epsilon = \frac{5 - 6q^2 - 2q^6 + 3q^8}{3q^2 + q^6} \tag{13}$$

and a zig-zag instability at

$$\epsilon = \frac{1 - (1+2\eta)q^4}{(1-s)q^2} \tag{14}$$

The resulting stability domain of normal rolls is shown in figure 2. It shows, for example, that, on increasing ϵ for a roll pattern locked in the first unstable wavevector ($\vec{q} = \vec{1}_x$), the first instability occurs at $\epsilon = 0.5$ and is of the zig-zag type, while for normal rolls with the optimal wavevector, the first instability occurs at $\epsilon = 1.68$ and is of the Eckhaus type . In any case, since the phase stability limits of normal rolls form a closed domain, this pattern always becomes unstable on increasing ϵ.

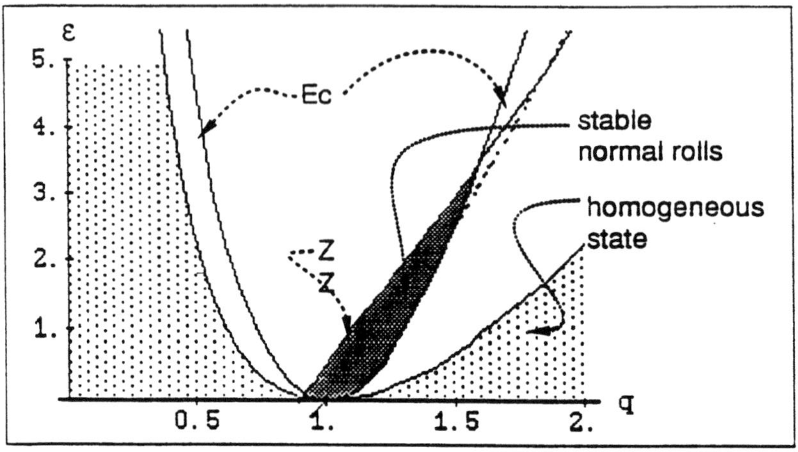

Figure 2 : Stability diagram for normal roll solutions of the amplitude equations (9).

3.2. UNDULATED ROLLS

Near a phase diffusive instability, since the diffusion coefficients become small or even negative, one has to include higher order space derivatives in the phase dynamics. For the normal rolls described in the preceeding section, one obtains :

$$\partial_t \phi = D_x \partial^2_{xx} \phi + D_y \partial^2_{yy} \phi - \partial^4_{xxxx} \phi - 2(1+\eta) \partial^4_{xxyy} \phi - (1+\tau) \partial^4_{yyyy} \phi \qquad (15)$$

One sees on the diagram displayed in figure 2 that the phase instability that a roll pattern encounters on increasing ϵ is of the zig-zag type. Hence the system will develope phase pertubations of wavevectors k perpendicular to the x axis and of linear evolution given by :

$$\partial_t \phi_k = -[D_y k^2 + (1+\tau) k^4] \phi_k \qquad (16)$$

and, for $D_y < 0$, there exists a whole band of unstable wavevectors k, the most unstable corresponding to

$$k_0 = \sqrt{-\frac{D_y}{2(1+\tau)}} \tag{17}$$

In the immediate neighbourhoog of the instability, the system will thus develope transient undulated rolls which are intermediate states between the normal rolls and the new bifurcating structure corresponding to oblique rolls.

3.3. OBLIQUE ROLLS

Above the zig-zag instability of normal rolls, one expects structures with non vanishing y components of their wavevectors. Since unstable modes with wavevectors $q_1 = q_x \vec{1}_x + q_y \vec{1}_y$ and $q_2 = q_x \vec{1}_x - q_y \vec{1}_y$ $(q_1^2 = q_2^2)$ have the same linear growth rate, the order parameterlike variable may be written as :

$$\begin{aligned}\sigma(x,y,t) =& A(x,y,t)e^{i(q_x x+q_y y)} + \bar{A}(x,y,t)e^{-i(q_x x+q_y y)} \\ &+ B(x,y,t)e^{i(q_x x-q_y y)} + \bar{B}(x,y,t)e^{-i(q_x x-q_y y)}\end{aligned} \tag{18}$$

The corresponding amplitude equations are given by :

$$\begin{aligned}\partial_t A =& [\epsilon(q_x^2 + s q_y^2) - (q^2-1)^2 - 2\eta q_x^2 q_y^2 - \tau q_y^4]A \\ &- 2iq_x[\epsilon + 2 - 2q_x^2 - 2(1+\eta)q_y^2]\partial_x A \\ &- 2iq_y[\epsilon s + 2 - 2(1+\eta)q_x^2 - 2(1+\tau)q_y^2]\partial_y A \\ &+ [4q_x^2 + 2(q^2-1) + 2\eta q_y^2 - \epsilon]\partial_{xx}^2 A + 8(1+\eta)q_x q_y \partial_{xy}^2 A \\ &+ [4q_y^2 + 2(q^2-1) + 2\eta q_x^2 + 6\tau q_y^2 - \epsilon s]\partial_{yy}^2 A \\ &- 3q_1^4|A|^2 A - 2[q_1^2 q_2^2 + 2(\vec{q}_1\vec{q}_2)^2]|B|^2 A - [q_1^2\nabla^2\bar{A} + 2\vec{q}_1.\vec{\nabla}(\vec{q}_1.\vec{\nabla}\bar{A})]A^2 \\ &+ [12iq_1^2(\vec{q}_1\vec{\nabla}A + 4\vec{\nabla}(\vec{q}_1.\vec{\nabla}A) + 2q_1^2\nabla^2 A]|A|^2 \\ &+ 2[q_2^2\nabla^2 A + 2iq_2^2(\vec{q}_1.\vec{\nabla}A) + 2(2i\vec{q}_1.\vec{q}_2 + .\vec{q}_2.\vec{\nabla})(\vec{q}_2.\vec{\nabla}A)]|B|^2 \\ &+ 2[i(q_1^2 + 2\vec{q}_1.\vec{q}_2)(\vec{q}_1 + \vec{q}_2).\vec{\nabla}B + 2\vec{q}_1.\vec{\nabla}(\vec{q}_2.\vec{\nabla}B) + (\vec{q}_1.\vec{q}_2)\nabla^2 B]A\bar{B} \\ &+ 2[i(q_1^2 + 2\vec{q}_1.\vec{q}_2)(\vec{q}_1 - \vec{q}_2).\vec{\nabla}\bar{B} - 2\vec{q}_1.\vec{\nabla}(\vec{q}_2.\vec{\nabla}\bar{B}) - (\vec{q}_1.\vec{q}_2)\nabla^2\bar{B}]AB\end{aligned} \tag{19a}$$

$$
\begin{aligned}
\partial_t B = & [\epsilon(q_x^2 + s q_y^2) - (q^2-1)^2 - 2\eta q_x^2 q_y^2 - \tau q_y^4] B \\
& - 2i q_x [\epsilon + 2 - 2q_x^2 - 2(1+\eta) q_y^2] \partial_x B \\
& + 2i q_y [\epsilon s + 2 - 2(1+\eta) q_x^2 - 2(1+\tau) q_y^2] \partial_y B \\
& + [4q_x^2 + 2(q^2-1) + 2\eta q_y^2 - \epsilon] \partial_{xx}^2 B - 8(1+\eta) q_x q_y \partial_{xy}^2 B \\
& + [4q_y^2 + 2(q^2-1) + 2\eta q_x^2 + 6\tau q_y^2 - \epsilon s] \partial_{yy}^2 B \\
& - 3q_2^4 |B|^2 B - 2[q_1^2 q_2^2 + 2(\vec{q}_1 \vec{q}_2)^2] |A|^2 B - [q_2^2 \nabla^2 \bar{B} + 2\vec{q}_2.\vec{\nabla}(\vec{q}_2.\vec{\nabla}\bar{B})] B^2 \\
& + [12 i q_2^2 (\vec{q}_2 \vec{\nabla} B + 4\vec{\nabla}(\vec{q}_2.\vec{\nabla} B) + 2q_2^2 \nabla^2 B] |B|^2 \\
& + 2[q_2^2 \nabla^2 B + 2i q_2^2 (\vec{q}_1.\vec{\nabla} B) + 2(2i\vec{q}_1.\vec{q}_2 + \vec{q}_2.\vec{\nabla})(\vec{q}_2.\vec{\nabla} B)] |A|^2 \\
& + 2[i(q_1^2 + 2\vec{q}_1.\vec{q}_2)(\vec{q}_1 + \vec{q}_2).\vec{\nabla} A + 2\vec{q}_1.\vec{\nabla}(\vec{q}_2.\vec{\nabla} A) + (\vec{q}_1.\vec{q}_2)\nabla^2 A] B\bar{A} \\
& + 2[i(q_1^2 + 2\vec{q}_1.\vec{q}_2)(\vec{q}_1 - \vec{q}_2).\vec{\nabla} \bar{A} - 2\vec{q}_1.\vec{\nabla}(\vec{q}_2.\vec{\nabla} \bar{A}) - (\vec{q}_1.\vec{q}_2)\nabla^2 \bar{A}] AB
\end{aligned}
\tag{19b}
$$

These equations admit stationary solutions corresponding to patterns of uniform amplitudes which may be of

- the oblique roll type :

$$A = R_0 expi\phi_0 \quad , \quad B = 0 \quad , \quad \text{or} \quad B = R_0 expi\phi_0 \quad , \quad A = 0$$

with

$$R_0^2 = \frac{\epsilon(q_x^2 + s q_y^2) - (q_1^2 - 1)^2 - 2\eta q_x^2 q_y^2 - \tau q_y^4}{3q_1^4} \quad , \quad \phi_0 = \text{cst} \tag{20}$$

- the bimodal type :

$$A = S_0 expi\phi_0 \quad , \quad B = S_0 expi\psi_0$$

with

$$S_0^2 = \frac{\epsilon(q_x^2 + s q_y^2) - (q_1^2 - 1)^2 - 2\eta q_x^2 q_y^2 - \tau q_y^4}{3q_1^4 + 2[q_1^2 q_2^2 + 2(\vec{q}_1 \vec{q}_2)^2]} \quad , \quad \phi_0, \psi_0 = \text{cst} \tag{21}$$

The oblique rolls are stable for $3q_1^4 < 2[q_1^2 q_2^2 + 2(\vec{q}_1\vec{q}_2)^2]$ or $cos^2\theta > 1/4$, θ being the angle between the wavevectors q_1 and q_2. Bimodal structures are stable for $3q_1^4 > 2[q_1^2 q_2^2 + 2(\vec{q}_1\vec{q}_2)^2]$ or $cos^2\theta < 1/4$. Hence, since q_y has a tendency to increase and q_x to decrease for increasing ϵ (cf. eq. (5)), we expect first the formation of zig-zag structures corresponding to alternate domains of oblique rolls of wavevectors q_1 and q_2 . The phase dynamics and stability domains of these structures may be determined as usual **[11]**. However, when the angle between q_1 and q_2 exceeds $\pi/3$, a transition to bimodal structures should occur.

3.4. BIMODAL STRUCTURES

As seen in the preceeding section, uniform bimodal structures are stable when their underlying wavevectors q_1 and q_2 make an angle θ such that $\pi/3 < \theta < 2\pi/3$. Since this angle increases with ϵ, the expected final structure corresponds to squares built on orthogonal wavevectors symmetric versus the easy axis 0x. The phase dynamics of this pattern may be derived along the usual lines [11], and one finds :

$$\begin{aligned}\partial_t \phi =& D^1_{XX}\partial^2_{XX}\phi + D^1_{XY}\partial^2_{XY}\phi + D^1_{YY}\partial^2_{YY}\phi \\ &+ D^2_{XX}\partial^2_{XX}\psi + D^2_{XY}\partial^2_{XY}\psi + D^2_{YY}\partial^2_{YY}\psi \\ \partial_t \psi =& D^1_{YY}\partial^2_{XX}\psi + D^1_{XY}\partial^2_{XY}\psi + D^1_{XX}\partial^2_{YY}\phi \\ &+ D^2_{YY}\partial^2_{XX}\phi + D^2_{XY}\partial^2_{XY}\phi + D^2_{XX}\partial^2_{YY}\phi \end{aligned} \tag{22}$$

where $X = (x+y)/\sqrt{2}$ and $Y = (x-y)/\sqrt{2}$. The diffusion coefficients are analytical functions of the various parameters of the dynamics and have been explicitely computed [11]. It turns out that the square pattern is only stable in a closed domain as shown in figure 3.

Hence, the square patterns are destabilized via a phase instability which induces the spontaneous nucleation of topological defect. As a result, they should naturally experience the experimentally observed defect mediated disorganization [12]. Furthermore, non variational contributions to the dynamics, which have been omitted in the present description, should strongly influence defect motion and take care of some aspects of the spatio-temporal complexity observed in this system [13]

4. Conclusions

We showed that the pattern formation phenomena that occur in the electrohydrodynamic instability of nematics can be described qualitatively with a reduced dynamical which takes essentially into account the intrinsic anisotropy of the system and the gradient dependence of the nonlinear couplings. This model provides an alternative to the too complex nematohydrodynamics for the study of the generic properties of pattern formation in anisotropic systems. The sequence of patterns, going from rolls to bimodal structures, results from the competition between linear anisotropy effects and the structure of the nonlinear couplings. Furthermore, it appears that the latter are responsible of the fact that bimodal structures are only stable in a closed domain of the parameter space. This is consistent with the defect mediated disorganization of the system which is observed for increasing field intensity. Of course, a quantitative descritpion of these phenomena requires the numerical analysis of the model, the fitting of some of its parameters, and eventually the introduction of non variational effects. We nevertheless think that the understanding of pattern formation in liquid crystal instabilities could benefit much from this description and that the proposed model could play here the same role that the Swift-Hohenberg equation played for Rayleigh-Bénard convection.

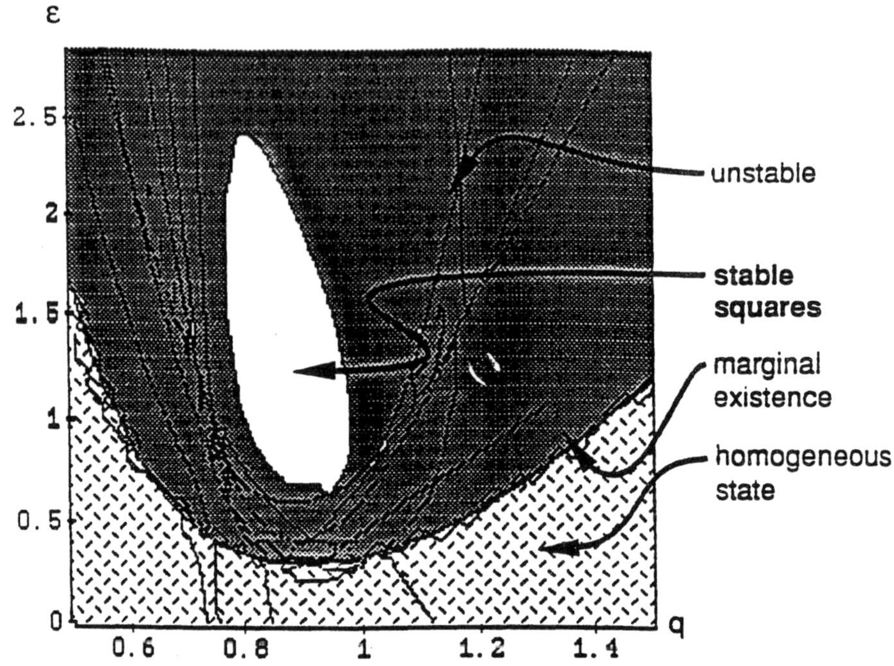

Figure 3 : Stability diagram for square patterns solutions of the amplitude equations (9) for $s = 2$, $\eta = 0.25$ and $\tau = 1.5$.

Acknowledgments. D.W. is grateful to the Fundacion Andes (Chile) for its support during the realization of this work.

References

1. A.Joets and R.Ribotta, J.Physique **47** (1986), p. 595.
2. S.Kai and W.Zimmerman, Prog.Theor Phys. **99** (1989), p. 458.
3. E.Dubois-Violette, P.G.De Gennes and O.Parodi, J.Physique **32** (1971), p. 305.
4. W.Pesch and L.Kramer, Z.Physik **B63** (1986), p. 121.
5. E.Bodenschatz, W.Zimmerman and L.Kramer, J.Physique **49** (1988), p. 1875.
6. D.Walgraef, in "Nematics," J.M.Coron et al., eds., Kluwer, Dordrecht, 1991, p. 391.
7. J.Swift and P.C.Hohenberg, Phys.Rev. **A15** (1977), p. 319.
8. D.Walgraef and C.Schiller, Physica **D27** (1987), p. 423.
9. D.Walgraef, Solid state Phenomena **3-4** (1988), p. 77.

10. E.Guazzelli, G.Dewel, P.Borckmans and D.Walgraef, Physica **D35** (1989), p. 220.
11. P.De Brouwer, "Pattern Formation in Nematics subjected to an Oscillating Electrical Field," Licenciaat Thesis, Vrije Universiteit Brussel, 1991.
12. J.Toner and D.Nelson, Phys.Rev. **B23** (1981), p. 316.
13. D.Walgraef, *Pattern Evolution in Extended Nonequilibrium Systems*, preprint, Free University of Brussels, 1992.

TURING STRUCTURES IN THE PRESENCE OF GRADIENTS

A. De Wit *, P. Borckmans †, G. Dewel †
Service de Chimie Physique and
Centre for Nonlinear Phenomena and Complex Systems
Université Libre de Bruxelles - CP 231
1050 Bruxelles
Belgium

ABSTRACT. To understand recent experimental observations of localized Turing structures in the presence of gradients of concentrations, the patterns observable in two and three-dimensional homogeneous model reaction-diffusion systems are first reviewed. The influence of several types of ramps on the selection and orientation of these Turing structures are then numerically analyzed.

1. Introduction

Major experimental results [1,2] have recently induced a renewal of interest in the study of spatial structure formation in open chemical reactors [3]. This spontaneous spatial self-organization in chemistry, proposed theoretically by Turing in 1952 [4], had never been observed experimentally until the conception these last years of new kinds of reactors. These consist in a flat piece of gel fed differently along two opposite sides by controlled reservoirs containing the reactants of the chlorite-iodide-malonic acid reaction. The gel suppresses parasitic convection effects while preserving all diffusion processes. This setup has allowed the first observation of Turing structures, stationary periodic variations of concentrations in space that arise solely through the coupling of diffusion and nonlinear chemical reactions. They thus belong to the class of dissipative structures arising from interactions between non linear local processes and transport phenomena far from thermodynamical equilibrium [5]. The observed structures are two or three-dimensional (2D, 3D) according to the importance of the gradients generated by the feeding. These results raise questions about the influence of gradients of concentrations on the selection and the orientation of patterns.
In this paper, we study numerically this problem in 2D and 3D. We first review in Section 2 the situation in reactors with homogeneous feeding of reactants. Section 3 analyses the effect of several types of ramps on 2D pattern selection. We study the behavior of 3D ramped structures in Section 4.

2. Homogeneous feeding of reactants

We study the pattern selection problem in chemistry on a non linear reaction diffusion model: the Brusselator [6] whose equations of evolution take the form:

---

* I.R.S.I.A. (Belgium) Fellow
† Senior Research Assistant with the F.N.R.S. (Belgium)

E. Tirapegui and W. Zeller (eds.), Instabilities and Nonequilibrium Structures IV, 247–258.

$$\frac{\partial X}{\partial t} = A - (B+1)X + X^2Y + D_x \nabla^2 X$$

$$\frac{\partial Y}{\partial t} = BX - X^2Y + D_y \nabla^2 Y \tag{1}$$

A and B are control species the concentration of which are kept constant and uniform in space and time. B is chosen as the control parameter. The reference homogeneous steady state $(X_0, Y_0)=(A,B/A)$ loses its stability for $B=B_c=[1+A\sqrt{D_X/D_Y}]^2$ when $D_Y > D_X A^2/[\sqrt{1+A^2} - 1]^2$ giving rise to patterns with an intrinsic wavelength $2\pi/k_c$ where $k_c^2=A / \sqrt{D_X D_Y}$.

The concentrations of the new states appearing beyond the critical point can thus be approximated by linear superpositions of M critical modes:

$$C(\mathbf{r},t) = C_o + \sum_{i=1}^{M} (A_i e^{i\mathbf{k}_i.\mathbf{r}} + A_i^* e^{-i\mathbf{k}_i.\mathbf{r}}) \qquad |\mathbf{k}_i| = k_c$$

Using standard bifurcation techniques [7], we derive from (1) the equation of evolution for the amplitudes A_i:

$$\frac{\partial A_i}{\partial t} = \left[\frac{B - B_c}{B_c} + \left(\frac{\partial}{\partial x_i} - \frac{i}{2k_c}\frac{\partial^2}{\partial y_i^2}\right)^2\right] A_i + v(B) \sum_j^M \sum_k^M A_j^* A_k^* \delta\, (\mathbf{k}_i+\mathbf{k}_j+\mathbf{k}_k) - \sum_j^M \gamma_{ij} |A_j|^2 A_i$$

$$- \sum_j^M \sum_k^M \sum_l^M \beta_{jkl} A_j^* A_k^* A_l^* \delta\, (\mathbf{k}_i+\mathbf{k}_j+\mathbf{k}_k+\mathbf{k}_l)$$

$(B-B_c)/B_c$ measures the relative distance from the bifurcation point. All other coefficients v, γ and β are functions of the parameters of our model. The last term represents the contributions of noncoplanar quadrangles of $\mathbf{k}_i$ that may arise in 3D. The possible patterns in 2D and 3D homogeneous reactors (i.e. where the concentration of A and B are uniform in space) result from the competition between the solutions of these equations.

2.1. TWO-DIMENSIONAL HOMOGENEOUS SYSTEMS

Near B_c, the possible competing structures are stripes, characterized by one (M=1) wavevector $\mathbf{k}_s$ and patterns of hexagonal symmetry when M=3 with $\mathbf{k}_1+\mathbf{k}_2+\mathbf{k}_3=\mathbf{0}$. There exist also mixed-modes solutions that are however unstable in our case.

To test these analytical predictions, we have numerically solved the reaction-diffusion equations of the Brusselator model using an explicit Euler scheme complemented by finite difference methods. The predicted patterns are indeed observed in a 2D numerical reactor (see fig.1):

- stripes

- depending on the sign of $v(B_c)$, one has either Hπ hexagons where the maxima of concentrations form a honeycomb lattice or H0 hexagons where this lattice is triangular. These two hexagonal patterns differ only through a phase relation.

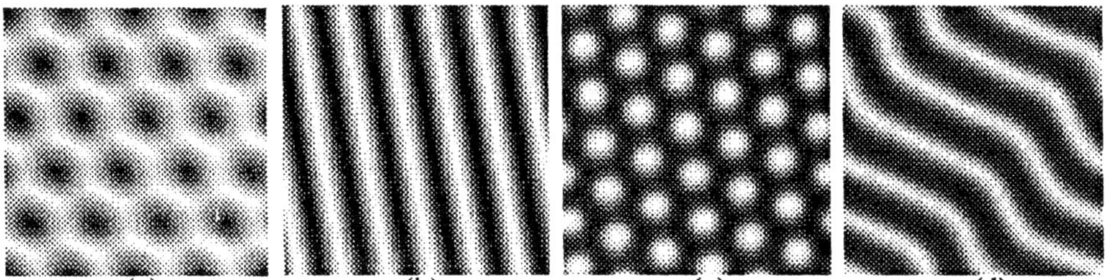

(a) (b) (c) (d)

fig.1: 2D Turing structures for the concentration of species X in the Brusselator. The integrations were performed on a square grid of length 64x64 with periodic boundary conditions. The gray scale corresponds to the concentrations between the absolute minimum (black) and maximum (white). They thus represent relative concentrations and not absolute ones.
a) Hπ
b) Stripes
c) H0
d) Zigzag structure

Increasing B, we have detected two scenarios:

-when $v(B_c)>0$, H0 is the only hexagonal pattern observed and no Hπ exist. Stripes may appear in a small region in the vicinity of B_c.

-when $v(B_c)<0$, the behaviour obtained is summarized on the bifurcation diagram of fig.2. Hπ is the first structure to appear when B is larger than B_c. These Hπ become unstable with respect to stripes for intermediate values of B. Reversing the variation of the bifurcation parameter, we come back to Hπ tracing an hysteresis loop. We recover thus near the bifurcation point the possibility of spatial bistability of the standard hexagons-rolls competition in hydrodynamics [8]. Spatial bistability means that either hexagons or rolls may occur for a same value of B depending on the initial conditions. At still higher B's, the stripes become unstable with respect to H0 and a second hysteresis loop is associated with this transition to reentrant hexagons. The bifurcation diagram drawn here describes the stability of the different patterns for the parameters chosen: A=4.5, DY/DX=8 (and thus B_c=6.71). We will keep this set of parameters throughout the rest of this paper to use fig.2 as a guide to understand the situation in the presence of a gradient.

Moreover, we have also detected a branch of stable zigzag stripes in the vicinity of the straight ones (fig.1d). There exists thus a domain of B where the Brusselator exhibits spatial tristability between straight stripes, zigzag stripes and hexagons. The two phases of hexagonal structures and the zigzag stripes obtained for the Brusselator were also obtained in the numerical study of the Schnackenberg model [9]. In both models [10,11] the stability domain of the different structures and the existence of reentrant hexagons may be explained in terms of nonlinear couplings.

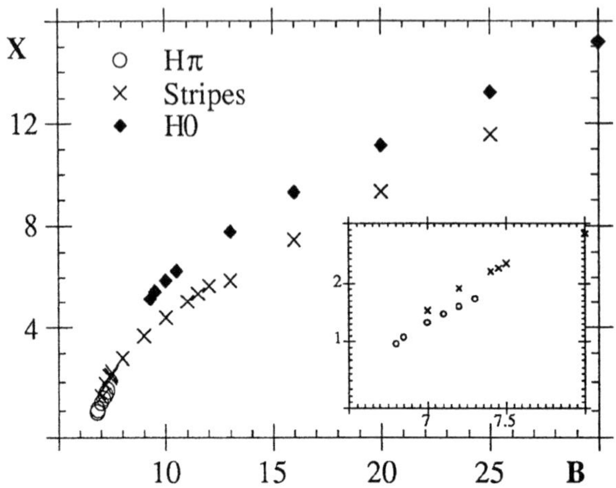

fig.2: Bifurcation diagram obtained from numerical integration of the 2D homogeneous Brusselator when A=4.5 and DY/DX=8. We draw the maximum concentration of X, from which we substracted the concentration of the uniform reference state (X_0=A), as a function of the bifurcation parameter B. The inset presents an enlarged view of the region close to B_c.

The stripes and both hexagonal phases have been observed in recent experiments [3]. Furthermore, the fact that these chemical patterns are identical to those that have been observed for a long time in hydrodynamics [14] reinforces the interest for the universal characteristics at the base of spatial self-organization.

2.2. THREE-DIMENSIONAL HOMOGENEOUS SYSTEMS

Theoretical and numerical analysis of the Brusselator model in a 3D reactor [15] predict the emergence of the following sequence of patterns when B is increased slightly above B_c: a structure with body centered cubic (bcc) symmetry (M=6) is the first to appear subcritically. It is followed, also subcritically, by hexagonal prisms (M=3) and then by supercritical lamellae (M=1). Phase related reentrant structures may also be present at higher values of B.

3. Two-dimensional ramped Turing structures

Knowing the situation in homogeneous systems, let us investigate how the presence of a gradient modifies this pattern selection problem.
In hydrodynamics, several works have already tried to understand how the variation in space of the control parameter affects the formation and selection of patterns. In Rayleigh-Bénard convection, non uniform heating has been showed to modify the amplitude equations as well as the nature of the bifurcation [16]. Furthermore, spatially varying external parameters can also influence wavelength selection mechanisms [17].
In chemistry, studies of chemical ramped structures have been initiated during the seventies in 1D [18]. Nowadays, study of 2D or even 3D non linear chemical models with a spatially varying control parameter is directly relevant to the experimental situation. Before starting

such an analysis, we have searched numerically for the occurence of generic behaviors in Turing structures submitted to gradients of concentrations [19].
The results of this analysis can be classified according to the type of ramps used.

3.1. LINEAR RAMPS

We have integrated the evolution equations of the Brusselator model on a rectangular domain of size Lxl, maintaining A uniform and constant in time and imposing at all times a linear ramp of the control parameter B along the side L. The direction of the gradient is denoted **G**. At L/3, we fix the value of B at its critical value B_c (see fig.3).

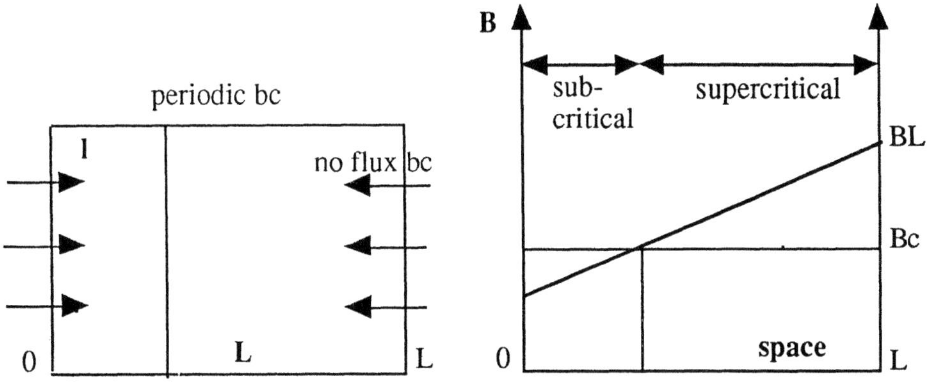

fig.3: Schematic drawing of a rectangular reactor Lxl used in numerical simulations of 2D ramped Brusselator. We impose a linear ramp of the control parameter B between the edges 0 and L where no-flux boundary conditions are imposed. Periodic boundary conditions are used along the other sides.

To change the slope of the ramp, we vary BL, the value of B at the edge L of the domain. When $BL>B_c$, the first third of the reactor is subcritical.
The diffusion coefficients and A are kept equal to the value for which the bifurcation diagram under uniform conditions of fig.2 has been drawn (thus $B_c = 6.71$). Imposing BL=8.0 and starting from random initial conditions, we observe coexistence in the same domain of $H\pi$ with wavevector $\mathbf{k_1} \perp \mathbf{G}$ $(\mathbf{k}_1+\mathbf{k}_2+\mathbf{k}_3=\mathbf{0})$ and stripes with $\mathbf{k_s}$ // $\mathbf{G}$ (see fig.4). Roughly, each type of pattern develops in the region of space where the local value of B allows it to be stable in the homogeneous case. We observe thus the unfolding in space of the bifurcation diagram.
Let us remark that $H\pi$ appear in the subcritical part of the reactor down to values of B much lower then the smallest B for which they are observed in homogeneous reactors. Indeed, the gradient plays the role of an imperfection providing a characteristic rounded off bifurcation. Compared to the homogeneous case, the presence of a linearly varying control parameter may thus modify the stability domain of a given pattern and the nature of the bifurcation. This is confirmed by the following numerical experiment in which we have raised the value of BL in small steps, thereby progressively increasing the slope of the ramp.
For BL=7.0, the supercritical region lies completely in the stability domain of the $H\pi$ that are indeed the only observed pattern. Using a random initial condition an hexagonal structure,

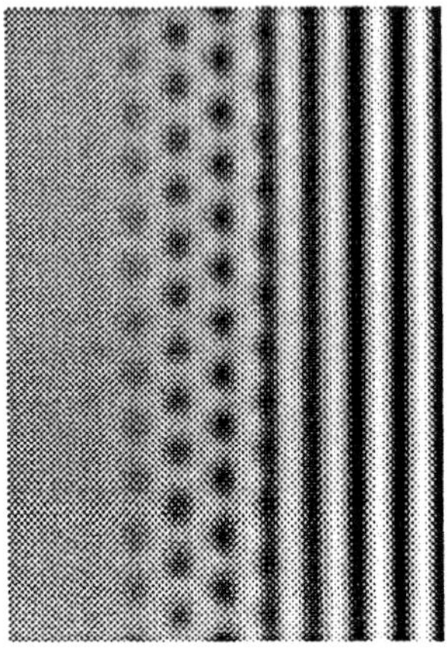

fig.4: Coexistence of $H\pi$ and stripes in a domain Lxl: 90x128 with A=4.5, DX=3.5, DY=28.0 (DY/DX=8). B_c=6.71 at L/3 and BL=8.0. The gradient **G** is thus directed from the left to the right. The initial condition was the local reference state $(X_o,Y_o(B))$ on which a small random noise has been added. The direction of the stripes is perpendicular to **G** and consequently, their wavevector $\mathbf{k_s}$ is // **G**. In the case of the π hexagons, $\mathbf{k_1} \perp \mathbf{G}$ (with $\mathbf{k_1}+\mathbf{k_2}+\mathbf{k_3}=0$) which means that one side of the hexagon (in white) is // **G**.

builds up (fig.5a) oriented in such a way that $\mathbf{k_1} \perp \mathbf{G}$ $(\mathbf{k_1}+\mathbf{k_2}+\mathbf{k_3}=\mathbf{0})$. Starting from this structure BL is increased to 8.0. The right part of the reactor now lies in a region of parameter space where $H\pi$ are unstable with respect to stripes under uniform conditions. However, fig.5b shows only somewhat deformed hexagons and no stripes. The stability limit of the $H\pi$ has thus obviously been shifted. Furthermore, remembering fig.4 that was obtained for the same values of the parameters (BL=8.0) by using random initial conditions, we notice that we have multistability between different ramped structures.

On increasing BL to 8.2 in the experiment of fig.5c, the $H\pi$ filled region finally snaps back to its righteous place in space and a set of stripes invades the remainder of the system. The transition takes place in such a way that $\mathbf{k_s}$ // $\mathbf{k_2}$ (or $\mathbf{k_3}$) rather than $\mathbf{k_s}$ // $\mathbf{k_1}$ where the reorganization would have to occur fully in the face of the gradient. As a consequence of this choice of orientation of the stripes, there is a parity breaking (choice between $\mathbf{k_s}$ // $\mathbf{k_2}$ or $\mathbf{k_s}$ // $\mathbf{k_3}$) with respect to the direction of the gradient. This induces a drift in time perpendicular to **G** of the whole pattern, including the $H\pi$ that are entrained because of the existence of their domain boundary with the stripes.

Increasing BL further, the slope of the ramp grows and the region where the $H\pi$ may develop shrinks concurrently until it becomes smaller than one wavelength. The hexagons are finally pinched out (even in the subcritical domain). As soon as this occurs, the stripes are freed because the blocking domain boundary also vanishes and a new set of stripes ($\mathbf{k_s}$ $\perp$ **G**) spontaneously invades the system (see fig.5e,f).

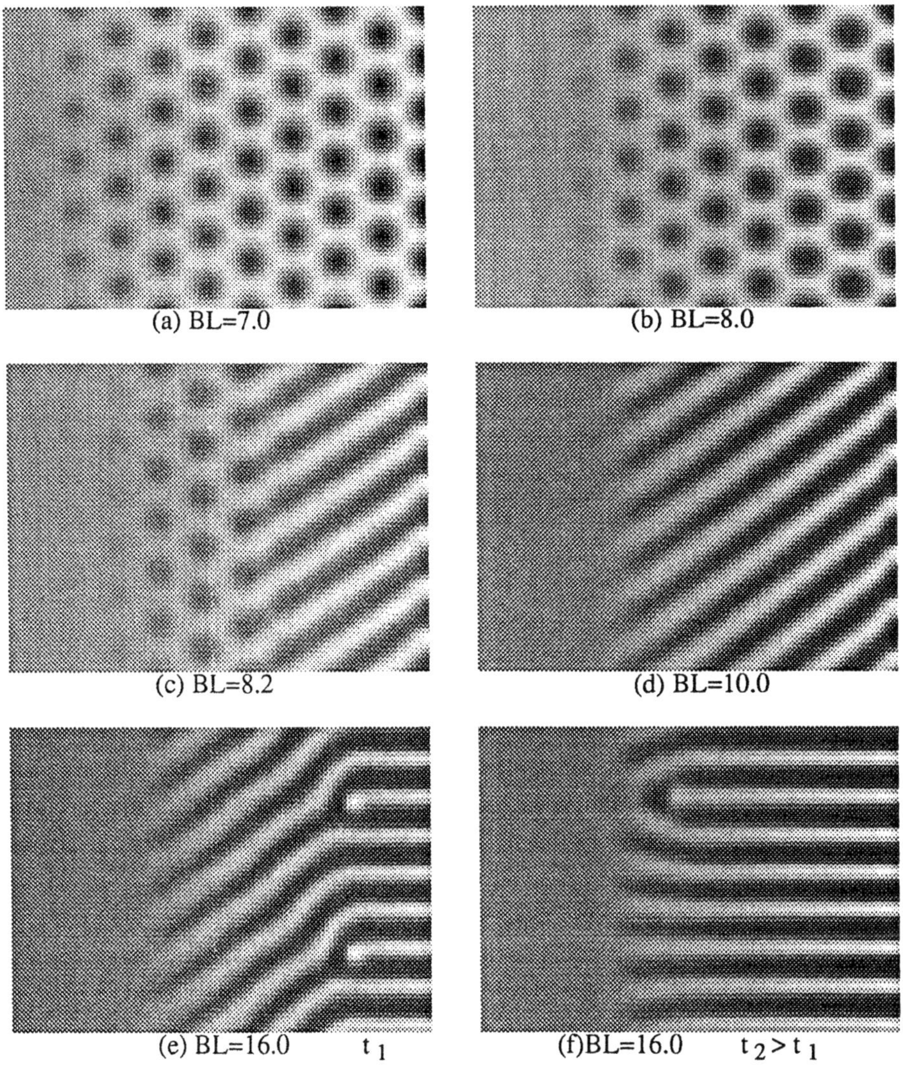

fig.5: Series of ramped structures obtained for the same values of parameters A, DX, DY as in fig.4 but in a domain Lxl: 90x64. We start from random initial conditions with BL=7.0 and obtain Hπ shown in a). After that, we start from the previously obtained pattern and increase BL quasistatically to go from b) to e). Each situation represented is the one obtained asymptotically after the change of BL except for e).

a) BL=7.0.

b) BL=8.0.

c) BL=8.2. Stripes appear suddenly such that $\mathbf{k_s}$ // $\mathbf{k_2}$ or $\mathbf{k_3}$. The pattern is drifting in the direction perpendicular to **G**.

d) BL=10.0. Subcritical Hπ are still present.

e) BL=16.0 at time t1.

f) BL=16.0 at time t2 much longer than time t1.

The nucleus of this new stripe system was already present for BL=8.2 as arising from the anchoring of the original Hπ to the BL boundary. Indeed, as may be seen from fig.5c,d the $\mathbf{k}_s$ // $\mathbf{k}_2$ (or $\mathbf{k}_3$) stripes exhibit a bending near this BL boundary to satisfy the imposed no-flux condition.

Therefore, the stripes that occur in fig.4 ($\mathbf{k}_s$ // $\mathbf{G}$) must result from a forcing due to the boundary condition at BL. This is confirmed by three new experiments.

In the first we follow the same routine as that presented in fig.5 but instead of increasing BL taking the structure obtained for its previous value, we always start from random initial condition. Then as already noted, we obtain fig.4 for BL=8.0. Starting with higher BL, the domain of stripes with $\mathbf{k}_s$ // $\mathbf{G}$ occupies the whole system and the Hπ are likewise pinched out. If the region near BL enters the domain of stability of H0 hexagons, the stripes in the bulk of the system are again freed of the BL boundary forcing and situations such as depicted in fig.6 result. Here we have spatial coexistence of stripes ($\mathbf{k}_s$ again $\perp$ $\mathbf{G}$), H0 oriented in the same way as were the Hπ and a lonely boundary stripe.

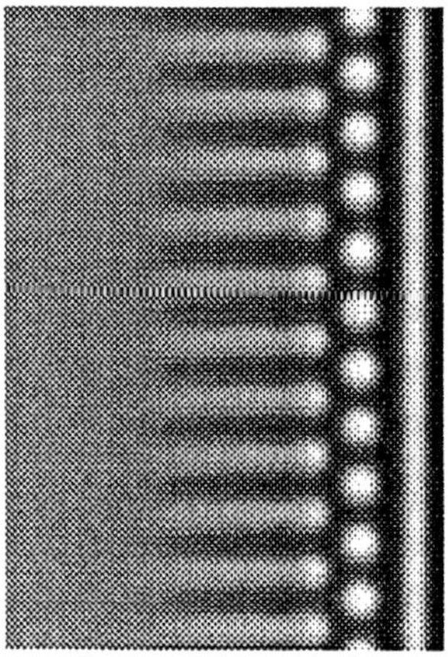

fig.6: Coexistence of stripes and H0 in a domain Lxl: 90x128 with BL=16.0, A=4.5, DX=7.0, DY=56.0. In this asymptotic state obtained from random initial conditions, $\mathbf{k}_s \perp \mathbf{G}$. This situation is bistable with that of fig.5f resulting from another initial condition.

The second experiment uses a completely supercritical system that lies in its totality in the region of parameters where, under uniform conditions, stripes are the only stable structures, i.e. we take B0=7.5 and BL=8.5. We now vary the slope by changing the size L of our system. In a "long" domain, thus with small slope, $\mathbf{k}_s$ is parallel $\mathbf{G}$ near the boundaries and appears directed in any direction in the middle of the system. We remark that many results lead to patterns containing bend stripes and defects because of the ubiquitous presence of zigzag stripes that are also stable for a wide range of values of B. For intermediate size domains, the anisotropic effect of the slope begins to play a role and in the middle of the domain, $\mathbf{k}_s \perp \mathbf{G}$ while $\mathbf{k}_s$ is still parallel to $\mathbf{G}$ near the boundaries (grain boundaries appear). Finally in very small domains, the influence of the boundaries prevails and only stripes with $\mathbf{k}_s$ // $\mathbf{G}$ are observed (fig.7a).

The forcing and orientational effect of the boundary giving rise to stripes with $\mathbf{k_s}$ // $\mathbf{G}$ can be linked to the fact that the ramped basic solution (X_0,Y_0) does not satisfy the imposed no-flux boundary condition on X and Y [20]. Let's therefore take other boundaries with no-flux of X (because $X_0=A=ct$) and a flux S/A of Y for a linear ramp in B of slope S (because $Y_0=B(x)/A=Sx/A$ so $\partial Y/\partial x|_{0,L}=S/A$). We see in fig.7b the spontaneous change of orientation of the stripes from $\mathbf{k_s}$ // $\mathbf{G}$ to $\mathbf{k_s} \perp \mathbf{G}$. So this $\mathbf{k_s} \perp \mathbf{G}$ orientation seems the natural one in the presence of the linear ramp.

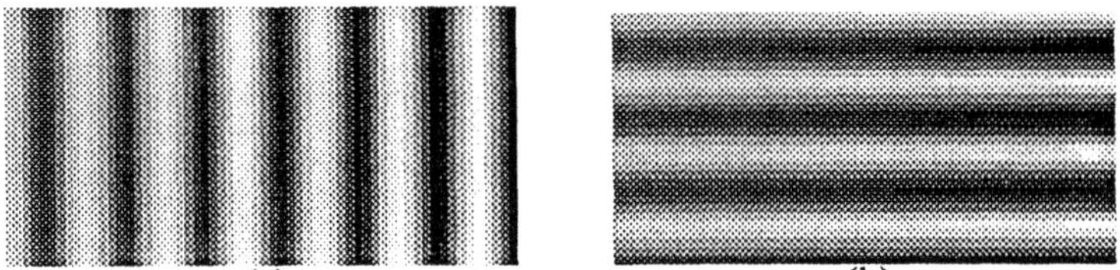

(a) (b)

fig.7: Stripes in a domain Lxl: 60x30 with B0=7.5 and BL=8.5.
a) no-flux boundary conditions are applied at all edges of l and L for both variables X and Y.
b) no-flux boundary conditions exist at all edges for the variable X while for Y,we take no-flux at the borders of l and a flux=S/A at the borders of L along which is directed the ramp.

The different numerical experiments performed here using a linear ramp of the control parameter confirm thus *coexistence* in a same reactor of several structures as observed in real experiments. *The gradient plays a role on the stability domain* of each pattern leading to *multistability* between different ramped structures and to a *imperfect bifurcation.* The *linear ramp influences also the orientation of the structures* favoring wavevectors directed perpendicularly to the direction of the gradient. *Orientational effects of the boundaries* have also to be taken into account.

3.2. CHAIN-LIKE RAMPS

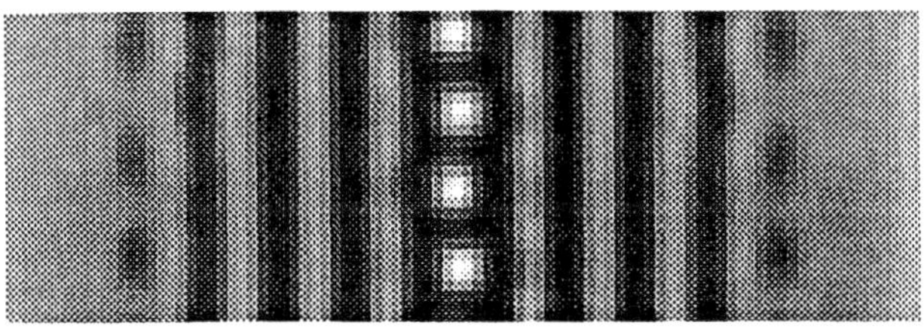

fig.8: Coexistence of H0, stripes and Hπ in a domain Lxl: 90x30 submitted along L to a chain-like profile of A, solution of the diffusive equation:

$$\delta A/\delta t = - k_a A + D_a \nabla^2 A$$

As initial condition, we fix at each point X and Y equal to the local reference state. On the line L/2, we multiply X by 2. The parameters are A(0)=A(L)=10, k_a=0.01, D_a=10.0, DX=3.5, DY=28.0 whereas B is constant and equal to 12.0.

This time, the reactor Lxl is fed at its two ends in such a way that B is constant but A has a diffusive, chain-like profile with a minimum A_{min} located at L/2. At each local value of A corresponds a local $B_c=B_c(A)$. If B is maintained fixed at a value higher then $B_c(A_{min})$, the central part of the reactor is supercritical.
Coexistence of different structures is also observed as shown in fig.8. These *localized* structures determine their own region of existence but may invade subcritical lateral parts of the reactor.
For that kind of profile, the *bifurcation is perfect.* The *delay of the threshold* calculated by the linear stability analysis increases with the curvature of A's profile. We have not studied the influence of the ramp on the direction of patterns in that case.

4. Three-dimensional ramped Turing structures

Some of the coexisting experimental patterns are in fact 3D as their wavelength is smaller then any side of the gel. We have integrated Eq (1) on a 3D domain lxlxL ($l<L$) in the presence of a gradient to see if the 2D results can be generalized to 3D. We again impose a linear ramp of B between the edges 0 and L of the side of length L and maintain A constant. We observe then coexistence in the same domain of bcc patterns, prisms and lamellae separated by complex boundaries (see fig.9) [21].
Further studies of the influence of the gradient on the selection and orientation of the 3D structures are under way.

5. Conclusions

The study of Turing structures in homogeneous systems show that some of the experimental results recently obtained may already be described by the classical approach of pattern selection. Our numerical study to ramped structures shed some light on the coexistence of different patterns in a same reactor and show that the influence of the ramp and the borders on the selection and orientation of structures may be quite complicated.
Besides, several questions raised by the experiments remain still unanswered, as for instance, the way by which time reappears in the dynamics of stationary patterns when a parameter is changed. Moreover, the 3D aspects of these structures is still triggering the attention and asks for more detailed analysis.
All the questions related to these chemical dissipative structures show the total originality and interest of patterns in chemistry compared to those obtained in hydrodynamics. Indeed, the usual presence of a privilegied direction in hydrodynamical instabilities prevents there the possibility of true 3D symmetry breaking. Moreover, the ease of obtaining hexagons in chemistry gives a good tool to study this pattern hardly controled in hydrodynamics.

Acknowledgements

We thank J. Boissonade, P. De Kepper and V. Dufiet for stimulating discussions. This work was sponsored by Twinning Contract #SC1*-CT91-0706(TSTS) of the EC Science Program.

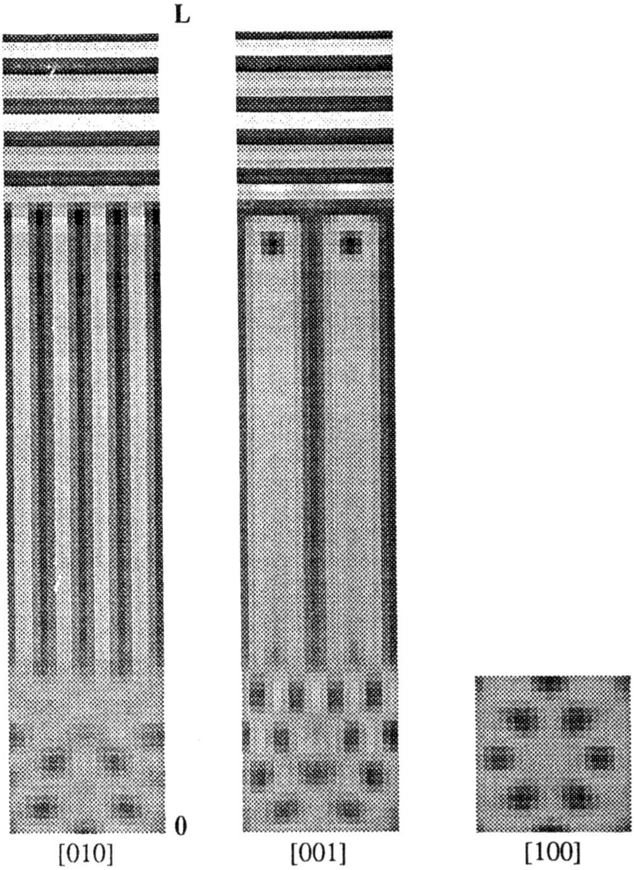

fig.9: 3D ramped Turing structures in a domain lxlxL:20x20x100. A linear ramp of B is fixed between edges 0 and L where no-flux boundary condition are applied. Periodic boundary conditions are used in the two other directions. The parameters are A=4.5, DX=1.0, DY=8.0 (and thus B_c=6.71), B0=6.7, BL=7.3. Visualisation of the pattern is made possible by showing three orthogonal planes noted by their Miller indices. Coexistence of three structures is observed: near the edge 0, we are just above B_c and a bcc pattern is seen. Lamellae appear at edge L for higher values of B. Hexagonal prisms impose themselves in the middle of the domain for intermediate values of B.

References

[1] V. Castets, E. Dulos, J. Boissonade, P. De Kepper, *Phys. Rev. Lett.*, **64**, 2953, 1990.

[2] Q. Ouyang, H. Swinney, *Nature*, **352**, 610, 1991.

[3] Proceedings of the NATO ARW:"Nonequilibrium Chemical Dynamics: From Experiment to Microscopic Simulations", (Brussels, 2-5 December 91), to appear in a special issue of *Physica A*.

[4] A. Turing, *Phil. Trans. R. Soc. Lond.*, **237B**, 27, 1952.

[5] G. Nicolis, I. Prigogine, "Self-organization in Nonequilibrium Systems", (Wiley, New-York, 1977).

[6] I. Prigogine, R. Lefever, *J. Chem. Phys.*, **48**, 1695, 1968.

[7] P. Manneville, "Dissipative Structures and Weak Turbulence", (Academic Press, Boston, 1990).

[8] F.H. Busse, *Rep. Prog. Phys.*, **40**, 1929, 1978.

[9] V. Dufiet, J. Boissonade, *J. Chem. Phys.*, **96**, 664, 1992.

[10] J. Verdasca, A. De Wit, P. Borckmans, G. Dewel, *Phys. Lett. A.* (to appear 1992).

[11] S. Metens, G. Dewel, P. Borckmans, submitted to *J. Chem. Phys.*

[12] R.D. Vigil, Q. Ouyang, H.L. Swinney, see ref. [3].

[13] G. Ertl, *Science*, **254**, 1750, 1991.
J. Falta, R. Imbihl, M. Henzler, *Phys. Rev. Lett.*, **64**, 1409, 1990.

[14] H. Bénard, *Revue Gén. Sci. Pur. Appl.*, **11**, 1261, 1900.

[15] A. De Wit, P. Borckmans, G. Dewel, D. Walgraef, *Physica D* (to appear 1992).

[16] P.M. Eagles, *Proc. R. Soc. Lond.*, **A371**, 359, 1980.
I.C. Walton, *SIAM*, **67**, 199, 1982.

[17] L. Kramer, E. Ben-Jacob, H. Brand, M.C. Cross, *Phys. Rev.Lett.*, **49**, 1891, 1982.
H. Riecke, H.-G. Paap, *Phys. Rev. Lett.*, **59**, 2570, 1987.

[18] M. Herschkowitz-Kaufman, G. Nicolis, *J. Chem. Phys.*, **56**, 1890, 1972.
J.F.G. Auchmuty, G. Nicolis, *Bull. Math. Biol.*, **37**, 323, 1975.

[19] J. Boissonade, *J. Phys.* (France), **49**, 571, 1988.
P. Borckmans, A. De Wit, G. Dewel, see ref. [3].

[20] I.C. Walton, *J. Fluid. Mech*, **131**, 455, 1983.

[21] P. Borckmans, G. Dewel, A. De Wit, D. Walgraef, *Entropie*, **164**, 83, 1991.

Stability Limits of Defects and Spatio-Temporal Chaos in Nonequilibrium Media

Igor S. Aranson
Racah Institute of Physics, Hebrew University of Jerusalem,
91904 Jerusalem, Israel

Lorenz Kramer and Andreas Weber
Physikalisches Institut der Universität Bayreuth,
8580 Bayreuth, Postfach 101251, Fed. Rep. Germany

Abstract

We show that the Eckhaus instability for traveling waves is of convective nature and does not lead directly to absolute instability. As a consequence spiral waves and hole solutions remain stable in a larger range than expected previously and the transition to spatio-temporal chaos can be delayed in some parameter range even beyond the Benjamen-Feir limit. We calculate the onset of absolute instability using the complex Ginzburg-Landau equation ($CGLE$) and verify the results by detailed simulations. For the 1d CGLE we show that stable localized oscillating hole solutions exist in a region of parameters preceding the appearance of spatio-temporal chaos. These solutions oscillate in the depth and position of the hole. They can be described in terms of a supercritical Hopf bifurcation of the stationary hole solutions. The analytical results have been verified numerically by direct simulations of the $CGLE$.

1 Introduction

Oscillatory nonequilibrium media have been the objective of intense study recently. In such media defect solutions are especially interesting because they typically emit waves which may invade the system. One may have phase slips in 1d, (topological) point defects in 2d (spirals) and line defects in 3d extended media, known from chemical reactions (e.g. Belousov-Zhabotinskii reaction) [1], electrohydrodynamic convection of liquid crystals [2], convection in binary mixtures [3], multimode lasers [4]. Many more examples in physics, chemistry and biology can be found [5]. Defects are also often responsible for the formation of spatio-temporal disorder and the appearence of turbulence. Moreover many important characteristics of this turbulence can be explained within the framework of defect dynamics. Although nonequilibrium media are studied intensively many unsolved problems still persist. In this article we concentrate on the questions of stability of defects and the transition to spatio-temporal chaos influenced by complex dynamics of the ensembles of defects.

The basic equations and solutions are presented in the rest of this section. In section II we investigate the stability boundaries of waves emitted by defects. It turns out that the usual Eckhaus instability is of convective nature and may therefore be surpassed, depending on the imposed boundary conditions until the absolute stability limit is reached. Section III is devoted to implications for spatio-temporal chaos. The comparison with results of numerical simulations performed by different authors is presented there. In section IV a

E. Tirapegui and W. Zeller (eds.), Instabilities and Nonequilibrium Structures IV, 259–272.

new class of oscillating defect solutions is studied. Such solutions can preceed the spatio-temporal chaos. Some problems of future interest are considered in the conclusion. Most of the results have been presented before in short publications [6, 7, 8].

The simplest description of oscillating nonequilibrium media is provided by the complex Ginzburg-Landau equation

$$\partial_t A = [(1+ib)\nabla^2 + 1 - (1+ic)|A|^2]A\ , \tag{1}$$

which is strictly valid in the weakly nonlinear region slightly above a supercritical bifurcation to the oscillatory state. The complex amplitude A describes the slowly varying envelope of the wave pattern. It exhibits traveling plane wave solutions

$$A = F exp[i(\vec{Q}\vec{r} - \omega t)] \tag{2}$$

where $F^2 = 1 - Q^2$ and $\omega = c + (b-c)Q^2$. They exist for $Q^2 < 1$. In order to test their stability one usually considers the complex growth rate λ of the modulational modes. Restricting oneselves to the most dangerous longitudinal perturbations proportional to $exp(\pm i\vec{k}\vec{r})$ with $\vec{k} \parallel \vec{Q}$ one easily finds.

$$\lambda(k) = -k^2 - 2iQbk - F^2 \pm \sqrt{(1+c^2)F^4 - (bk^2 - 2iQk + cF^2)^2}\ . \tag{3}$$

In the long-wavelength limit ($k \to 0$) one may expand (3) leading to

$$\lambda(k) = iv_g k - D_{\|}k^2 + \mathcal{O}(k^3) \tag{4}$$

$$\text{with} \quad v_g = 2(c-b)Q,\ \ D_{\|} = 1 + bc - 2(1+c^2)Q^2/(1.-Q^2)\ . \tag{5}$$

Traveling waves are long-wavelength stable for $D_{\|}(Q) > 0$ which designates the Eckhaus stable range.

Two-dimensional defect solutions of Eq. (1) with simple zeros in A are spiral waves. An isolated spiral has the form

$$A = F(r)\, exp(i[-\omega t - m\varphi + \psi(r)]). \tag{6}$$

Here (r, φ) are polar coordinates, $m = \pm 1$ is the topological charge or circulation, $F(r)$ and $\psi(r)$ are real functions with $F(0) = 0$ and asymptotics $\lim_{r\to\infty} F(r) = \sqrt{1-Q^2}$, where $Q = \lim_{r\to\infty} \psi'(r)$ is the asymptotic wavenumber which is a unique function of b and c that can in general not be determined analytically [12], and $\omega = c + (b-c)Q^2$ is the frequency of rotation.

The 1d-analog of the defect solutions of the *CGLE* are the "stationary hole solutions" presented by Nozaki and Bekki [9], which are of the form

$$A_h = \sqrt{1-Q^2}\tanh(\frac{x - x_0}{p})\exp(i(\Phi(x - x_0) - \omega t + \varphi_0))\ , \tag{7}$$

where $\partial_x\Phi = Q\tanh((x - x_0)/p)$, x_0 and φ_0 are arbitrary constants characterizing the position and the phase of the hole and p is given by

$$(c-b)p^4 - (4(c-b) + 9b(1+bc))p^2 + (4(c-b) + 18b(1+b^2)) = 0\ . \tag{8}$$

In this case the asymptotic wavenumber Q is determined analytically from the relation

$$Q = \tfrac{2-p^2}{3bp} \tag{9}$$

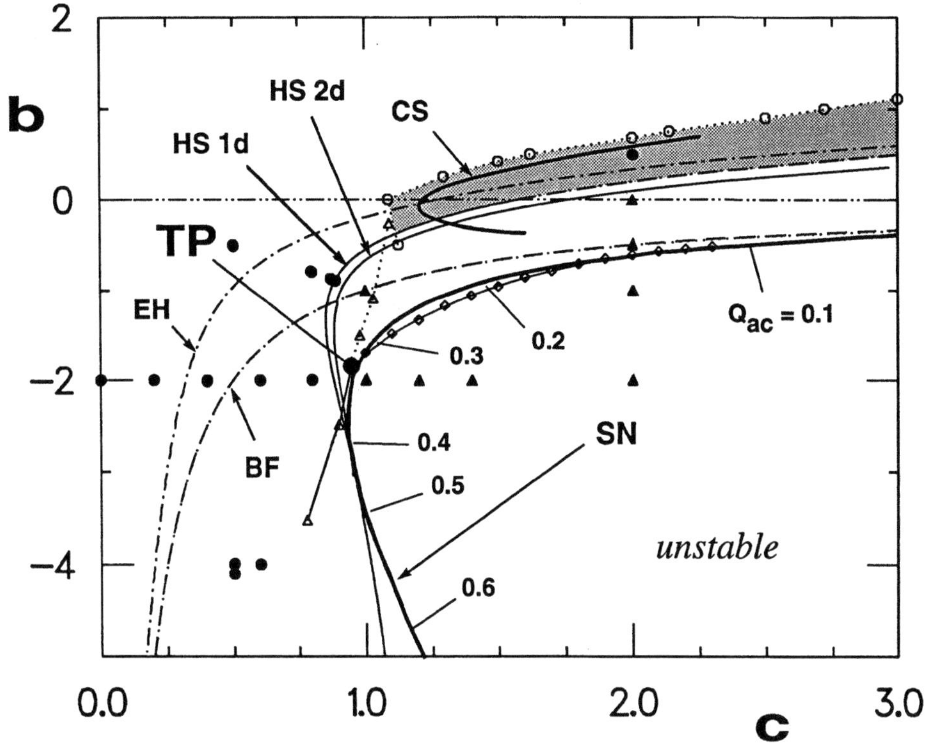

Figure 1: *Stability limits for 2d and 1d defects in the b-c-plane. Presented are: B-F-limit* $1 + bc = 0$ *(curve BF); long wavelength Eckhaus limit* $(1+bc)/(3+bc+2c^2) = Q^2$ *with* $Q(b,c)$ *corresponding to 2d defects (curve EH); absolute instability limits according to (14) for 1d defects (curve HS1d) and 2d defects (curve HS2d). Solid circles (stable defects) and solid triangles (defect turbulent states) give the results of the 2d numerical simulations of Eq. (1) with small random initial conditions. The curve SN gives the limit up to which convectively stable wavenumbers exist. Some values for the corresponding wavenumbers* Q_{ac} *along SN are indicated. The lines marked with* $\triangle$, $\diamond$ *and* $\bigcirc$ *are taken from recent 1d numerical investigations [10, 11]. The analog of defect turbulence exists to the right of the curve defined by the triangles [11]. In the shaded region stable hole solutions were found numerically [10]. The curve (CS) gives the location of the Hopf bifurcation where stationary hole solutions loose stability.*

Thus the defects provide a wavenumber selection mechanism. They emit asymptotically plane waves with a unique wavenumber and a necessary condition for its stability would seem $D_{||}(Q) > 0$. In Fig. 1 the boundary of the Eckhaus-stable range of 2d defect solutions is shown in the b-c-plane (EH, dash-dotted; the plot can be continued to negative c by noting the symmetry $(b, c) \to -(b, c)$). In order to determine the emitted wavenumber in the b-c-plane we first took values along the lines $b = 0$ or $b = -c$ from numerical simulations and then filled the plane by use of the similarity transformation [12, 13]

$$Q'(b', c') = \frac{Q(b, c)}{\sqrt{(1 - Q^2(b, c))\,|z'/z| + Q^2(b, c)}} \tag{10}$$

$$\text{where} \qquad z = \frac{1 + ib}{1 + ic}\ , \quad z' = \frac{1 + ib'}{1 + ic'} \tag{11}$$

and (b, c) and (b', c') have to fulfil the relation

$$\frac{b - c}{1 + bc} = \frac{b' - c'}{1 + b'c'} \tag{12}$$

Also shown in Fig. 1 is the Benjamin-Feir (B-F) limit (dash-dotted) where the solution with $Q = 0$ becomes unstable - all other plane-wave solutions being then unstable already [14].

2 Absolute Instability of Waves

The usual picture for the generic long-time behavior of sufficiently large systems starting from random initial conditions and b, c of order one is roughly as follows: In the B-F unstable range one has mainly "defect-mediated turbulence", i.e. a disordered state [15] with random creation, motion and annihilation of defect pairs and a well-defined average defect density [16]. For this state the dimension of chaos is proportional to the size of the system [17]. In the B-F stable but Eckhaus unstable range one has considerable sensitivity on initial conditions and one may have various defect densities up to a maximum with the defects either disordered and sometimes rather motionless or ordered in a lattice [13, 16]. Finally, in the Eckaus stable range, the density of defects would usually vanish. However, with appropriate initial conditions defect lattices can be formed [13]. Their minimal spacing appears to diverge when $b - c$ becomes too small.

Here we wish to provide analytic and numerical evidence that spirals and holes, even if they are essentially isolated, remain stable in the Eckhaus- and, for appropriate parameters, even in the B-F unstable range. This has a profound influence on the turbulent regime and explains some of the observed features. Some numerical evidence in this direction was already found earlier [13]. In fact, since defects emit traveling waves with a nonzero group velocity $v_g = \partial_Q \omega = 2(b-c)Q$ directed outward the Eckhaus criterion can be taken only as a test for convective instability where a localized initial perturbation $S_o(x)$ of the asymptotic plane wave, although amplified in time drifts away and does not necessarily amplify at a fixed position [18]. To test for absolute instability one has to consider the time evolution of the localized perturbation which is in the linear range given by

$$S(x, t) = \int_{-\infty}^{\infty} dk/(2\pi)\hat{S}_o(k)exp(ikx + \lambda(k)t) \tag{13}$$

where $\hat{S}_o(k)$ is the Fourier transform of $S_o(x)$[†]. The integral can be deformed into the complex k-plane. In the limit $t \to \infty$ the integral is dominated by the largest saddle point of $\lambda(k)$ (steepest descent method [19]) and the test for absolute instability is

$$Re[\lambda(k_o)] > 0 \quad \text{with} \quad \partial_k \lambda(k_o) = 0, \quad Re[\partial_k^2 \lambda(k_o)] < 0 \tag{14}$$

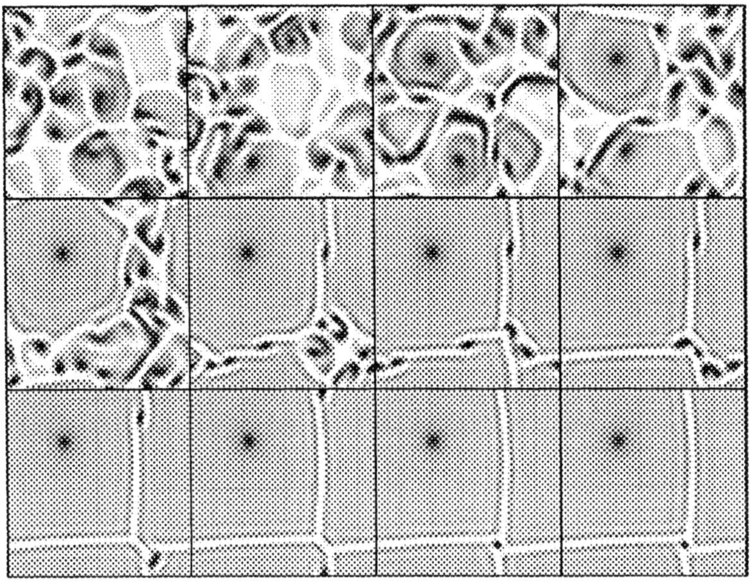

Figure 2: *Snapshots of a simulation with small random initial conditions and periodic boundary conditions in the convectively unstable range* ($b = -2$, $c = 0.6$, *system size 100x100*). $\mid A \mid^2$ *is coded in the grey scale (white:* $\mid A \mid^2 = 1$, *black:* $\mid A \mid^2 = 1$*). The first picture is taken at* $t = 225$, *the others follow at time intervals of* 25.

The long-wavelength expansion (4) indicates that at the Eckhaus instability, where $D_{\|}$ becomes negative, the system can remain stable in the above sense. When $D_{\|}$ vanishes and $Q \neq 0$ the main contribution comes from the linear term that then can suppress instability.

Using the general expression (3) together with the wavenumber emitted by the spirals as before we obtain the curve $HS2d$ in Fig. 1 as the stability limit according to the conditions (14)[‡]. Also included is the corresponding curve for the hole solution (curve $HS1d$ in Fig. 1). We have tested these results systematically by direct numerical simulations of 1d holes in a large interval and were able to verify the stability limit (see next section). Some points were also checked for the 2d case (see Fig. 1).

Clearly in a finite system the stability can be achieved only if the reflection at the boundaries is not too strong, so in principle one has to expect the system to lose stability before the exact limit is reached. In the CGL the interaction of the emitted waves with the boundaries leads to shocks, that are also formed when waves of different defects collide. These shocks are very strong perturbations to the plane wave solutions but it turns out that they absorb the incoming perturbations. Also, the convectively unstable state is very susceptible to noise, so noise induced transitions to turbulence should occur.

3 Boundaries of Spatio-Temporal Chaos

It turns out that for $b > -1.2$ the new stability limit is more restrictive than the B-F-instability and less restrictive otherwise. In the first region earlier simulations [16] along the line $c = 2$ (actually -2 was used, but we make use of the symmetry) exhibited existence of a disordered state until about $b = 0$ when coming from the B-F unstable regime. This

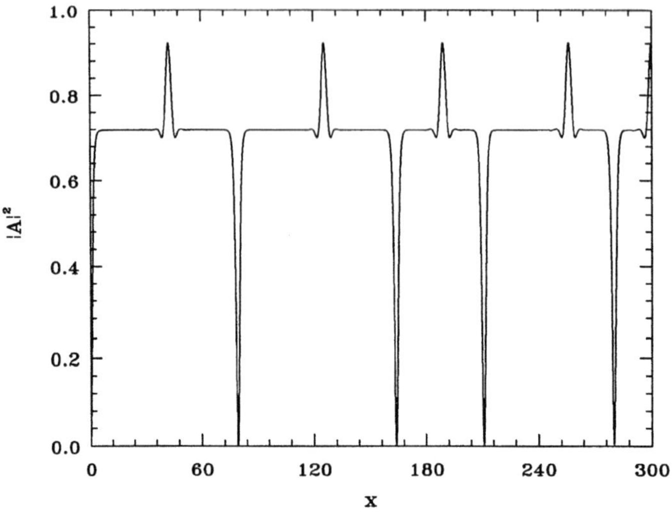

Figure 3: *Plot of* $|A|^2$ *for a stationary, stable, nonperiodic 1d solution with zeros of* $|A|^2$ *and shocks inbetween in the slightly absolutely unstable range* ($b = 0$, $c = 1.62$). *In 1d the stability limit for* $b = 0$ *is at* $c = 1.567$.

boundary is rather near to the new stability curve so one is tempted to conclude that the curve has some significance as a limit of existence of defect-mediated turbulence. Our stability boundary is also consistent with the boundary for persistent turbulence determined numerically for $b = -1$ by the criterion of the Lyapunov exponent dropping to zero in a discretized version of the CGL [20]. Of course the region $b < -1.2$ is especially interesting. We have performed numerical simulations of Eq. (1) with small random initial conditions along the line $c = 2$ and also along $b = -2$. The final state of the system is indicated in Fig. 1 (solid circles: stable defects, solid triangles: defect turbulence). Initially always many defects are formed. In the convectively unstable range eventually one defect becomes dominant pushing away the other ones which ultimately annihilate. However, with periodic boundary conditions, for topological reasons another defect with oppostite circulation and without the far field typical for spirals survives inside the shock regime. In Fig. 2 an example of this evolution is shown for a simulation with periodic boundary conditions. Note the shocks where the waves of spirals from neighboring periodicity regions collide. Such defect lattices exist also in the Eckhaus stable regime and were observed before [13]. In fact defect lattices exist also in the (slightly) absolutely unstable range, however, with an upper bound on their spacing. Actually there is some evidence that also stationary, spatially nonperiodic arrangements exist. In Fig. 3 we show $|A|^2$ for a 1d analog of such a solution which developed after long transients.

We conclude that defect turbulence can exist in 2d only on the unstable side of $HS2d$. This explains the results of previous simulations [15, 16, 20, 13]. More generally we have the following scenario for the stability of travelling waves in the b-c-plane: On the BF unstable side traveling waves with $|Q|$ sufficiently small are absolutely unstable. Sufficiently near to the BF line there exists a "stable" (actually convectively unstable) band of wavenumbers $0 < Q_{a1} < Q < Q_{a2}$ (we restrict ourselves to $Q > 0$ since there is a symmetry $Q \rightleftharpoons -Q$).

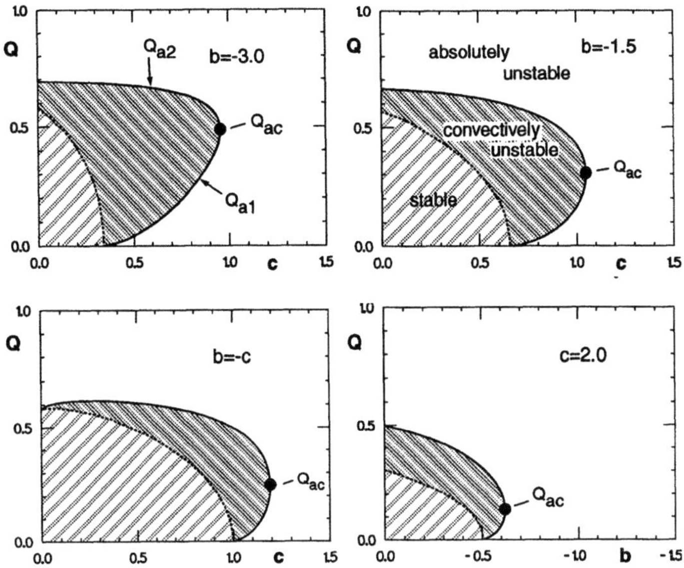

Figure 4: *Four cuts in the b-c-Q-space are shown. The upper bound Q_{a2} for convectively unstable solutions merges with the lower bound Q_{a1} at Q_{ac}. Q_{ac} lies on the curve SN in Fig. 1. Q_{a1} goes to zero at the BF limit.*

Q_{a1} goes to zero at the BF line as $(1+bc)^{3/2}$ as can be seen from the long-wavelength expansion (4) with the fourth-order term included. Moving away from the BF line (into the unstable regime) Q_{a1} increases and Q_{a2} decreases until they come together in a saddle-node-type process at a value which we denote by Q_{ac}. In Fig. 4 the scenario is demonstrated for four cuts in the b-c-Q-space. The saddle node curve is also plotted in Fig. 1 (curve SN) together with some values of Q_{ac}. Thus convectively unstable waves exist up to this curve.

As mentioned above we have verified the above concept in particular by testing the stability of the plane waves emitted by the hole solutions (7). The absolute instability limit (curve $HS1d$ in Fig. 1) corresponds to Q_{a2} for $|b|\lesssim 3$ and to Q_{a1} otherwise. Actually the hole solutions are unstable in the core regime except (roughly) below and to the right of the curve designated by ○ (Fig. 1, shaded region), see Sakaguchi [10] and next section. This instability can be suppressed by imposing the zero as a boundary condition. The stability limit of the emitted plane waves as determined by direct numerical simulation coincides with $HS1d$. Thus it appears that the phase-slip solutions are fully stable in the shaded regime.

The 1-d analog of defect turbulence is turbulence with permanent occurrence of phase slips. Its range of existence and properties were investigated very recently in great detail numerically [11, 21] (some parts of the scenario were investigated previously by Sakaguchi [22]). In Fig. 1 the border line of existence of "1-d defect turbulence" is included (△) [11]. When approaching this curve from the right the average rate of phase slips appears to vanish continuously. Between this curve and the BF line one has pure "phase turbulence", i.e. a complex dynamical state with the amplitude $|A|$ remaining near 1. The authors of ref. [11] also find phase turbulence above the line corresponding to the diamonds (and below the BF

line) so that here one has a region of multistability with different turbulent attractors. The bistable range terminates at the "tri-critical point" TP where the two curves (diamonds and triangles) meet.

We now suggest that the "stable" (actually convectively unstable) band of wavenumbers that encloses the absolutely unstable range centered around zero (see Fig. 4) can actually stabilize phase turbulence, i.e. prevent the occurrance of phase slips and restrict the system to a dynamics with small local wavenumber (if it was initialized properly). This seems quite plausible since large amplitude depressions go in parallel with large local wavenumbers which presumably get swept away in the convectively unstable range. Apparently the fluctuations are then annihilated by collisions from neighboring regions. The mechanism presumable works as long as Q_{a1} is sufficiently small, which is the case essentially all the way up to Q_{ac} above the tri-critical point TP. Then the border line of phase turbulence ($\diamond$) coincides essentially with $SN1d$. The re-entrant behavior of Q_a suggests that there may also exist a defect-turbulent attractor with large local wavenumbers on the average, as observed numerically [11]. Below TP Q_{ac} becomes rather large and phase turbulence is apparently not sustained all the way up to Q_{ac}. Also the unstable region $1 > Q > Q_{ac}$ appears to become too small to support amplitude turbulence. Hopefully these arguments can be made more rigorous in the future. We point out that preliminary simulations show that the phase turbulence exists also in two dimensions. A systematic study analogous to the one presented in ref. [11], for two dimensions would be very useful.

Since convectively unstable states cause a spatial amplification of noise [23] the above scenarios may change drastically when small amounts of noise are present. It should be interesting to study such effects.

4 Oscillating Hole Solutions

Numerical simulations by Sakaguchi [10] indicate that the hole solutions are stable in a relatively narrow "belt" in the $b-c$ plane (see Fig. 1, shaded region). The instability of the hole solution can be qualitatively separated into two classes, the instability of the emitted waves and the core instability [10]. The first class is characterized by the continuous spectrum of extended eigenfunctions. Because of the asymptotic behavior for $x \to \pm\infty$, those eigenfunctions tend to plane waves and the stability analysis reduces to the analysis of the (absolute) stability of the plane wave with given $Q(b,c)$ discussed above (see Fig. 1, curve $HS1d$).

The core instability is related to localized eigenmodes, corresponding to a discrete spectrum of the linearized problem. The spectrum always includes two marginally stable eigenfunctions, corresponding to the gauge and translation symmetries of Eq. (1) [††]. Other eigenfunctions have to be computed numerically. For convenience of the analysis it is useful to change from the variable A to

$$W = A\exp(-i(\Phi(x-x_0) - \omega t + \varphi_0)) \, , \tag{15}$$

where the function $\Phi(x-x_0)$, ω and φ_0 are defined in Eq. (7). Equation (1) then becomes (x_0 is chosen zero)

$$\begin{aligned} \partial_t W &= W\left[1 + i\omega - (1+ib)Q^2\tanh^2\frac{x}{p} + \frac{iQ}{p}(1+ib)\cosh^{-2}\frac{x}{p}\right] - \\ &\quad -(1+ic) \mid W \mid^2 W + (1+ib)(\partial_x^2 W + 2iQ\tanh\frac{x}{p}\partial_x W) \, . \end{aligned} \tag{16}$$

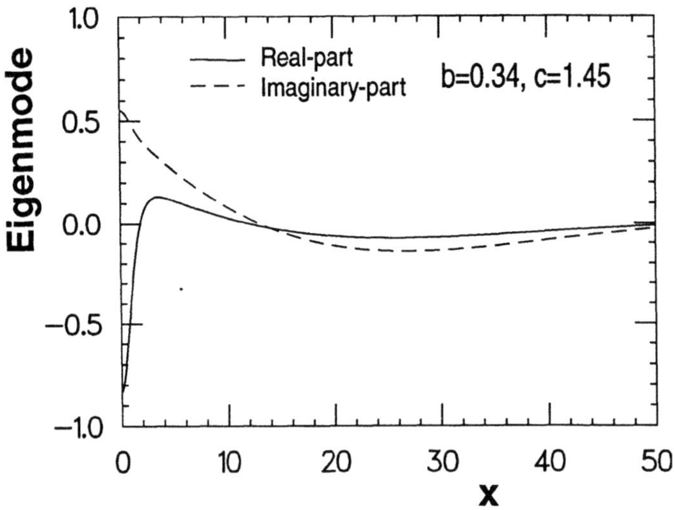

Figure 5: *The real and imaginary parts of* $\xi = C\eta + C^*\eta'$ *for* $b = 0.34$, $c = 1.45$, *are shown for the most unstable modes. The corresponding eigenvalues are* $\lambda = 0.0358 \pm i * 0.0611$.

The hole solution Eq. (7) now has the form

$$W_0 = \sqrt{1-Q^2}\tanh\frac{x}{p}\,. \tag{17}$$

The perturbations of this solution are described by the linear equation ($W = W_0 + \xi$), obtained from Eq. (16) using expression (17) :

$$\begin{aligned}\partial_t\xi = \hat{L}\xi = \cosh^{-2}\frac{x}{p}[1 - i\omega + \frac{iQ}{p}(1+ib)]\xi - \\ -(1+ic)(1-Q^2)\tanh^2\frac{x}{p}(\xi^* + \xi) + 2iQ(1+ib)\tanh\frac{x}{p}\partial_x\xi + (1+ib)\partial_x^2\xi\,.\end{aligned} \tag{18}$$

The eigenmodes of Eq. (18) are of the form $Ce^{\lambda t}\eta(x)$ if λ is real and $Ce^{\lambda t}\eta(x) + C^* e^{\lambda^* t}\eta'(x)$ if λ is complex, C is an arbitrary constant (generally η' does not coincide with η^*). The marginally stable eigenfunctions ($\lambda = 0$) connected with gauge and translation symmetry are, respectively

$$\eta_1 = i\tanh\frac{x}{p}\ ,\ \eta_2 = \frac{1}{p\cosh^2\frac{x}{p}} + iQ\tanh^2\frac{x}{p} \tag{19}$$

These eigenmodes can be obtained easily by differentation of the hole solution, Eq. (7), with respect to φ_0 and x_0.

It is known that for $b = c = 0$ Eq. (18) has one unstable localized eigenmode $\eta = i\cosh^{-1}(x/\sqrt{2})$ corresponding to the eigenvalue $\lambda = 1/2$ [12].

Our numerical computation has shown that for some values of b and c the eigenvalue decreases and becomes negative. More precisely this pertains to the real part of λ, because before the transition occurs the eigenvalue becomes complex.

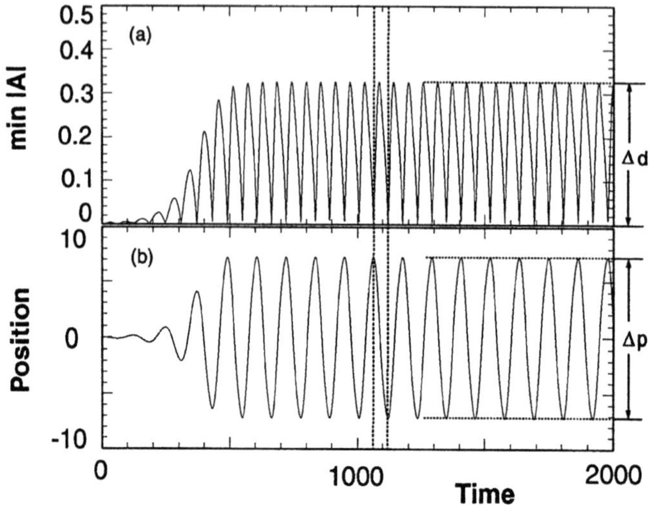

Figure 6: *The evolution of an oscillating hole solution for* $b = 0.34$, $c = 1.45$ *and system length* $l = 400$. *The initial condition was a hole with* $|A| = 0$ *at the center* $x_0 = 200$. *In (a) the minimum* $|A|$ *of the hole is shown, (b) shows the position of the minimum of the hole. The frequency of the oscillations of the position are in good agreement with the theoretical value which is near the threshold equal to* $Im(\lambda)$.

The neutral curve (corresponding to $Re(\lambda) = 0$) in the plane of the parameters (b, c) is shown as curve CS in Fig 1. The eigenmode η corresponding to a complex unstable λ is shown in Fig. 5. We conclude that the neutral curve $Re(\lambda) = 0$ is the boundary of a *Hopf bifurcation.* The condition $Re(\lambda) < 0$ (right of curve CS in Fig. 1) together with the stability condition for the continuous spectrum (above curve $HS1d$ in Fig. 1) provides necessary (and apparently also sufficient) conditions for the stability of the stationary hole solution.

On the left side of the neutral curve CS there is the possibility that the instability of the localized eigenmode is saturated by the nonlinearities so that a forward bifurcation results. We have verified numerically that this is indeed the case and find solutions where the hole oscillates periodically in depth and in the position of the center (Figs. 6, 7).

This behavior can be understood in the framework of a weakly nonlinear approach. Slightly beyond the Hopf bifurcation ($0 \leq Re(\lambda) \ll 1$) we can look for a solution in the form of a "superposition" of the slightly perturbed hole and the localized unstable eigenmode

$$\begin{aligned} A(x,t) &= \left[\sqrt{1-Q^2}\tanh\left(\frac{x-x_0(t)}{p}\right) + (C(t)\eta(x-x_0) + C^*(t)\eta'(x-x_0)) + z(x,t)\right] \times \\ &\times exp\,(\,i\,(\Phi(x-x_0(t)) - \omega t + \varphi_0(t))) \end{aligned} \qquad (20)$$

where $x_0(t)$ and $\varphi_0(t)$ are now slowly varying functions of time, $C(t)$ is the complex amplitude of the unstable localized mode and $z(x,t)$ is the correction term. From the condition of the boundedness of z we obtain the equation for $C(t)$, $x_0(t)$ and $\varphi_0(t)$. The solution (20) obviously describes the hole with varying depth and position because $|\eta(0)| \neq 0$. $\varphi_0(t)$ and $x_0(t)$ describe corrections to the frequency and the position of the hole (see below). [‡‡]

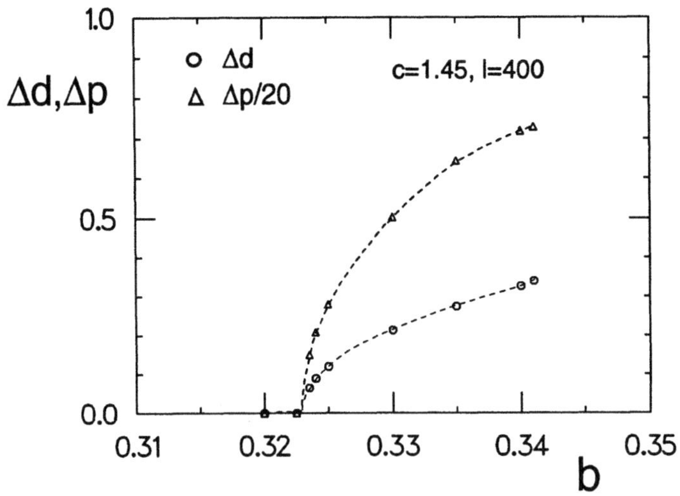

Figure 7: *The maximal variations of the depth of the hole Δd and of the position of the minimum Δp are shown as functions of b for $c = 1.45$ and $l = 400$. Below $b \simeq 0.323$ the nonoscillating hole solution is stable; at $b \simeq 0.342$ the transition to a defect-free solution sets in.*

Substituting (20) into (1) (neglecting higher order terms in z) one obtains

$$\hat{L}z = -\dot{x}_0\eta_2 + \dot{\varphi}_0\eta_1 + ((\dot{C} - \lambda C)\eta + (\dot{C} - \lambda C)^*\eta') + (1 + ic)N(C) + O(|\, z\,|^2)\,, \tag{21}$$

$N(C)$ are the terms nonlinear in C providing the interaction between modes,

$$\begin{aligned} N \;=\; & \left|\sqrt{1-Q^2}\tanh\frac{x}{p} + \zeta(x)\right|^2 \left(\sqrt{1-Q^2}\tanh\frac{x}{p} + \zeta(x)\right) \\ & - \left(\sqrt{1-Q^2}\tanh\frac{x}{p}\right)^3 - (1-Q^2)\tanh^2\frac{x}{p}\,[2\zeta(x) + \zeta^*(x)]\,, \end{aligned} \tag{22}$$

where the abbriviation $\zeta(x) = C\eta(x) + C^*\eta'(x)$ is used. The condition of boundedness of z is the orthogonality of the r.h.s. of the Eq. (21) to the "dangerous" eigenmodes η_1 and η_2 (more precisely: the corresponding eigenmodes of $\hat{L}^+$). This condition can be fulfilled by adjusting $\dot{x}_0(t)$ and $\dot{\varphi}_0(t)$. The condition of boundedness of $C(t)$ is that on the average the "projection" of the r.h.s. of (21) on the mode η is zero. These conditions give us the equations for $\dot{C}(t)$, $\dot{\varphi}_0$, $\dot{x}_0$.

Performing the projections leads to the following equations

$$\begin{aligned} \dot{C} &= \lambda C - \beta\,|\,C\,|^2\,C \\ \dot{x}_0 &= Re(\gamma C)\,|\,C\,|^2 + h.o.t. \\ \dot{\varphi}_0 &= \delta\,|\,C\,|^2 + h.o.t. \end{aligned} \tag{23}$$

where β, γ, δ are constants defined by the scalar product of the modes η_1, η_2, η and N.

The system (23) has stable periodic solutions for $Re(\beta) > 0)$ in the form

$$C(t) = C_0 e^{i\Omega t}\;,\; x_0(t) = \frac{|\gamma||C_0|^3}{\Omega}\cos(\Omega t + \mu) \tag{24}$$

where $|C_0|^2 = Re(\lambda)/Re(\beta)$, $\Omega = Im(\lambda) - |C_0|^2 Im(\beta)$, $\mu = const$. These solutions describe the periodically oscillating hole. In contrast to phase slip solutions in superconductivity [24], here not only the depth but also the position of the hole oscillates.

The solutions exist in a narrow belt arround the neutral curve. As one moves away from the neutral curve the oscillations of the depth become higher and for some critical value of (b,c) this solution breaks down and a transition to a plane wave occurs. In Fig. 7 the amplitudes of the oscillations of min $|A|$ and the position of min $|A|$ are shown for fixed c as functions of b (these quantities are denoted by $\Delta d, \Delta p$ and marked as $\bigcirc, \triangle$ correspondingly, see also Fig. 6). Above $b \simeq 0.323$ the amplitude Δd grows continuously from zero to a rather large value and at $b \simeq 0.342$ the transition to a defect-free solution sets in. The supercritical nature of the process can clearly be seen.

5 Conclusion

We have shown that the Eckhaus instability for defects in oscillatory unstable media is of convective nature. Due to the effect of the formation of shocks the defects remain stable even for periodic boundary conditions in the Eckhaus unstable range. We expect the interaction of well-separated 2d defects (spirals) to be exponentially screened in the convectively unstable region, as was shown to be the case essentially in the Eckhaus-stable regime [25, 26]. The evidence comes from the fact that localized static perturbations decay exponentially towards the spiral core in both regimes. The decay length $2\pi/Im(k)$ is obtained from Eq. (3) with $\lambda = 0$ and it coincides with the interaction length obtained before for $Q^2 \ll 1$ [25, 26].

We found that in the one-dimensional *CGLE* oscillating hole solutions exist stably in a certain parameter range. The solutions can be understood in terms of a Hopf bifurcation from the non-oscillating hole solution and the existence has been shown numerically. Earlier simulations [10] had indicated that the oscillations in position would grow until the holes were finally absorbed by the boundaries. Possibly the rather small parameter range for stable oscillations was missed. Preliminary simulations of hole solutions extended in two dimensions indicate that these solutions are unstable against transversal perturbations and break up into spirals that exist stably in that parameter range.

Another interesting question is the interaction of the oscillating hole solutions. Our numerical results show that the oscillating holes annihilate for small distances. For wider separations they form bound states that are characterized by a synchronization of the frequency and the phase of oscillations, see Fig. 8.

Our results can also explain the spontaneous appearance of a stable stationary lattice of phase slips (hole solutions, see Fig. 3) slightly below the boundary of the absolute instability (see [10, 7]). In this region the waves emitted by isolated hole loose stability while the core persists to be stable. One can then expect that the large scale instability of the emitted waves will be suppressed by the nucleation of new phase-slips. Then modes with a wavelength of more than about one half the phase slip separation are suppressed. Presumably the boundary of spatio-temporal chaos for this region is defined by the line of core instability CS (see Fig. 1) inside the region of the absolute instability of emitted waves.

6 Acknowledgments

Financial support by Deutsche Forschungsgemeinschaft (Sonderforschungsbereich 213 and Graduiertenkolleg) and the German-Israeli Foundation (G.I.F. Research Project No. I0130-091.07/89) is gratefully acknowledged. The simulations were performed on the CRAY YMP's of the HLRZ Jülich and LRZ München.

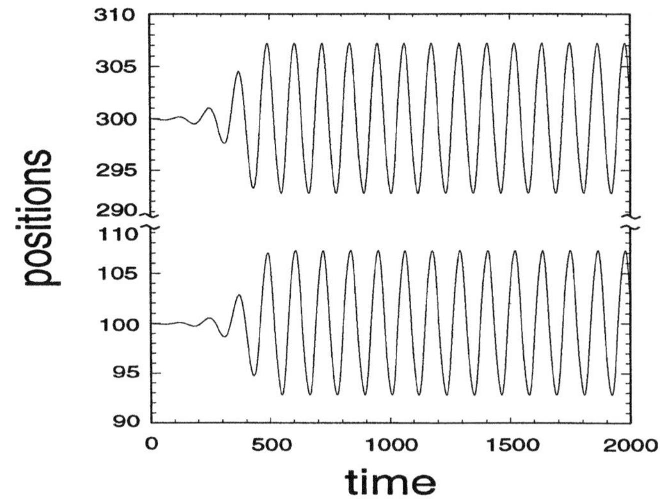

Figure 8: *The time evolution of the positions of two interacting oscillating holes are shown for* $b = 0.34$, $c = 1.45$ *and* $l = 400$.

References

[†] It can be shown strictly that the destabilization occurs at first for purly longitudinal perturbations ($\vec{Q} \parallel \vec{k}$)

[‡] To obtain this solid curve one needs values of Q as function of b and c inside the B-F unstable region which can not be obtained with the transformation Eq. (10) from the stable range. So we reproduced the dependence of Q for $c = 0.5$ and $b < -2$, and then used the transformation.

[††] The modes of the continuous spectrum satisfy the symmetry $\xi(x) = -\xi(-x)$, which correspond to the boundary condition $\xi(0) = 0$

[‡‡] The main contribution to the variation of the position indeed comes from the term with η and η'.

[1] S. C. Müller, T. Plesser, and B. Hess, Physica **24 D**, 71 (1987).

[2] I. Rehberg, B. L. Winkler, M. de la Torre Juarez, S. Rasenat, and W. Schöpf, in *Festkörperprobleme: Advances in Solid State Physics*, Vol. 29, Vieweg, Braunschweig 1989.

[3] R.W. Walden, P. Kolodner, A. Passner, C.M. Surko, Phys. Rev. Lett. **55**, (1985); E. Moses, V. Steinberg, Phys. Rev. **A 34**, 693 (1986); I. Rehberg, G. Ahlers, Phys. Rev. Lett. **55**, 500 (1985).

[4] F. T. Arecchi, G. Giacomelli, P. L. Ramazza, and S. Residori, Phys. Rev. Lett. **65**, 2531 (1990); Phys. Ref. Lett. **67**, 3749 (1991).

[5] see for example *Nonlinear Processes in Excitable Media*, A.V. Holden, M. Markus, H.G. Othmer, NATO ASI Series B, Plenum Press, 1990.

[6] I. S. Aranson, L. Kramer and A. Weber, "Localized Oscillating Solutions in the 1d Complex Ginzburg-Landau Equation", preprint 1992.

[7] I. Aranson, L. Aranson, L. Kramer and A. Weber, "Stability limits of spirals and traveling waves in nonequilibrium media", preprint 1991.

[8] A. Weber, L. Kramer I.S Aranson and L.B. Aranson, Physica **D** (in press).

[9] K. Nozaki and N. Bekki, J. Phys. Soc. Jpn. **53**, 1581 (1984).

[10] H. Sakaguchi, Prog. Theor. Phys. **85**, 417 (1991).

[11] B.I. Shraiman, A. Pumir, W. van Saarloos, P.C. Hohenberg, H. Chate, M. Holen, "Spatiotemporal chaos in the one-dimensional complex Ginzburg-Landau equation", Physica **D** (in press).

[12] P. S. Hagan, SIAM J. Appl. Math **42**, 762 (1982).

[13] E. Bodenschatz, M. Kaiser, L. Kramer, W. Pesch, A. Weber and W. Zimmermann, in *New Trends in Nonlinear Dynamics and Pattern Forming Phenomena: The Geometry of Nonequilibrium*, edited by P. Coullet and P. Huerre, NATO ASI Series, Plenum Press, 1990, p. 111. E. Bodenschatz, A. Weber, and L. Kramer, in *Nonlinear Processes in Excitable Media*, edited by A. V. Holden, M. Markus, and H. G. Othmer, NATO ASI Series B, Plenum Press, 1990.

[14] A. C. Newell, Lect. Appl. Math. **15**, 157 (1974).

[15] Y. Kuramoto and S. Koga, Prog. Theor. Phys. **66**, 1081 (1981). Y. Kuramoto, *Chemical Oscillations, Waves and Turbulence*, Springer Series in Synergetics, Berlin, 1984.

[16] P. Coullet, L. Gil, and J. Lega, Phys. Rev. Lett. **62**, 1619 (1989).

[17] I. S. Aranson, A. V. Gaponov-Grekhov, and M. I. Rabinovich, Zh. Eksp. Teor. Fiz. **89**, 92 (1985) [Sov. Phys. JETP **62(1)**, 52 (1985)].

[18] L. D. Landau and E. M. Lifshitz, *Fluid Mechanics* (Pergamon Press, London, 1959), p.111.

[19] P. M. Morse and H. Feshbach, *Methods of Theoretical Physics, Part I*, (McGraw-Hill, 1953), p.437.

[20] T. Bohr, A.W. Pedersen and M.H. Jensen, Phys. Rev. **A42**, 3626 (1991).

[21] A. Pumir, B.I. Shraiman, W. van Saarloos, P.C. Hohenberg, H. Chate, M. Holen, Physica **D** (in press).

[22] H. Sakaguchi, Theor. Phys. **84**, 792 (1990).

[23] R.J. Deissler, Physica **25D**, 233 (1987).

[24] L. Kramer and R.J. Watts-Tobin, Phys. Rev. Lett. **40**, 1041 (1978); R.J. Watts-Tobin, Y. Krähenbühl and L. Kramer, J. LowTemp. Phys. **42**, 459 (1981).

[25] I. S. Aranson, L. Kramer, and A. Weber, Physica **D** **53**, 376 (1991).

[26] L. N. Pismen and A. A. Nepomnyashchy, Physica D (in press).

The fragmentation of a drop falling in a fluid: geometry and velocity

A. Garcimartín, H.L. Mancini[#] and C. Pérez-García[*]

Departamento de Física y Matemática Aplicada,
Facultad de Ciencias,
Universidad de Navarra,
31080 Pamplona, Spain.

Abstract

In this paper we provide experimental data on the fragmentation of a drop that falls in a miscible fluid. We have accurately measured the velocities and the trajectories of the drop and the fragments. This allowed us to establish several relations between velocities and the geometry of the process.

The study of instabilities of fluid motion, that very often appear in nature, has been the aim of many works and papers since the beginning of this century. Some of them had been described even earlier, but it has not been possible to explain thoroughly all the situations, mainly because mathematics are quite involved and in some cases the problems are only partially understood using numerical simulations. This is just the case of a seemingly simple instability: the fragmentation of a drop that falls in another liquid. It was described by J.J. Thompson and H.F. Newall as early as 1885 [1]. A description of the circulation pattern inside an undeformable drop can be found in [2]. The shapes and trajectories of the fragments resemble the anatomical constitution of some living organisms, as stressed by D'Arcy Thompson [3]. It is also an interesting problem because fragmentation phenomena occur in some process of industrial or technological importance, such as mixing. It is one of the simplest open flows that show a pattern of instability. Nevertheless, the theoretical explanation of the problem is far from being complete, despite several recent efforts [4].

In this work we provide several measurements of the features of the phenomenon

---

[#] Also at CITEFA/CONICET (1603 Villa Martelli) and Universidad de San Luis, Chacabuco y Pedernera, 5700 San Luis, Argentina.

[*] Also at Departament de Física Fonamental, Facultat de Física, Universitat de Barcelona, Avgda. Diagonal 647, 08028 Barcelona (Catalonia), Spain.

E. Tirapegui and W. Zeller (eds.), Instabilities and Nonequilibrium Structures IV, 273–279.

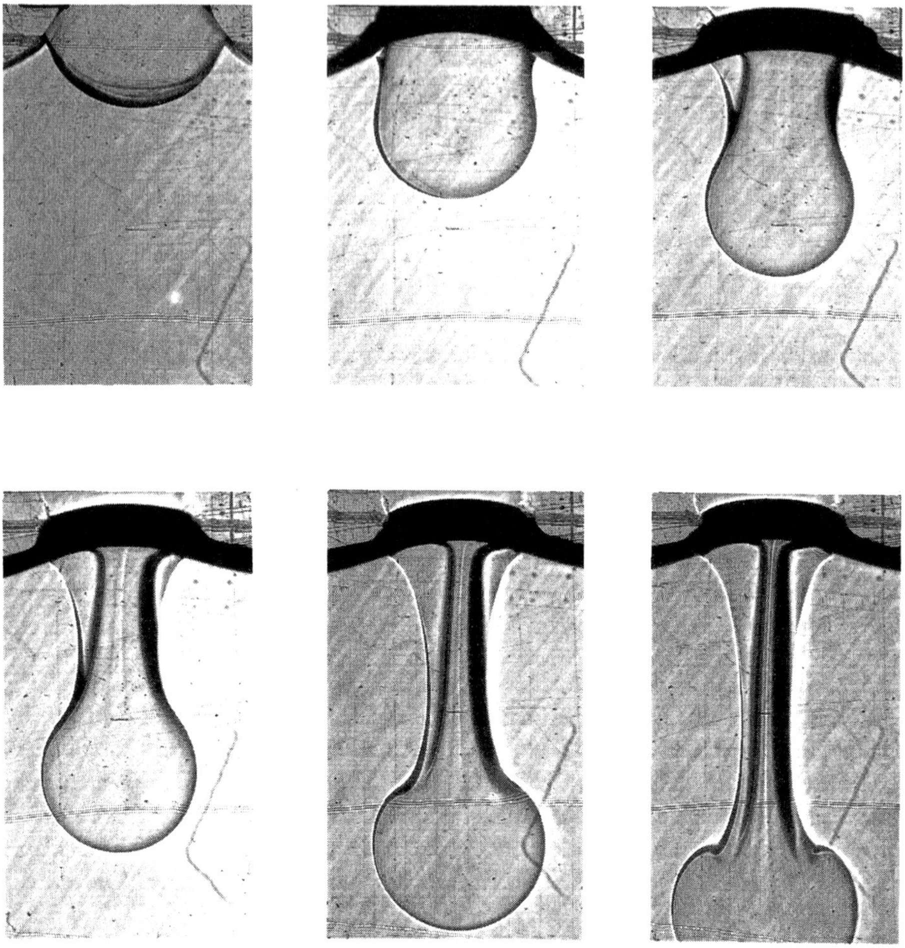

Fig.1. Formation of the drop near the free surface of the liquid. The radius of the drop is about 1 mm and the interval between the first and last photographs is approximately 3 sec.

that could help testing future theoretical works. We will first briefly discuss the general features of the phenomenon. For the sake of simplicity we will only deal with two miscible liquids. Surface tension is a stabilizing factor and so it makes the observation of the fragmentation more difficult. Moreover some features are even more surprising if we consider that both liquids are miscible and so interfacial effects are not present. It must be said that the existence of interfacial dynamical forces has been conjectured [5]. A study of those more complicated cases has been performed in recent papers [6].

A drop of the heavier fluid is carefully deposited on the free surface of a lighter fluid. The system is then in a situation that leads to the Rayleigh-Taylor instability: the upper fluid selects a wavelength and begins to fall in the other. If the quantity of heavy liquid is the appropriate, only one drop will be formed. It must be pointed that this instability has a characteristic time much lower than the following process, and that the form of the drop is indeed close to the sphere (see Fig. 1). It is known that the fragmentation will also take place if a drop of a liquid lighter than the fluid is released inside the liquid, the drop then moving upward [1][7]-[9].

Once the drop begins falling, a circulation is developed inside it due to viscous shear in the walls. If the drop kept the spherical shape the circulation inside it would be the so-called Haddamard-Rybczynski flow [2]: the fluid goes down along the vertical axis and up along the walls. Such circulation is axisymmetric. A stagnation line is located at $r=a/\sqrt{2}$ where a is the radius of the drop. The stream function is given by the expression

$$\psi=\frac{1}{4}Ur^2\sin^2\theta\left[\left(\frac{a}{r}\right)^3-3\frac{a}{r}\right]$$

where U is the sedimentation velocity of the drop times $1/[3(2+3\ \lambda)]$; λ is the ratio between the drop and the liquid viscosities.

But the circulation is always unstable if the liquid in the drop and the surrounding fluid are miscible, and the drop begins to be deformed and mix with the external fluid. As a result the velocity decreases. Eventually the drop breaks up into a torus that expands radially. But this torus is in turn unstable, because the conditions for the Rayleigh-Taylor instability are present again: heavier liquid over lighter liquid. And so the torus itself suffers a break-up into several fragments. Each one of the fragments follows the same process again until the daughter drops are so small that they fail to develop an internal circulation.

Some photographs that beautifully illustrate the whole process are shown in [7][8]. In these papers the fragmentation is interpreted as a competition between diffusion and the transfer of velocity across the radius. The first one has a characteristic time $\tau_d=r^2/D$ while the second one is $\tau_t=r/v_\infty$. The quotient between them yields a non dimensional number, namely

$$F = \frac{r^3 \Delta\rho\ g}{D\ \mu}$$

which is used to predict if the drop will break up or not: if $F > F_c$ then the drop will break, but it will fail to do so if $F < F_c$. In a similar way, another non dimensional number S -the Schmidt number, which is the quotient between diffusion and viscous characteristic times- allows the classification after the number of fragments resulting from each break-up [8].

Numerical simulations that reproduce the initial stages of the drop deformation have been performed. Kojima *et al.* [4] have carried out a linear analysis assuming a perturbation of the drop radius as a series of Legendre polynomials. Pozrikidis [10] took a further step, adding nonlinear terms. While the former calculations are valid only in the first stages of the evolution of the drop, the latter reproduces quite well several features of the drop: the ejection of a tail at the rear and the entrance of liquid into it. Nevertheless the fragmentation is never fully reproduced in those theoretical studies. Moreover mixing, which cannot be neglected in the process of fragmentation, is not taken into account in those works.

In order to investigate the internal velocity field we restricted the geometry adopting a Hele-Shaw cell that provides a sort of vertical section of the process (a quasi 2D system). We checked that the dynamical parameters (such as velocities, distance and timing) change only in a small amount respect to the 3D case. The main difference is that the drop does not develop a torus but instead it always breaks into two fragments of approximately the same size (see Fig. 2). The liquids employed were mixtures of glycerin and water in different proportions.

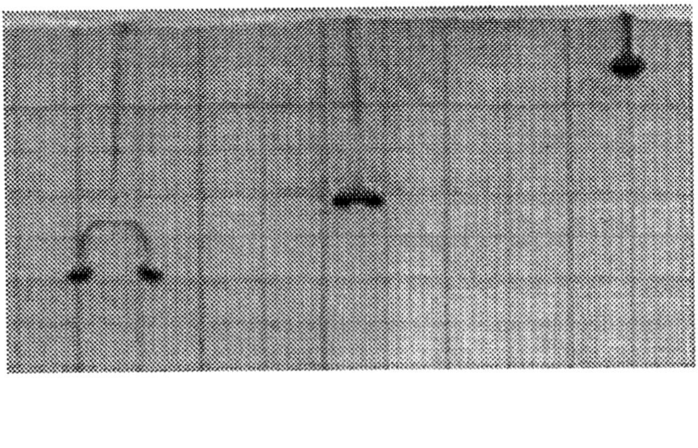

Fig. 2. Three drops falling in a Hele-Shaw cell. The one at the left has already broken up.

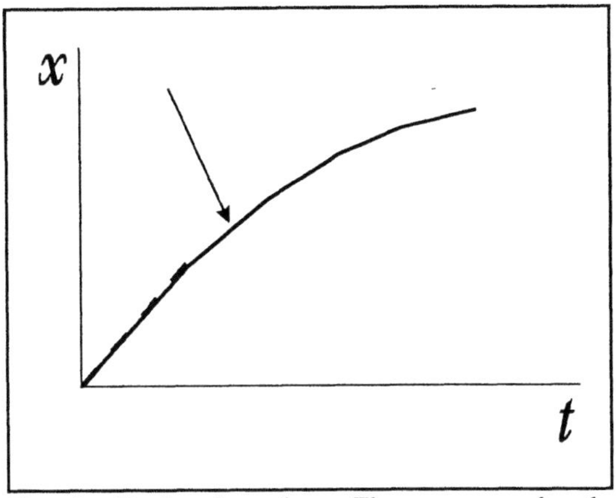

Fig. 3. Track of a drop. The arrow marks the fragmentation at about 20 mm from the surface.

We arranged a magnifying system that allowed us a very good resolution in space and time. We recorded the fragmentation process with a CCD camera and stored the images in a standard VCR. We then measured velocities and positions. This can be easily done by sweeping just one line of pixels that can be moved to follow the drop. In this fashion we get a dark line in an image whose coordinates are distance and time. The slope of that line is the velocity (see Fig. 3). This facts support the existence of a sedimentation velocity quite close to that obtained by the Stokes law for solid spheres (dashed line in Fig.3). Before the fragmentation the drop begins to grow and so it slows down, as can be readily seen. A new fragmentation velocity is achieved after fragmentation.

The velocity inside the drop can also be studied. We have obtained [9] a close agreement between break up times τ_{bu} and the characteristic time of velocity transfer across the radius (a/U). The results are plotted in Fig. 4. In a similar way velocities inside the drop can be obtained by tracking small latex particles. We have confirmed qualitative experiments [11] that suggested that the Haddamard-Rybczynski formula was a good approximation even for non-rigid spheres.

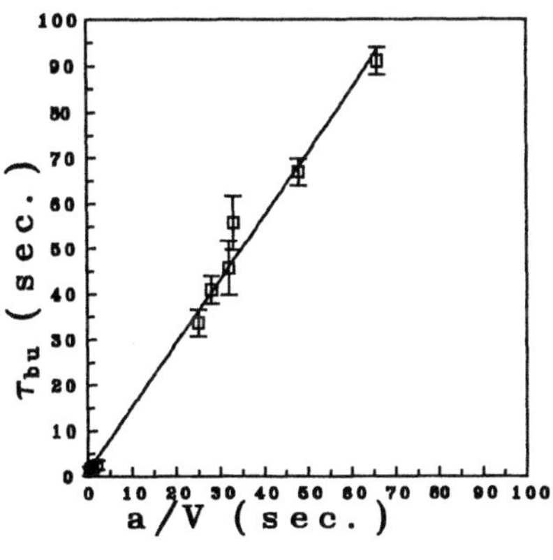

Fig. 4: Break-up times vs. characteristic time of velocity transfer across the radius (from [11]).

An interest feature of the fragmentation that deserves a closer investigation is the drastic separation between the daughter drops (see Fig. 5). Typically, in two or three seconds the distance between them can be as big as five times the radius. During this time the fragments rotate only in one direction, much like a rigid body. After that they develop again the same circulation pattern that the drop had before fragmentation. We have obtained a representation of the trajectories from the records. Photographs like that shown in Fig. 2 do not show the path followed by the fragments because the tail is falling along with the drop. In fact the fragments describe nearly parabolic paths. This indicates the existence of a horizontal force responsible for the first stage of the separation of the fragments whose nature remains unclear. In Fig. 6 we plot the velocities of the drop versus the separation achieved by the fragments. It is difficult to explain the seemingly linear relation between those two parameters.

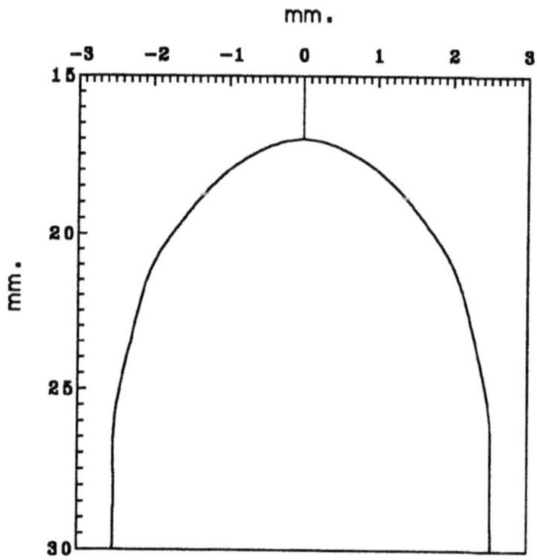

Fig. 5. Paths followed by the fragments.

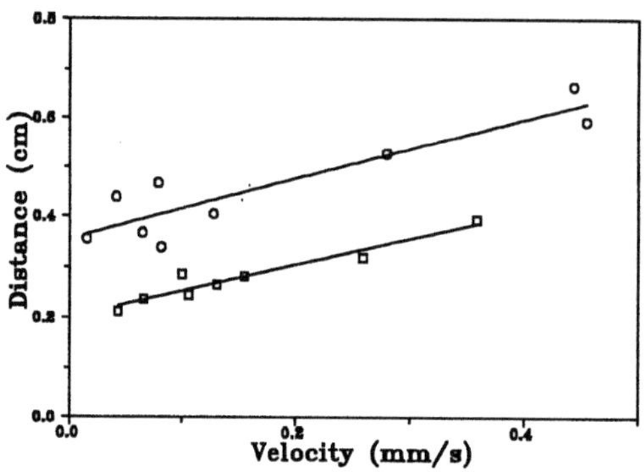

Fig. 6: Distance between fragments vs. drop velocity.

We have then shown that the dynamics of the drop in 2D, while being quite similar to the 3D case, allow increased precision in the measurements and the study of the internal circulation. A cascade of fragmentations is also evident in our case. The Stokes law, valid for rigid spheres in 3D, is almost valid in the present situation. We have not been able to explain the drastic separation of the fragments, but we provided experimental data on the path followed by them. Other interesting points of this phenomenon remain unexplained theoretically. It would be interesting to perform a stability analysis not by perturbating velocity but vorticity. The last one has a simple expression, with only one component $\omega_{\varphi}=5\ \frac{U\ r\ \sin\theta}{a^2}$. Besides this, income of mass leading to a mixing process has not been taken into account in theoretical works. The existence of transient interfacial forces is also an open question. Finally, this study could be completed by determining the critical fragmentation number and the scaling of the main instability parameters when the gap of the Hele-Shaw cell is varied [12].

* * *

We are indebted to Prof. Arecchi (Florence, Italy) and Dr. Buah-Bassuah (CapeTown, Ghana) for their comments and suggestions. We also thank Pedro Elizalde for technical helping. Two undergraduate students, L. Quesada and T. Traviesa, helped us in the data gathering and processing. Financial support has been received from DGICYT (Spanish Government) through contract OP90-0098 and a sabbatical year for one of us (H.L.M.). This work has also been supported by the Gobierno Foral de Navarra (OF 725/91).

REFERENCES

[1] J.J. THOMPSON and H.F. NEWALL, *Proc. R. Soc. London* **39** (1885) 447

[2] J. HAPPEL and H. BRENNER, **"Low Reynolds Number Hydrodynamics"**, Noordhoff International Publishing, Leyden (1965).

[3] D'ARCY W. THOMPSON, **"On Growth and Form"**, Cambridge University Press (1963), pages 393 and ss.

[4] M. KOJIMA, E.J. HINCH and A. ACRIVOS, *Phys. Fluids* **27** (1984) 19.

[5] S.E. MAY and J.V. MAHER , *PhyS. Rev. Lett* **67** (1991) 2013.

[6] D.D. JOSEPH, *Eur. J. Mech B/ Fluids*, **9** (1990) 565; and N. BAUNMANN, D.D. JOSEPH , P. MOHR and Y. RENARDY, *Phys. Fluids A* **4 (3)** (1992) 567.

[7] F.T. ARECCHI, P.K. BUAH-BASSUAH, F. FRANCINI, C. PÉREZ-GARCÍA and F. QUERCIOLI, *Europhys. Lett.* **9** (1989) 333.

[8] F.T. ARECCHI, P.K. BUAH-BASSUAH and C. PÉREZ-GARCÍA, *Europhys. Lett.* **15** (1991) 429.

[9] A. GARCIMARTÍN, H.L. MANCINI and C. PÉREZ-GARCÍA, *Europhys. Lett.* **19** (1992) 17

[10] C. POZRIKIDIS, *J. Fluid Mech.* **210** (1990) 1.

[11] K.E. SPELLS, *Phys. Soc. Proc. B* **65** (1952) 541.

[12] F.T. ARECCHI, private communication.

THE ROLE OF DISPERSION IN THE GENERALIZED KURAMOTO SIVASHINSKY EQUATION

C. M. Alfaro, R. D. Benguria & M. C. Depassier
Facultad de Física
Universidad Católica de Chile
Casilla 306, Santiago 22, Chile

ABSTRACT. We study numerically the equation $u_t + uu_x + \delta u_{xxx} + u_{xx} + u_{xxxx} = 0$ with periodic boundary conditions in a small domain $[0, L]$. Increasing dispersion leads to the stabilization of the traveling waves that branch off the trivial solution $u = 0$. Beyond a certain value of δ, for all L, some initial conditions lead to solitary like traveling pulses. For lower values of δ we find ranges of L for which traveling waves are not stable, modulated traveling waves are stable in these regions.

Introduction

In the study of long wavelength oscillatory instabilities such as found in the problem of fluid flow along an inclined plane[3,6,10], in the study of surface waves driven by temperature gradients[5] and in the study of the Eckhaus instability of traveling waves in the Ginzburg Landau equation[4], it has been found that the equation that governs the evolution of disturbances is the dispersion modified Kuramoto-Sivashinsky equation $u_t + uu_x + \delta u_{xxx} + u_{xx} + u_{xxxx} + \lambda(uu_x)_x = 0$. The special case $\lambda = 0$ has been the subject of many studies[1,2,8,17,11,9,12] as it is one of the simplest models that includes nonlinearity, dissipation, dispersion and instability. In the absence of dispersion that is for $\delta = 0$, this is the well known Kuramoto Sivashinsky equation which, with periodic boundary conditions in a small domain, leads to depending on the value of L to steady states, rotating waves and homoclinic cycles which have been denominated pulsating waves[7]. These numerical results have been qualitatively and quantitatively explained by a reduction to a center unstable manifold near the second bifurcation from the trivial solution[13]. An essential feature is the $O(2)$ symmetry of the equation[15,16]. When dispersion is included, this symmetry is broken, it has been shown that the breaking of this symmetry destroys the homoclinic cycle, which is replaced by quasi-periodic flow on a 2-torus[14]. Numerical studies have shown that large dispersion leads to qualitatively different behavior, most initial conditions evolve to arrays of solitary like traveling pulses.

In this work we study a five mode truncated system which we expect that, as for the Kuramoto Sivashinsky equation, is qualitatively and quantitavely accurate in a small domain. We find that the inclusion of dispersion leads to the stabilization of traveling wave solutions, a value of δ is found beyond which for all L in the domain considered, a single or two pulse traveling wave is stable leading to different final states depending on initial conditions.

E. Tirapegui and W. Zeller (eds.), Instabilities and Nonequilibrium Structures IV, 281–287.

Mathematical Formulation

We consider the equation $u_t + uu_x + \delta u_{xxx} + u_{xx} + u_{xxxx} = 0$ subject to periodic boundary conditions $u(0,t) = u(L,t)$ with initial condition $u(x,0) = u_0(x)$.

We expand the solution for u in the Fourier series $u(x,t) = \sum_{n=-\infty}^{\infty} a_n(t)e^{ik_n x}$ where $k_n = 2n\pi/L$ and the coefficients satisfy $a_{-n}(t) = \overline{a}_n$. Here $\overline{a}$ denotes the complex conjugate of a. Since the spatial average of u is conserved we fix it to be zero by letting $a_0 = 0$. Replacing the series expansion in the equation and truncating at five modes we obtain the following system for the time evolution of the Fourier amplitudes:

$$\dot{a_1} + (\mu_1 - i\delta k^3)a_1 + ik(\overline{a}_1 a_2 + \overline{a}_2 a_3 + \overline{a}_3 a_4 + \overline{a}_4 a_5) = 0 \quad (1)$$

$$\dot{a_2} + (\mu_2 - 8i\delta k^3)a_2 + ik(a_1^2 + 2\overline{a}_1 a_3 + 2\overline{a}_2 a_4 + 2\overline{a}_3 a_5) = 0 \quad (2)$$

$$\dot{a_3} + (\mu_3 - 27i\delta k^3)a_3 + 3ik(a_1 a_2 + \overline{a}_1 a_4 + \overline{a}_2 a_5) = 0 \quad (3)$$

$$\dot{a_4} + (\mu_4 - 64i\delta k^3)a_4 + 2ik(a_2^2 + 2a_1 a_3 + 2\overline{a}_1 a_5) = 0 \quad (4)$$

$$\dot{a_5} + (\mu_5 - 125i\delta k^3)a_5 + 5ik(a_1 a_4 + a_2 a_3) = 0 \quad (5)$$

where $k = 2\pi/L$ and $\mu_n = k_n^4 - k_n^2$.

In the polar coordinate representation for the complex amplitudes $a_n = \rho_n \exp^{in\theta}$, traveling waves correspond to fixed points of ρ_n, and of $\theta_n - (n/n_o)\theta_{n_o}$ with constant time derivative of θ_{n_o}. The speed of the wave is then $c = -\frac{\dot{\theta}_{n_o}}{kn_o}$, with n_o the first nonvanishing Fourier component. Modulated traveling waves correspond to periodic solutions. We have studied numerically the solution to this system of equations. In the next section we describe the results.

Numerical Results

We have studied the bifurcation structure of the system in the range $0 < L < 6.5\pi$ for different values of dispersion δ. The third bifurcation from $u = 0$ at $L = 6\pi$ is not contained in our system, its main components being a_3 and a_6, therefore beyond this value of L we cannot expect qualitative agreement with results from the full equation. We have studied the stability of traveling waves for different values of δ. We refer to [7] for the case $\delta = 0$. In Figure 1 we show the branches of traveling waves for $\delta = .05$. We observe that now the solution that branches from $L = 2\pi$ (TW1) extends into the larger L region, there is a secondary solution (TW3) born from the solution that branches from $u = 0$ (TW2) at $L = 4\pi$. The solid lines show the stable region of each branch. The branches TW1 and TW3 lose stability through a Hopf bifurcation, the branch TW2 gains stability when two real eigenvalues cross zero and it becomes unstable at larger L through a Hopf bifurcation. There is a range of L between 4.34π and 4.78π where the only stable solutions are modulated traveling waves. We do not observe the homoclinic cycle which is no longer present as shown in [14]. As δ is increased, the region of stability of the traveling waves is increased. We show in Figure 2 the same diagram as in the previous figure, now for $\delta = .1$. We observe that TW2 loses stability at large L and regains it later, in this interval chaotic solutions can be found. As this occurs near $L = 6\pi$ the persistence of this behavior should be studied in a higher order system. Shown in Fig. 3 are the results for $\delta = 1$. Notice that now TW1 is stable into larger values of L; in the region where both TW1 and TW2 are simultaneously stable different initial conditions lead to different final states, with different number of pulses. The branch TW3 is unstable and exists in a smaller range of L. A new branch

TW4 exists now for larger values of L. It corresponds to a different two pulse solution for $u(x,t)$ which coexists stably with TW2. Once any of the branches has lost stability, different initial conditions lead to either traveling waves or to modulated traveling waves. On Figure 4 we show the results for $\delta = 1.3$ where we see the gradual disappearance of TW3, and the approach towards lower values of L of TW4, while finally in Figure 5 the results for $\delta = 3$ show that TW3 is no longer present and that over large regions of L different types of traveling waves are simultaneously stable. Which state is reached depends on initial conditions. In all cases where a stable traveling wave coexists with an unstable portion of a different traveling wave, some initial conditions lead to a TW, others to a modulated TW. In Figs 6 we show $u(x,t)$ as a function of time when the system has evolved into a modulated traveling wave. We see that humps in this case can be clearly identified but their speed reverses sign. Different types of behavior can be found for $u(x,t)$ for different values of L and δ when the system evolves into a modulated traveling wave, several cases can be found for which humps do not preserve their identity.

Summary

We have studied a five mode truncation of an equation that models oscillatory instabilities in an extended system. The five mode truncated system should be qualitatively accurate in a small domain of the horizontal periodicity interval L. We find that for small dispersion we can find ranges of L for which traveling waves are not stable. For large dispersion stable traveling waves exist for all values of L, and several of them can be simultaneously stable. For large dispersion one can show that the height of the bumps increase linearly with δ[12]. This truncated system reproduces qualitatively most of the features observed from integrations of the full partial differential equation.

Acknowledgements

This work was partially supported by FONDECYT. C. M. Alfaro acknowledges the support of Fundación Andes.

References

[1] T. Kawahara, Phys. Rev. Lett. **51** (1983) 381.

[2] S. Toh and T. Kawahara, J. Phys. Soc. Jpn. **54** (1985) 1257.

[3] J. Topper and T. Kawahara, J. Phys. Soc. Jpn. **44** (1978) 663.

[4] B. Janiaud, A. Pumir, D. Bensimon, V. Croquette, H. Richter and L. Kramer, Physica D **55** (1992) 269.

[5] H. Aspe and M. C. Depassier, Phys. Rev. A **41** (1990) 3125.

[6] A. Nepomnyaschy, Proceedings of Perm State (A. M. Gorki) University, **N 362**, (1976) 114.

[7] I. K. Kevrekidis, B. Nicolaenko and J. C. Scovel, Siam J. Appl. Math. **50** (1990) 760.

[8] T. Kawahara and S. Toh, Phys. Fluids **28** (1985) 1636.

[9] C. Elphick, G. R. Ierley, O. Regev and E. A. Spiegel, Phys. Rev. A **44** (1991) 1110.

[10] A. Oron and D. A. Edwards, *Stability of a falling liquid film in the presence of interfacial viscous stresses* preprint 1992.

[11] N. A. Kudryashov Phys. Lett. A **147** (1990) 287.

[12] R. D. Benguria and M. C. Depassier, Phys. Rev. A, **45** (1992) 5566.

[13] D. Armbruster, J. Guckenheimer and P. J. Holmes, SIAM J. Appl. Math., **49** (1989) 679.

[14] P. Chossat, *Forced reflectional symmetry breaking of an O(2)-symmetric homoclinic cycle*, preprint 1992.

[15] D. Armbruster, J. Guckenheimer and P. J. Holmes, Physica D **29** (1988) 257.

[16] M. R. E. Proctor and C. A. Jones, J. Fluid Mech. **188** (1988) 301.

[17] T. Kawahara and S. Toh, Phys. Fluids **31** (1988) 2103.

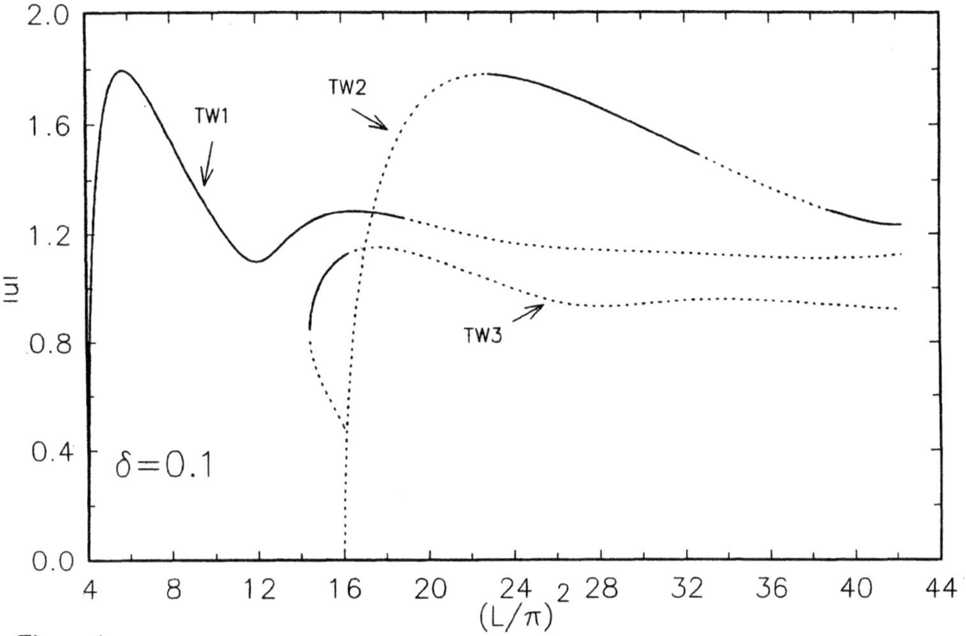

Figure 1.
The norm of u for the three branches of traveling waves as a function of $(L/\pi)^2$ for $\delta = .05$. The solid line indicates the stable regions.

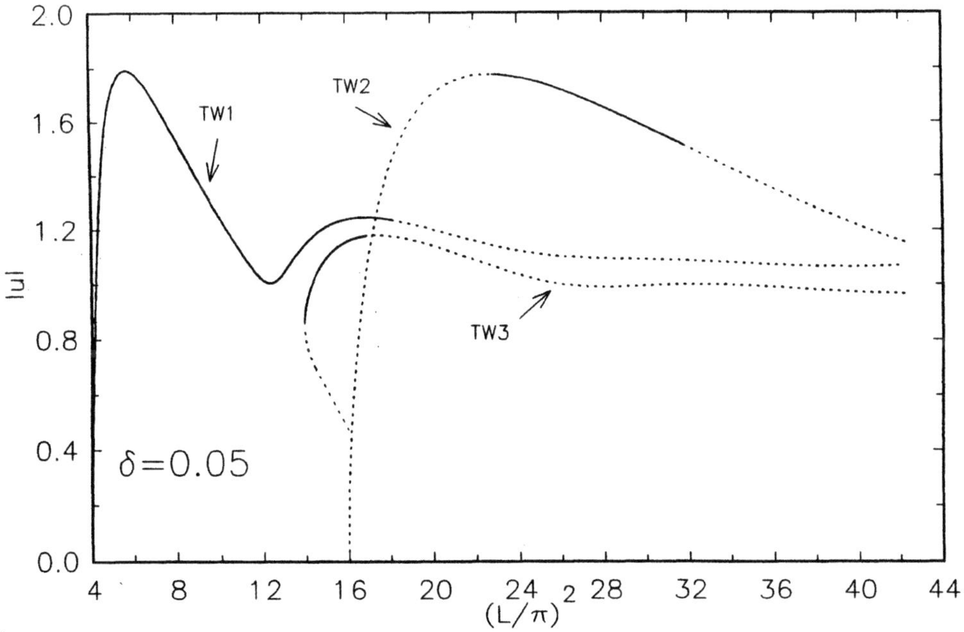

Figure 2.
At $\delta = .1$ the branch TW1 has a larger region of stability. The branch TW2 loses and regains stability at large L. In this window chaotic solutions are found.

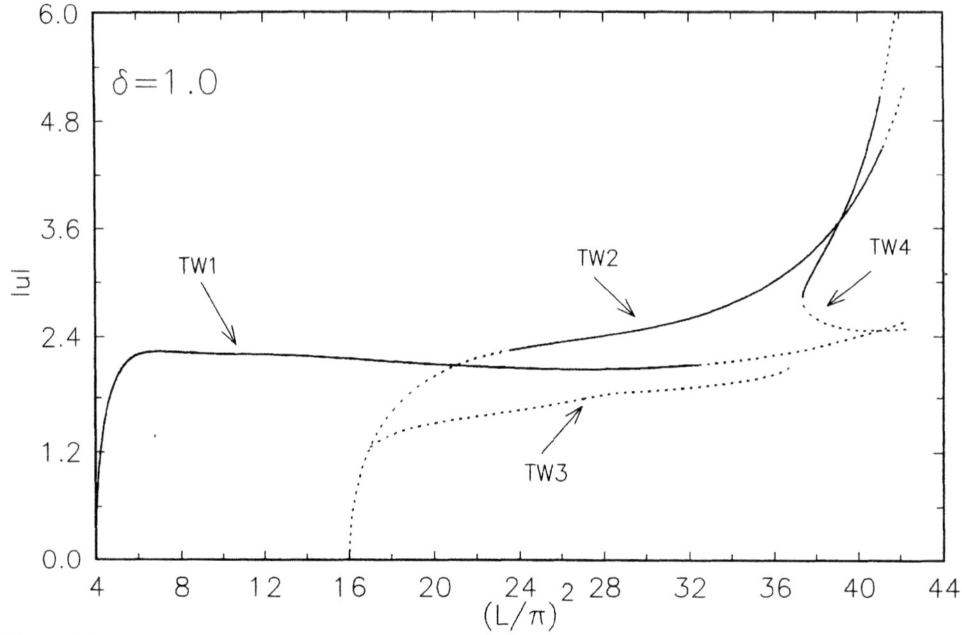

Figure 3.
At $\delta = 1$ a new branch TW4 exists. In the region where TW1 and TW2 or where TW2 and TW4 are simultaneously stable, different initial conditions lead to different traveling waves. TW3 exists in a smaller region.

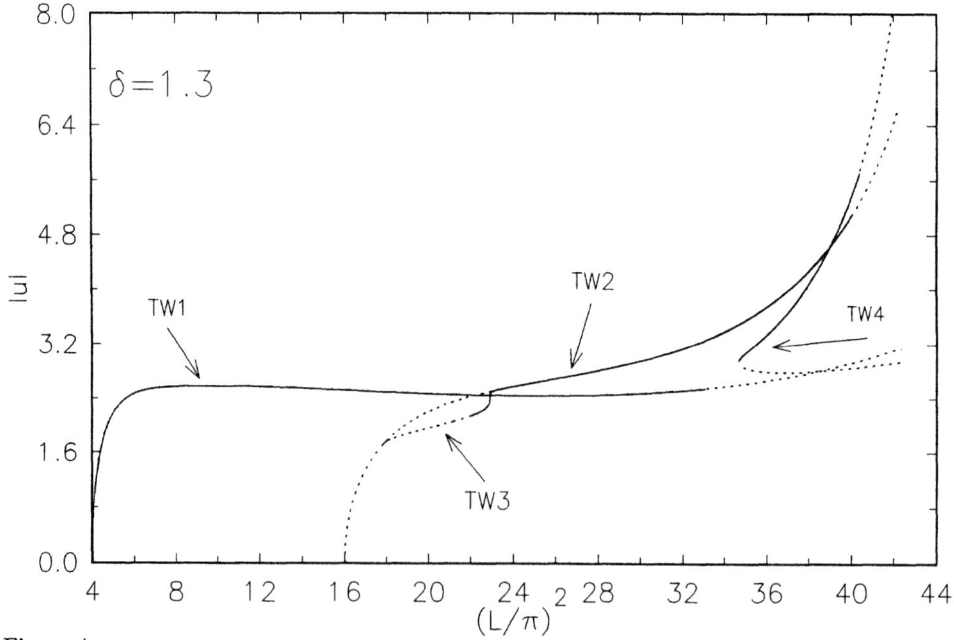

Figure 4.
At $\delta = 1.3$ TW1 and TW2 increase their region of stability, TW4 exists at lower values of L.

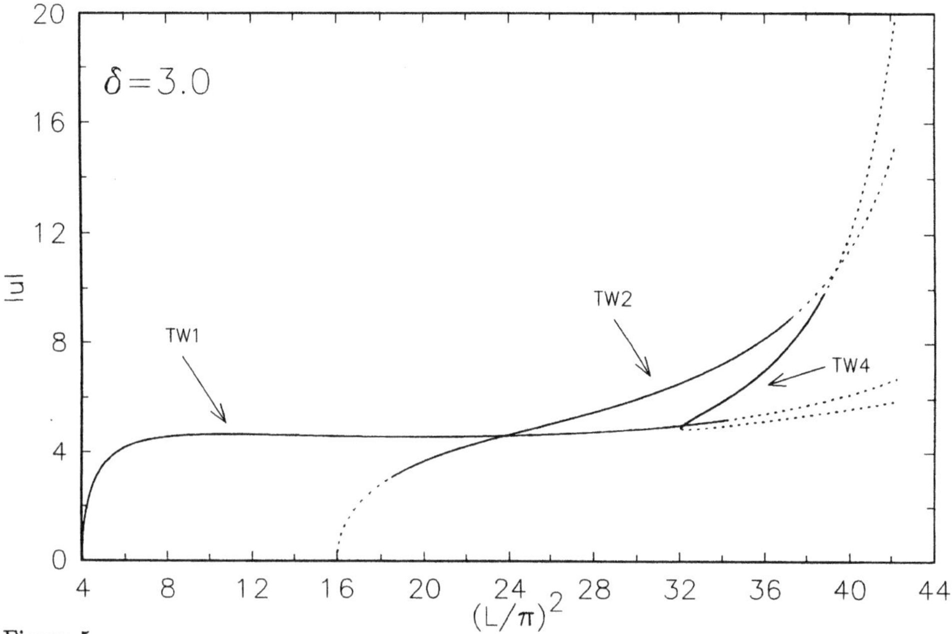

Figure 5.
At $\delta = 3$, TW3 no longer exists. TW1, TW2, and TW4 are simultaneously stable in a range of L.

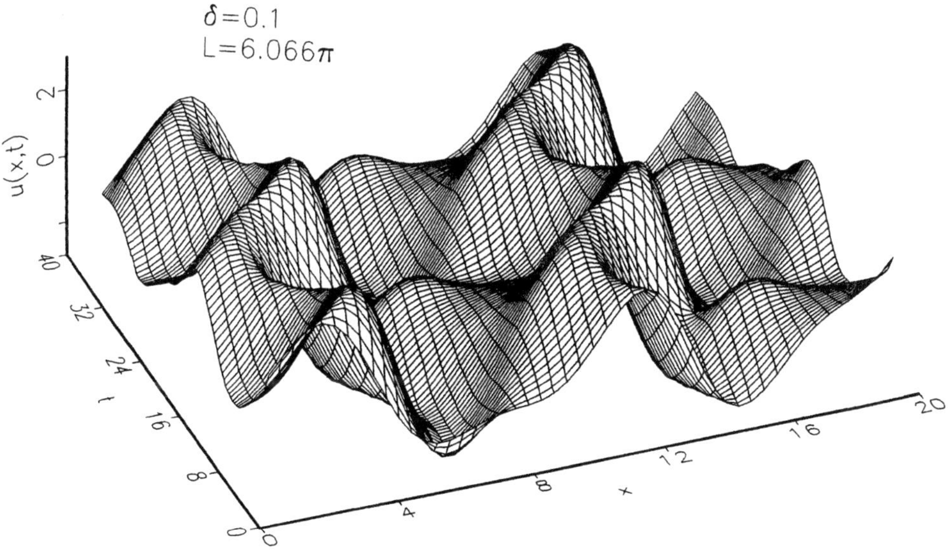

Figure 6.
We show $u(x,t)$ at $L = 6.066\pi$, $\delta = .1$ when the system has evolved into a modulated traveling wave. The pulses preserve their identity but reverse their direction of propagation periodically.

THREE-MODES NONLINEAR STATISTICAL DESCRIPTION FOR A GLOBAL DRIFT WAVE TURBULENCE

R. Montagne and A.C. Sicardi Schifino
Instituto de Física, Facultad de Ciencias,
Universidad de la República, CC 10773, CP 11200, Montevideo, Uruguay
and
Instituto de Física, Facultad de Ingeniería,
Universidad de la República, CC No 30, CP 11000, Montevideo, Uruguay

ABSTRACT. A nonlinear description of drift waves turbulence in a dissipative media with a driven force is built, using a statistic based in a three-modes interaction model. The statistical dynamics of the system is investigated on the basis of the existence of a most physically important triad (MPI triad). Then, the neglected modes are modeled as a white-noise contribution leading to a Langevin equation. A probability density function is obtained for the associated Fokker-Planck equation of the turbulent field. The spectrum cascades and the most physically important triad (MPI triad) are analysed.

I. INTRODUCTION

The electrostatic potential in a magnetized inhomogeneous of the drift wave turbulence plasma obeys a two dimensional equation known as the Hasegawa-Mima equation[1]. We will work with a reduced one-field adiabatic model. We conserve the dissipation, and in order to achieve a nontrivial stationary state we introduce a driven force. This approach was already done in a bidmensional Navier-Stokes fluid[2]. We calculate the spectrum of the ideal invariants of our problem, the energy[1,2] $W = \frac{1}{2}\int[(\nabla\phi)^2+\phi^2]dV$ and the generalized enstrophy $U = \frac{1}{2}\int[(\nabla^2\phi)^2+(\nabla\phi)^2]dV$.

In this work, we will develop an approximation based on a generic three-modes interaction model This construction is achieved imposing the preservation of the feature of quadratic nonlinear interaction and the selection of the triad which plays the most physically relevant role in the developed turbulence. The choice of this triad, that is the most physically important triad (MPI triad) has its starting point of approach of ref [2]. However, very different and surprising results are obtained here. Then, the effects of the remaining couplings are modeled as a white-noise contribution leading to a Langevin equation. A probability density function is obtained for the associa-

E. Tirapegui and W. Zeller (eds.), Instabilities and Nonequilibrium Structures IV, 289–296.

ted Fokker-Planck equation of the turbulent field. Then using this model, we obtain the whole spectrum cascade.

II. BASIC EQUATIONS

It is well known in the literature that the electrostatic potential of the drift waves turbulence in plasma physics obeys a nonlinear partial differential equation, the Hasegawa-Mima (H-M) equation(1). Adding a dissipation term and a driven force term to the basic equations, from which the H-M equation is derived, it can be proved that this equation can take the form

$$\frac{\partial}{\partial t}(\nabla^2\varphi - \varphi) + (\nabla\varphi \times \hat{z}\cdot\nabla)(\ln n_o - \nabla^2\varphi) = \frac{\nu}{n_o}\nabla^4\varphi + \nabla^2 f. \qquad \text{(II.1)}$$

Where we have used a dimensionless form for the system and taked the one-field fluid model for the potential where the electrons behave adiabatically.

This equation is a modified version of the H-M equation. A dissipative term and a forcing term were added in order to study the stationary turbulence. Here $\varphi \equiv e\phi/T$, with ϕ the electrostatic potential, n_o the ion density, ν the kinematic viscosity and $f = f(x,y,t)$ the driven force. Also it was used an unit system in which $m = \omega_{ci} = \rho_L = 1$ (where m is the ion mass, ω_{ci} is the ciclotron frequency and ρ_L is the Larmor radius).

When the right hand of the equation (II.1) is null, the original Hasegawa-Mima equation(1) is reobtained.

We write the function φ and the driven force f in terms of their corresponding Fourier amplitudes. Then, the Fourier transformation of equation (II.1) is

$$\frac{\partial}{\partial t}\varphi_k(t) + i\omega_k\varphi_k(t) + \nu_k\varphi_k(t) = \frac{k}{k^2+1}f_k + \sum_{\substack{k',k'' \\ k=k'+k''}} \Gamma^k_{k'k''}\varphi_{k'}(t)\varphi_{k''}(t), \qquad \text{(II.2)}$$

where $\omega_k = -\mathbf{k}\times\hat{z}\cdot\ln n_o/(k^2+1)$, $\nu_k = (\nu/n_o)k^4/(k^2+1)$ and $\Gamma^k_{k'k''} = \hat{z}\cdot\mathbf{k}\times\mathbf{k}'(k''^2-k'^2)/2(k+1)$.

III. TRIVIAL EQUILIBRIUM, LINEAR STABILITY AND APPROXIMATED EVOLUTION EQUATIONS

In order to solve the equation (II.2) we have to choose a specific form for the driven force $f_k = f\exp(-i\omega_{ko}t)\delta_{kko}$, $f = -\mathrm{f}$, finding the trivial solution to be

$$\varphi_{ko} = \frac{-1}{(\nu/n_o)k_o^2}f \equiv \mathrm{f} \quad , \quad \varphi_k = 0 \qquad \forall\ k \not\equiv k_o . \qquad \text{(III.1)}$$

It have been also assumed the frecuency matching $\omega + \omega - \omega \cong 0$ (3).

However, it is expected that the situation where the turbulence is developed this trivial equilibrium would be unstable and the system reach a new stationary solution, with a strong energy transfer between the different modes.

So we proceed to study the linear stability of this trivial solution. Arriving to an equation of first order in the perturbation $\delta\varphi$ of the solution (III.1)

$$\delta\varphi = A\dot{\delta}\varphi, \tag{III.2}$$

where the matrix A contains the characteristic three modes coupled interaction. The explicit form of the matrix is the same as the one found by A. Costa et al.[2], but instead of k^2 appears ν_k and instead of f appear $\mathbf{f}$. So the analysis is the same, and we will only outline the results. The eigenvalues of the matrix A are, for each triad $(\mathbf{k}\ ,\mathbf{k},\mathbf{k}')$

$$\lambda_o = -1 \quad , \quad \lambda_{\pm} = -\beta \pm \sqrt{\beta^2 - \gamma} \ , \tag{III.3}$$

where

$$\beta = \frac{\nu_k + \nu_{k'}}{2} \quad , \quad \gamma = \nu_k \nu_{k'} \left[1 - |\mathbf{f}|^2 \frac{\Gamma^k \Gamma^{k'}}{\nu_k \nu_{k'}} \right] , \tag{III.4}$$

with eigenvectors φ_+,φ_- and $\delta\varphi_{ko}$, of the same form as the one obtained in ref. [2] (just changing k^2 by ν_k). The only eigenvalue that could become positive (unstable mode) is λ_+. That is, when λ_+ becomes positive the trivial equilibrium will be unstable and this would be the beginning of the turbulent regime.

It can be shown[2] that the "unstable triangles" $(\mathbf{k},\mathbf{k}',\mathbf{k}_o)$, corresponding to a matrix block with positive λ_+, must necessarily have their vertices inside the region of the (k_x,k_y) plane for which the product $\Gamma^k\Gamma^{k'}$ is positive.

The sign of the eigenvalue λ_+ can be controlled through the value of the driven force f. So we find that $\mu=|\mathbf{f}|$ is the control parameter of the problem. When μ surpasses its critical value μ_c ($\mu>\mu_c$) the trivial solution become unstable.

We will study the evolution of the system after the trivial equilibrium has becomes unstable. Then, to analyse the nonlinear evolution, we consider the approximate coupled equations for each direct interaction triad represented by $\mathbf{k}\ ,\mathbf{k},\mathbf{k}'$

$$\dot{\varphi}_{ko} + \nu_{ko}\varphi_{ko} = \frac{k_o^2}{k^2+1} \mathbf{f} + \sum_{\substack{k',k'' \\ ko=k'+k''}} \Gamma^{ko}_{kk'}(\varphi^*_{k'}\varphi_k - \varphi_{k'}\varphi^*_k) \ , \tag{III.5a}$$

$$\dot{\varphi}_k + \nu_k \varphi_k = \Gamma^{k'} \varphi_{k'}\varphi^*_{ko} \tag{III.5b}$$

$$\dot{\varphi}_{k'} + \nu_{k'}\varphi_{k'} = \Gamma^{k'} \varphi_k \varphi^*_{ko} \tag{III.5c}$$

For the derivation of this set of equation we make use of resonance condition for the three vectors $\mathbf{k}=\mathbf{k}'+\mathbf{k}_o$, $\omega_k=\omega_{k'}+\omega_{ko}$ and $\bar{\mathbf{k}}=\bar{\mathbf{k}}'-\mathbf{k}_o$, $\omega_{\bar{k}}=\omega_{\bar{k}'}+\omega_{ko}$[3], as well as the same symmetry hypothesis on the modes as one used by A. Costa et al.[2].

The equation (III.5a) for the $\mathbf{k}_o$ mode is exact. That is, the interaction with all the modes is conserved. This is not the case for the equation (III.5b-c) where the only modes preserved are the one of direct interaction with the $\mathbf{k}_o$ mode. We have neglected the other modes but they will be modeled as a white noise later. Here, we will analyse the nonlinear saturation. For doing so, we assume that there is a leading term in th sum of equation (III.5a). Then, we will continue working with the evolution equations for each triad of direct interaction, conserving only this leading term. That is, we change the summation by a nonlinear term that implies the direct interaction of the most physically important triad. We will explain later the criteria for choosing that triad.

So, when $\mu > \mu_c$ a new set of solutions arise. We are facing at bifurcation from one node to another node place for each triad. This bifurcation gives rise to breaking of symmetry in the phases of the modes.

IV. NONLINEAR SATURATION, NOISE AND LYAPUNOV FUNCTION

The goal is to find the temporal evolution of the system for values of μ slightly greater than the critical value μ_c. In order to do so we write the set of equations for $(\varphi_{ko},\varphi_k,\varphi_{k'})$ in terms of the eigenvectors $(\varphi_+,\varphi_-$ and $\delta\varphi_{ko})$ and making use of the so called "slaving principle"[4], for each triad we proceed to solve the set of differential equations. As φ_- and $\delta\varphi_{ko}$ are the damped modes (negative eigenvalues) we make use of the adiabatic approximation $\dot{\varphi}_- \approx 0$ and $\delta\dot{\varphi}_{ko} \approx 0$. This tell us that φ_- and $\delta\varphi_{ko}$ are slaved modes and can be written as function of the unstable mode φ_+ leading to the evolution equation for $\varphi_+ = |\varphi_+|e^{i\phi_+}$

$$\frac{\partial}{\partial t}|\varphi_+| = \lambda_+|\varphi_+| - b|\varphi_+|^3 \quad , \qquad \frac{\partial}{\partial t}\varphi_+ = 0 \tag{IV.1}$$

where b stands for $b = 4\Gamma^{ko}\Gamma^{k'}_{k}\nu_{ko}/\nu_{ko}(\nu_k+\nu_{k'})^3$. The solution of equation (IV.1) is well known in the leterature[2,4]. Its stationary solution is $|\varphi_+(\infty)|^2=\lambda_+/b$.

We are interested now in obtaining the evolution for φ_+, but also taking into account the effect of the other modes on the triad as a noise effect. In order to do so two kind of terms have to be added to the simplified equations (III.5). There are certain terms in corresponding to the interaction of the triad of interest with the nearby triads and there are other terms which corresponds to the feedback of the following triad. It was found that the noise which corresponds to the feedback was negligible compare with the noise corresponding to the interaction of the nearby triads. Then the terms

$\Gamma^{k}_{k-ko}\varphi^{*}_{ko}\varphi_{\tilde{k}}$, $\Gamma^{k}_{k'ko}\varphi_{ko}\varphi_{\tilde{k}'}$, corresponding to the modes $\varphi_{k},\varphi_{\tilde{k}}$ and $\varphi_{k'},\varphi_{\tilde{k}'}$, which verifies $\tilde{\mathbf{k}}=\mathbf{k}+\mathbf{k}_{o}$ (with $\omega_{\tilde{k}}=\omega_{k}+\omega_{ko}$), $\tilde{\mathbf{k}}'=\mathbf{k}'-\mathbf{k}'_{o}$ (with $\omega_{\tilde{k}'}=\omega_{k'}-\omega_{ko}$), were added to the equations (III.5b) and (III.5c) respectively; in addition with the condition of white noise

$$\langle\varphi_{\tilde{k}}(t)\rangle = \langle\varphi_{\tilde{k}}(t)\rangle = 0 \tag{IV.2}$$

$$\langle\varphi_{\tilde{k}}(t)\varphi^{*}_{\tilde{k}'}(t)\rangle = \langle\varphi_{\tilde{k}'}(t)\varphi^{*}_{\tilde{k}}(t)\rangle = Q\delta(t-t').$$

This condition for the reduced approximation scheme resembles the Markovian closure model, as it is often shown in the literature[5]. Then, in the Fourier space, the modes out of the triad are modeled as a white noise.

Then, the equation of the temporal evolution of φ_{+} is modified, leading to a Langevin equation

$$\frac{\partial}{\partial t}\varphi_{+} = \lambda_{+}\varphi_{+} - b\varphi_{+}^{3} + \zeta(t) , \tag{IV.3}$$

where $\zeta(t)$ is a stochastic variable which verifies $\langle\zeta(t)\rangle = 0$, $\langle\zeta(t)\zeta^{*}(t)\rangle = \tilde{Q}\delta(t-t')$. Here $\tilde{Q} = [\nu^{2}_{k'}|\mu_{c}|^{2}(\Gamma^{k}_{k-ko})^{2} + |\mu_{c}|^{4}(\Gamma^{k'}_{kko})^{2}]Q$.

We can associate a Fokker-Planck equation to the Langevin equation (IV.3) for the probability distribution $f(\varphi_{+},t)$ defined on the phase space of φ_{+}

$$\frac{\partial}{\partial t} f = - \frac{\partial}{\partial\varphi_{+}}\left[[\lambda_{+}\varphi_{+} - b|\varphi_{+}|^{2}\varphi_{+}] f\right] + \frac{1}{2}\frac{\partial^{2}}{\partial\varphi_{+}\partial\varphi^{*}_{+}}(\tilde{Q}f) . \tag{IV.4}$$

Its stationary solution ($t\to\infty$) is[4] $f = N\exp[2S/\tilde{Q}]$, where N is the normalization factor and $S(|\varphi_{+}|) = (\lambda_{+}/2)|\varphi_{+}|^{2} - (b/4)|\varphi_{+}|^{4}$. In absence of noise S is a Lyapunov function (it can be shown that $\dot{S}\geq 0$[2]), whose maxima correspond to the stable equilibrium configuration. Using this information theory, S can be thermodynamically interpreted as the entropy associated to the probability distribution $f(\varphi_{+},t)$.

It can be shown[2] that, near the bifurcation point, the entropy S is a linear combination of the energy W and the generalized enstrophy U. It can also be seen that the distribution probabilitym when $t\to\infty$, is a canonical distribution on the ideal invariants W and U. Once again, it is important to remark that this distribution could be deduced, in the proximity of the bifurcation point, without making any hypothesis of thermodynamic equilibrium and in despite of the viscosity of the fluid.

V. DISCUSSION

We try, now, to obtain the spectrum cascade. We are interested in obtaining the direction of the energy $W_{k} = \frac{1}{2}(k^{2}+1)|\varphi_{k}|^{2}$ and the enstrophy $U_{k} =$

$\frac{1}{2}(k^2+1)k^2|\varphi_k|^2$cascades on the onset of the turbulence and extrapolate those results to the fully developed turbulence. As it was already said the energy transfer is done through a nonlienar interaction that involves triads which verifies $\mathbf{k}=\mathbf{k'}+\mathbf{k}_o$. That is, the energy is injected in the k_o mode and then, this energy is distributed to the $\mathbf{k}$ and $\mathbf{k'}$ modes. But, the triads doesn't contribute all with the same strength to the develop turbulence. The stable mode decay on time, so they do not contribute to the fully developed turbulence. Also the unstable triads have not all the same relevance. As we shall see there are only a few of them that have physical relevance and those are just the ones that five the direction of the cascades in the fully developed turbulence.

The unstable triads correspond to the points lying in the regions of the space (k_x,k_y) mentioned in section III. Our procedure is to compare the $(W_k,W_{k'},W_{ko})$ [or $(U_k,U_{k'},U_{ko})$] for every triad in the unstable region to decide the direction of the energy (enstrophy) cascade. Then, once the subregions where different directions of the energy and the enstrophy cascades were compute we try to find in which region lay the MPI triad. We take the MPI triad as the one which will give us the direction of the cascades on the fully developed turbulence.

In previous works[1,2] this MPI triad was taken as the one which grows faster. Actually Hasegawa[1] use the triad that makes the product $\Gamma^k\Gamma^{k'}$ maximum. Later for a 2-D Navier-Stokes fluid A. Costa et al.[2] used the triad that makes the growth rate (λ_+) maximum. But in the present work we use the triad that satures higher, that is $\frac{\lambda_+}{b}$ maximum, to be the MPI triad.

From the strictly linear analisis one should take into account only the triads that grows faster (the modes just grow exponentially) but once nonlinear effects are taken into account one has to consider the saturation ofthe growing modes. That is, if the triad which grows faster is looked for, but it could happened that for t long enough it simply stop growing and another triad with a slower growing rate satures higher becomming more importnat than one with faster growing rate.

The MPI triad is numerically computed for different values of the control parameter, the driven force μ. It was found that the MPI triad has a very weak dependece with the control parameter μ for the allowed values. These values of μ are the ones that makes $\mu - \mu_c$ to be small enough for the validity of the slaving principle.

FIGURE. The arrows indicates the corresponding cascade direction. Triads in the central region contribute with inverse cascade of energy and the direct cascade of enstrophy. The dark region corresponds to direct cascade of both energy and enstrophy. In the dotted region both cascades are inverse. The spectrum for ko « 1 is shown; the MPI triad lies completely in the region where the energy cascade is inverse and the enstrophy cascade is direct. This was done for ko = 0.1.

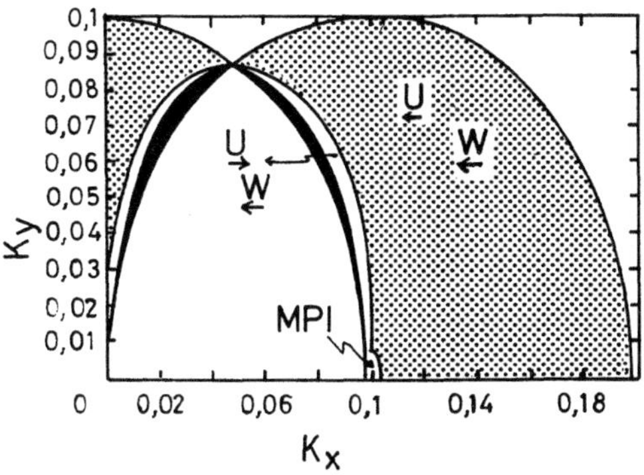

In the adjoining figure the cascade direction of the energy and enstrophy for each triad and the MPI triad are shown. The region in the **k**-plane where a triad can become unstable is found divided in sub-regions. Each sub-region corresponds to a given direction of the enstrophy and energy cascading. Obviously the physical sub-region is the central one, where the inverse cascade of the energy and direct cascade of enstrophy take place, as it is well known[6]. Then, as it is expected, the triad with physical meaning appears in this region. In the figure, k is given in a unit system such that the Larmor radius is one. The characteristic behaviour of the spectrum for $k_o \ll 1$ is shown. Actually this spectrum was found for $k_o = 0.1$. The physical sub-region is narrow. It was also found that the asymptotic value of $|\varphi+|^2$, considered as a function of the triad, presents a sharp maximum, the effect of other triads than the MPI triad can be neglected.

Finally,it was found that the MPI triad tends to condense at a certain value of k near to the x-axis, that is in the neighborhood of $k_c=(k_c,0)$, predicting the formation of zonal flows in the direction orthogonal to the z-axis.

VI. CONCLUSIONS

Starting from the Hasegawa-Mima equation for the drift waves turbulence in a dissipative media with a driven force, an approximation based on a generic three-modes interaction model was developed. This construction was achieved imposing a set of physical constrains. We have chosen a triad which plays a relevant role in the developed turbulence. For this system it was found that when μ (the driven force which acts as a control parameter of the problem) is increased, the trivial equilibrium solution (III.1) turns unstable giving rise to turbulence. This transition to turbulence is a consequence that in each representative triad a node-to-node bifurcation to a new stable equilibrium has taken place. That is, in this bifurcation, the energy is transferred from each unstable mode to the next mode producing an energy cascade. We were interested in the evolution of the unstable solution which was obtained through the slaving principle. Them, the effects of the remaining couplings are modeled as white-noise contribution leading to a Langevin equation and then the associated Foller-Planck equation of the turbulent field was obtained. The distribution is function of S, a Lyapunov function of the noiseless evolution, which can be identified with the non equilibrium entropy of the problem. Furthermore, in the proximity of the bifurcation point, it can be also shown that the distribution found, which could be derived without the hypothesis of thermodynamic equilibrium, resembles that of a canonical ensemble (with two ideal invariants). We have compute the most physically important (MPI triad), defined as the one which satures higher. And for this to be done we maximizes the asymptotic value for the MPI triad $|\varphi+|^2$, that is the quotient $\frac{\lambda+}{b}$. Several points must be remarked about the MPI triad.

- The MPI triad was calculated taking into account the nonlinear effects.
- The function $\frac{\lambda+}{b}$ has a sharp distribution arround the MPI traid. Giving auto-consistency for the three-modes interaction approximated model.
- This MPI triad predicts the same direction cascade as the fully developped turbulence.
- The MPI triad condense in the **k**-space. There is a $\mathbf{k}_c = (k_c, 0)$ that predicts the formation of zonal flows in the y direction.

ACKNOWLEDGMENTS
The authors would like to express their appreciation to R. Gonzâlez and D. Gómez (Argentina) for valuable discussions. And also acknowledge the support received from the PEDECIBA (Project URU/84/002 - UNDP).

REFERENCES

(1) Hasegawa, A. "Self-organization processes in continuous media", Adv.Phys. 34, 1-42 (1985).
(2) Costa, A., Gómez, D., Ponce Dawson, S., Ferro Fontán, C. and Sicardi Schifino, A. "Bifurcations in a three-mode model of the Navier-Stokes equation", Phys. Rev. A 38, 2037-2043 (1988).
(3) Sagdeev, R.Z. and Galeev, A.A. "Nonlinear Plasma Theory", Benjamin, New York (1969).
(4) Haken, H. "Advance Synergetics", Springer-Verlag, Berlin (1983).
(5) Ottaviani, M., Bowman, J.C. and Krommes, J.A. "Advances in the analytical theory of plasma turbulence and transport: Realizable Markovian statistical closures", Phys. Fluids B 3, 2186-2191 (1991).
(6) Kraichnan, R. "Inertial ranges in two-dimensional turbulence", Phys. Fluids 10, 1417-1423 (1967).

SPREADING OF A DROPLET ON A SOLID SURFACE AND THE HOFFMAN-TANNER LAW

V.A. KUZ
Instituto de Física de Líquidos y Sistemas Biológicos
Universidad Nacional de La Plata
C.C. 565, 1900 La Plata, Argentina

ABSTRACT. We present here an alternavite derivation to that given by De Gennes, of the Hoffman-Tanner law. By considering that the driving force acting upon the wedge-precursor film system is proportional to the surface tension gradient and by assuming that the flow of this system is of Couette type, we find the Hoffman-Tanner law.

1. INTRODUCTION

The knowledge of the spreading process of a droplet on a solid is important in many practical situation (paints, textile dying, lubrication, etc.). The spreading mechanism is not yet fully understood. Many old [1] and recent [2, 3] experimental and theoretical work [4, 5] contribute to understand better the subject. There is a relation, experimentally found, between the dynamic contact angle θ_d and the mean advancing velocity $\langle u \rangle$ of the precursor film [6]. This relation usually called Tanner's law, is $\theta_d^3 \cong C_a$, with $C_a = \eta\langle u \rangle/\sigma$ the capillary number, η the liquid viscosity and σ the liquid-vapor (L-V) interfacial tension. Earlier Hoffman proposed the same relation in capillary rise experiments. Experiments concerned with the time variation of the apparent radius R(t) of a spreading drop were compiled by A Marmur (1983). The scaling law is: $R(t) \cong t^{1/10}$, which can also derived [4] from the Hoffman-Tanner law. As it is known [9] almost always a primary film is present when the spreading occurs. It was suggested by Hardy in 1919 that a mechanism

E. Tirapegui and W. Zeller (eds.), Instabilities and Nonequilibrium Structures IV, 297–300.

of evaporation-condensation leads the motion of the primary film, later Bascon et al. suggested that film advances due to migration of liquid along the solid and they referred it, as a surface diffusion. R. Williams, studying the advancing front of a spreading liquid suggested that the drop was pulled in the perimeter by imbalance surface forces. By applying the thermodynamic theory of stability and the molecular-kinetic theory to polymolecular films, Frenkel (1946) found first the condition of spreading and showed later that the surface tension changes with a given power of the film depth's. Also Marmur (1988), when discussing the role of thin films in wetting, and the validity of Young's equation, indicated that interfacial tensions near the contact line region are different from those far away from it. This is essentially due to the effects of the long-range surfaces forces that the wall exert on the adjacent fluid Derjaguin (1987). Supported by these ideas, it will be assumed that the driving force of the system is the imbalance of the surface forces on the precursor L-V interface (from the triple line to the wedge). As this driving force acts on the surface of a liquid wedge, it will generate a rolling motion (as it was observed experimentally [16]), because the force can be split in two components, one parallel and one normal to the solid surface. Then the dynamic of the horizontal motion is simply done by assuming a pure shear flow of Couette type.

2. HYDRODYNAMIC EQUATION AND TANNER'S LAW

The dynamic problem of the spreading of a liquid droplet on a solid surface, will be considered in the regime where the effect of gravity is neglected. It is accepted that the apreading of a droplet is preceded by a thin precursor film. We write the balance equation for the momentum, at the (L-V) surface of this film; it reads:

$$\eta \left. \frac{\partial u}{\partial z} \right|_{z=l-}^{z=l+} + \frac{\partial \sigma}{\partial x} + (\bar{\kappa} + \bar{\eta}) \frac{\partial^2 u^s}{\partial x^2} = 0 \tag{1}$$

where $z = l+$ and $z = l-$ indicate the upper and the lower position of the interface, η and $(\bar{\kappa} + \bar{\eta})$ are the bulk and the surface viscosities and $\frac{\partial \sigma}{\partial x}$ is the surface tension gradient, u is the tangential component of the velocity and u^s indicates the velocity value at the (L-V) interface.

Let us consider the case where the fluid interface has a neglectable surface viscosity and $\eta_{gas} \ll \eta_{liq}$, then equation (1) becomes:

$$\eta \left. \frac{\partial u}{\partial z} \right|_{z=l-} + \frac{\partial \sigma}{\partial x} = 0 \ . \qquad (2)$$

To solve this equation we must know the velocity profile at the precursor film. Experiments suggest [15] a caterpillar motion of this film. The film advance rolling over the solid surface. The horizontal velocity will go from a maximum value $u(z = l) = u^s$ to $u(z = 0) = 0$; that is we assume a Couette type of flow for the precursor film. The velocity at this film will be given by:

$$u = \frac{z}{l} u^s \ . \qquad (3)$$

By inserting this equation into equation (2) we have:

$$u^s = \frac{l}{\eta} \frac{\partial \sigma}{\partial x} , \qquad (4)$$

where η indicates the viscosity of the film. The viscous stress exerted on the solid surface by the motion of the precursor film is:

$$\sigma_{xz} = \eta \left. \frac{\partial u}{\partial z} \right|_{z=0} = \frac{1}{l} \frac{\partial \sigma}{\partial x} \ . \qquad (5)$$

The viscous force per unit of the contact line is:

$$F_{vis} = \int_0^l \sigma_{xz} dx = \frac{\eta}{\theta_d} u^s , \qquad (6)$$

where the dynamic contact angle is $\theta_d = z/x$. This driving force, which is proportional to the surface velocity, must be balanced at the wedge of the droplet, by the total macroscopic capillary force. De Gennes has shown [4] that if θ_d is small, this force per unit length is equal to:

$$F_{cap} = \sigma\theta_d^2/2 \ . \qquad (7)$$

Then from equations (6) and (7) it results that:

$$\theta_d^3 = 4C_a , \qquad (8)$$

Here the capillary number is a function of the mean velocity $\langle u \rangle = u^s/2$, then $C_a = \eta\langle u \rangle/\sigma$. Equation (8) is the Hoffman-Tanner law.

3. COMMENTS

We show here and alternative way to derive the Hoffman-Tanner law. In this derivation the force on the solid surface is not singular and the advancing motion of the wedge-precursor system is of a rolling type.

REFERENCES

(1) Gould R.F. (ed.), Contact Angle, Wettability and Adhesion, Advances in Chemistry Series, 43, Washington D.C. (1964).

(2) Levinson P., Cazabat A.M., Cohen Stuart M.A., Heslot F. and Nicolet S., Revue Phys. Appl., 23, 1009 (1988).

(3) Leger L., Erman E., Guinet-Picard A.M., Aussure D., Strazielle C., Benattar J.J., Rieutor F., Daillant J. and Bosio L., Revue Phys. Appl., 23, 1047 (1988).

(4) a) De Gennes P.G., Rev. Mod. Phys., 57, 827 (1985).
b) De Gennes P.G., C.R. Acad. S.C. Paris t. 298, Serie II, Num. 4 (1984).

(5) Joanny J.F., J. Theor. and Appl. Mech., Num. Special 9, 249 (1986).

(6) Tanner L., J. Phys. D, 12, 1473 (1979).

(7) Hoffman R., J. Colloid and Interf. Sc., 50, 228 (1975).

(8) Marmur A., Adv. in Colloid and Interf. Sc., 19, 75 (1983).

(9) Bascom W.D., Cottington R.L., Singleterry C.R., and Gould R.F. (ed.) Contact Angle, Wettability and Adhesion; Advances in Chemistry Series, 43, Washington D.C. p. 381 (1964).

(10) Hardy H.W., Phil. Mag., 38, 49 (1919).

(11) Williams R., Nature, 226, 153 (1977).

(12) Frenkel J., Kinetic Theory of Liquids, Clarendon Pres, Oxford (1946).

(13) Marmur A., Rev. Phys. Appl., 23, 987 (1988).

(14) Deryaguin B.V., Langmuir, 3, 601 (1987).

(15) Dussan E.V., J. Fluid Mech., 65, 71 (1974).

CORIOLIS FORCE AND CENTRIFUGAL FORCE INDUCED FLOW INSTABILITIES

Innocent Mutabazi
Laboratoire de Mécanique des Fluides, Université du Havre
B.P. 265, F-76055 Le Havre Cedex, France
and
José Eduardo Wesfreid
Laboratoire de Physique et Mécanique des Milieux Hétérogènes,
URA CNRS 857, ESPCI, 10, rue Vauquelin, F-75231 Paris Cedex 05, France

ABSTRACT

Coriolis force and centrifugal force acting on flow induce longitudinal vortices in the main direction of the base flow. They couple either to give rise to oscillatory modes or to stabilize the flow when one of the forces is dominant with respect to the other. A comparative description of rotation and curvature effects is performed using the generalized Rayleigh criterion and linear stability theory.

1. Introduction

Laminar and turbulent flows are met in our everyday life (in natural and industrial flows), but the frontiers between the two flow regimes are not yet well established. The transition to turbulence in nonlinear systems such as hydrodynamic flows has become a challenging research activity in the last decades. It has been well understood that in many flow systems, transition to turbulence occurs via a successive steps of regular states in time and in space which loose continously their periodicity via bifurcations by increasing a control parameter of the system. There exist different physical parameters that can alter the bifurcation process and, in particular, they may reduce or increase the value of the control parameter for which the bifurcation occurs. Some of the mostly important parameters are : curvature, rotation, magnetic field, density stratification, temperature stratification,...

E. Tirapegui and W. Zeller (eds.), Instabilities and Nonequilibrium Structures IV, 301–316.

The understanding of the different steps towards turbulence may help the design of many industrial systems working with flow by enhancing their transfer properties. Rotation and curvature strongly affect a large variety of laminar and turbulent flows and concerns applications to engineering problems (turbomachiney, reciprocating engines with swirl or tumble effects, etc.) as well as to astrophysics and geophysics (atmospheric or oceanic flows). In many types of rotating machinery, the rotation and curvature play a significant role on the transition from laminar to turbulent regime (i.e. to the boundary layer development). Curvature plays an important role in technical applications such as flow along guide vanes, flow in heat exchangers and flow past airfoils. The fundamental mechanism involved in the formation of oriented vortices due to the presence of the rotation or curvature are not completely understood.

In this paper we will be concerned basically by the effect of rotation and curvature on the stability of the flow system; we will give a tentative unification scheme of their effects on simple flows such as Poiseuille channel flow.

The curvature of the flow systems can be realized either by the geometry (bounding the flow such as a concave wall) or the complex flow in 3-dimensional flow (curved streamlines). The most important known effect of curvature is the occurrence of centrifugal instability in curved geometries such as in the circular Couette flow known as *Taylor-Couette instability* [1], in curved Poiseuille channel flow known as *Dean instability* [2], and in the boundary layer over a concave wall known as *Görtler instability* [2,3]. All these instabilities occur at sufficiently low Reynolds number (Re $\approx$ 100 - 200) and exhibit longitudinal rolls superimposed on the base flow. The Taylor-Couette instability is the most investigated centrifugal instability because of its relative simplicity which allows mathematical modelling and easiest comparison with experimental results. General properties of the centrifugal instabilities have been recently formulated in [4].

The rotation of flow systems induces Coriolis force which in turn modifies the flow stability. It is commonly admitted that rotation has a stabilizing influence by inhibiting the amplification of perturbations developing in the flow. However, under some circumstances, the rotation can have a destabilizing role. The known example is the occurrence, for very low values of Reynolds number (Re $\approx$ 90), of longitudinal rolls in the plane Poiseuille channel flow rotating about its own axis [5].

The paper is organized as follows : the second section recalls the general equations governing the flow in rotating or curved configurations and the stability criterion of the base solutions. In the section 3, we discuss thoroughly the generalized Rayleigh discriminant obtained as an eigenvalue of an inertial operator, we recall the main result from linear stability analysis and the experimental state of art. The section 4 addresses the analogy between rotation and density stratification effects. Recapitulation and concluding remarks are formulated in section 5.

2. Governing equations in rotating frame and in curved flow configurations

2.1. FLOW IN ROTATING FRAME OF REFERENCE

The governing equations of the flow of an incompressible fluid in rotating frame of reference with a constant angular velocity $\mathbf{\Omega}$ have the following form [1]:

$$\frac{\partial \mathbf{v}}{\partial t} + (\mathbf{v}.\nabla)\mathbf{v} - 2\mathbf{\Omega}\ \mathrm{x}\mathbf{v} = -\nabla\Pi + \nu\,\Delta\,\mathbf{v} \qquad \text{(1-a)}$$

$$\mathrm{div}\ \mathbf{v} = 0 \qquad \text{(1-b)}$$

where the modified pressure $\Pi = [\frac{P}{\rho} - \frac{1}{2}|\mathbf{\Omega}\ \mathrm{x}\ \mathbf{r}|^2 + G]$, G is the potential of external forces, ρ and ν are the density and viscosity of the fluid, $\mathbf{v}$ and P are the velocity and the pressure fields. Using scaling factors as follows : $t = (d^2/\nu)\ t'$, $\mathbf{v} = V_0\ \mathbf{v}'$, $\Pi = (\nu V_0/d^2)\ \Pi'$, $\mathbf{x} = d\ \mathbf{x}'$where d and V_0 are the typical length and velocity respectively, one gets two flow control parameters entering the problem : the Reynolds number $Re = V_0 d/\nu$ and the rotation number $Ro = \Omega d/V_0$ (which is the inverse of the Rossby number).

In the following,we will consider the flow situations where the rotation axis is parallel to the flow system axis i.e. $\mathbf{\Omega} = \Omega\ \mathbf{e}_3$. The governing hydrodynamic equations then read (primes have been omitted for simplicity)

$$\frac{\partial \mathbf{v}}{\partial t} + Re[(\mathbf{v}.\nabla)\mathbf{v} - 2Ro\ \mathbf{e}_3\mathrm{x}\mathbf{v}] = -\nabla\Pi + \nu\,\Delta\,\mathbf{v} \qquad \text{(2-a)}$$

$$\mathrm{div}\ \mathbf{v} = 0 \qquad \text{(2-b)}$$

In that case, the rotation does not affect the base flow velocity $\mathbf{V} = (0, V(x), 0)$, but it introduces a normal to walls pressure gradient $\partial P/\partial x$ ($x\in[0,1]$). For example, the flow between two plates in relative motion (linear Couette flow) and rotating about its axis is given by $V(x) = x$, $\partial P/\partial x = 2\Omega V$; for a plane Poiseuille flow in a straight channel, one has $V(x) = 6x(x-1)$, $\partial P/\partial x = 2\Omega V$.

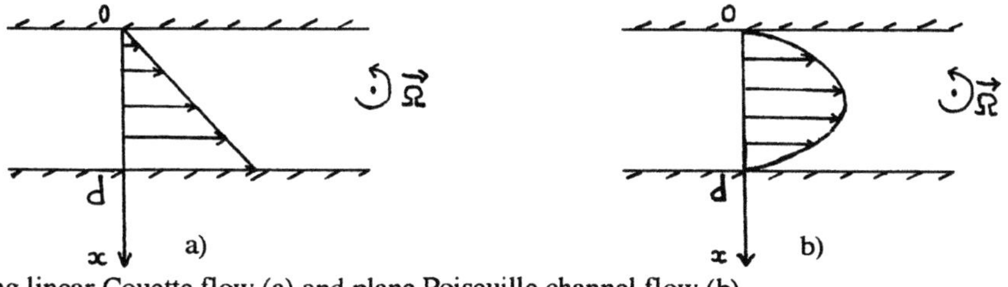

Fig.1 : Rotating linear Couette flow (a) and plane Poiseuille channel flow (b)

When the rotation axis is not parallel to the base flow shear vorticity, there is no unidirectional base flow solution, since the rotation modifies the parallel component and forces a second component in a perpendicular direction. The stability of such a flow situation has been investigated many years ago for plane Poiseuille channel [6] and recently in Couette system [7, 8] and will not be considered in the present work.

2.2. FLOW IN CURVED CONFIGURATIONS

The Navier-Stokes equations governing the flow in curved configuration of constant curvature are written in cylindrical coordinates (r,θ,z) as follows [9] :

$$\frac{\partial v_r}{\partial t}+v_r\frac{\partial v_r}{\partial r}+\frac{v_\theta}{r}\frac{\partial v_r}{\partial \theta}+v_z\frac{\partial v_r}{\partial z}-\frac{v_\theta^2}{r}=-\frac{\partial \Pi}{\partial r}+\nu\left(\Delta v_r-\frac{v_r}{r}-\frac{2}{r^2}\frac{\partial v_\theta}{\partial \theta}\right) \quad (3\text{-a})$$

$$\frac{\partial v_\theta}{\partial r}+v_r\frac{\partial v_\theta}{\partial r}+\frac{v_\theta}{r}\frac{\partial v_r}{\partial \theta}+v_z\frac{\partial v_\theta}{\partial z}+\frac{v_r v_\theta}{r}=-\frac{\partial \Pi}{\partial \theta}+\nu\left(\Delta v_\theta-\frac{v_\theta}{r}+\frac{2}{r^2}\frac{\partial v_r}{\partial \theta}\right) \quad (3\text{-b})$$

$$\frac{\partial v_z}{\partial t}+v_r\frac{\partial v_z}{\partial r}+\frac{v_\theta}{r}\frac{\partial v_z}{\partial \theta}+v_z\frac{\partial v_z}{\partial z}=-\frac{\partial \Pi}{\partial z}+\nu\Delta v_z \quad (3\text{-c})$$

$$\frac{\partial v_r}{\partial r}+\frac{v_r}{r}+\frac{1}{r}\frac{\partial v_\theta}{\partial \theta}+\frac{\partial v_z}{\partial z}=0 \quad (3\text{-d})$$

where the reduced pressure $\Pi = P/\rho + G$, G has been defined previously. The base flow velocity is assumed to be unidirectional $\mathbf{v} = V(r)\,\mathbf{e}_\theta$, and the radial momentum equation gives the radial pressure gradient generated by the curvature of the streamlines : $\frac{\partial \Pi}{\partial r}=\frac{V^2}{r}$. For a circular Couette flow, there is no azimuthal pressure gradient and the base flow velocity profile is given by : $V(r) = Ar+\frac{B}{r}$, while for a curved Poiseuille channel flow, it has the following form : $V(r) = C\left(r \ln\frac{r}{R}+Ar+\frac{B}{r}\right)$. The constant $C=\frac{1}{r}\frac{\partial \Pi}{\partial \theta}$ is the longitudinal pressure gradient, the constants A and B are given by corresponding boundary conditions. The expressions of these coefficients for the circular Couette and the circular Poiseuille flow can be found in literature [1, 2].

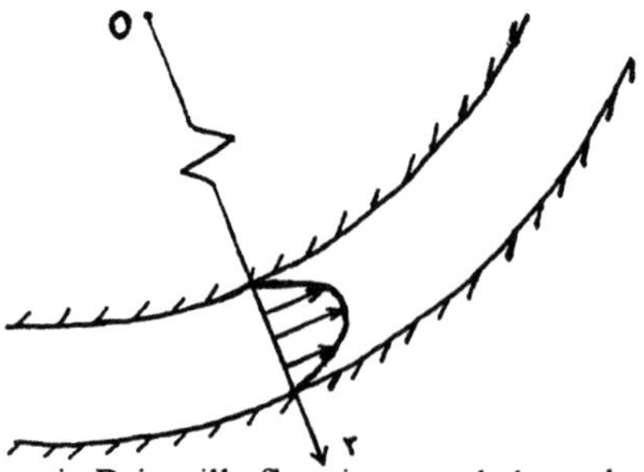

Fig. 2 : Schematic Poiseuille flow in curved channel

2. 3. INVISCID STABILITY CRITERION OF ROTATING OR CURVED FLOW SYSTEMS

The flow under the action of Coriolis force due to the rotation about the flow axis is potentially unstable to inviscid perturbations if the absolute vorticity is of opposite sign of that of the angular velocity of the system, i.e. the quantity

$$\Phi_1(x) = 2\Omega\left(\frac{dV}{dx} + 2\Omega\right) < 0 \tag{4}$$

Here the quantities are dimensional variables ($x \in [0,d]$, d is the width of the channel). The quantity $\xi_a = \frac{dV}{dx} + 2\Omega$ is the absolute vorticity of the rotating fluid in fixed frame of reference. This criterion was given to explain the occurence of longitudinal rolls observed in experiments by Lezius and Johnston [5]. The function $\Phi_1(x)$ is not invariant under the change of the rotation direction, but it is invariant under the inversion of rotation and velocity i.e. $\Phi_1(V, \Omega) \neq \Phi_1(V, -\Omega)$ and $\Phi_1(V, \Omega) = \Phi_1(-V,-\Omega)$

The flow under the action of the centrifugal force due the configuration curvature is potentially unstable and may give rise to longitudinal rolls if the *Rayleigh discriminant*

$$\Phi_2(r) = \frac{2V}{r}\left(\frac{dV}{dr} + \frac{V}{r}\right) < 0 \tag{5}$$

The Rayleigh discriminant is invariant under the inversion of the velocity : $\Phi_2(V) = \Phi_2(-V)$.

The quantity $\omega = V/r$ is the solid rotation velocity of fluid particles induced by the streamline curvature while the quantity $\zeta = \frac{dV}{dr} + \frac{V}{r}$ is the shear vorticity of the base flow. Thus, the Coriolis force and the centrifugal force have the same effect to induce longitudinal rolls superimposed on the base flow. Comparing expressions (4) and (5), the quantity Φ_1 may be called, by analogy, *Rayleigh discriminant* for flows in rotating frame of reference. The Rayleigh discriminant for both curved flows or flows in rotating systems is equal to the product of the angular velocity and the shear flow vorticity $\Phi = 2\omega\,\zeta$. The last relation can be considered as the *covariant definition of the Rayleigh discriminant* in any absolute frame of reference.

In case when the Coriolis force and the centrifugal force are acting together, for example in the curved channel Poiseuille flow rotating about its own axis, the covariance property of the Rayleigh discriminant allows to write:

$$\Phi(r) = 2\omega_a\,\zeta_a = 2\left(\frac{V}{r} + \Omega\right)\left(\frac{dV}{dr} + \frac{V}{r} + 2\Omega\right) \tag{6}$$

where $\omega_a = \omega + \Omega$ is the absolute angular velocity and $\zeta_a = \zeta + 2\Omega$ is the absolute vorticity of the shear flow [10]. The quantity $\Phi(r)$ has been called the *generalized Rayleigh discriminant* and it has been derived using the particle displacement argument [11]. In the next section, we will give an alternative derivation of it. The generalized Rayleigh discriminant is invariant under the simultaneous inversion of rotation and velocity : $\Phi(\Omega,V) = \Phi(-\Omega,-V)$, but it is not invariant under the rotation inversion or the velocity inversion separately.

3. Linear stability analysis of rotating curved flow configuration

In order to study the linear stability of a flow, the velocity and pressure fields are expanded into a base flow field ($\mathbf{V}(r)$, P) and a perturbation field $\mathbf{v}'(t,\mathbf{r})$, $p'(t,\mathbf{r})$:

$$\mathbf{v}(t, \mathbf{r}) = V(r)\, \mathbf{e}_\theta + \mathbf{v}'(t,\mathbf{r}), \quad p = P + p'$$

When both the centrifugal and Coriolis forces are acting in the system, the small perturbation velocity field $\mathbf{v}'$ obeys the following set of linearized equations :

$$\left(\frac{\partial}{\partial t} + \boldsymbol{M}\right) \mathbf{v}' = - \nabla \Pi + \nu\, \Delta \mathbf{v}' \qquad \text{(7-a)}$$

$$\mathrm{div}\, \mathbf{v}' = 0 \qquad \text{(7-b)}$$

where the *inertial operator* $\boldsymbol{M}$ has the following expression :

$$\boldsymbol{M} = \begin{pmatrix} \omega\frac{\partial}{\partial\theta} & 2(\omega+\Omega) & 0 \\ \frac{1}{r}\frac{d(r^2\omega)}{dr} + 2\Omega & \omega\frac{\partial}{\partial\theta} & 0 \\ 0 & 0 & \omega\frac{\partial}{\partial\theta} \end{pmatrix} \qquad (8)$$

Equations (7) admit normal modes solutions of the form $\mathbf{v}' = (u,v,w) \exp\{st + i\,(m\theta + kz)\}$. The modes with $m = 0$ correspond to axisymmetric rolls with spatial periodicity $k = 2\pi/\lambda$ along the axial direction, while the perturbations with $m \neq 0$ give rise to inclined rolls propagating both in the axial and azimuthal direction. These rolls occur in form of spirals in Taylor-Couette system [12] or in form of traveling inclined rolls in the Taylor-Dean system [13].

3.1. INVISCID NATURE OF THE GENERALIZED RAYLEIGH DISCRIMINANT

Following a well known procedure of linear stability theory [1,14], the generalized Rayleigh discriminant can be derived from the eigenvalues of the inertial operator for inviscid perturbations ($\nu = 0$). In general, the inertial operator $\boldsymbol{M}$ is not self-adjoint and therefore, it has complex eigenvalues :

$$s_{1,2} = i\, m\, \omega \pm \{-\Phi(r)\}^{1/2},\ s_3 = i\, m\, \omega \quad \textit{where} \quad \Phi(r) = 2[\,\Omega + \frac{V}{r}][2\Omega + \frac{dV}{dr} + \frac{V}{r}] \qquad (9)$$

For axisymmetric inviscid perturbations, the inertial operator is self-adjoint and its eigenvalues are real when $\Phi(r) < 0$: $s_1 = - s_2 = \{-\Phi(r)\}^{1/2}$, $s_3 = 0$. The quantity $\Phi(r)$, which is nothing else than the *generalized Rayleigh discriminant* , is the measure of the kinetic momentum stratification of the fluid particles. The instability will occur for a *negative stratification* i.e. when $\Phi(r) < 0$, while for a positive stratification ($\Phi(r) > 0$), the flow is stable. The marginal stratification $\Phi(r)=0$ defines the line separating the potentially unstable and potentially stable flow zones. To get a good physical insight of the generalized Rayleigh discriminant, we may rewrite it as follows

$$\Phi(r) = (1+\frac{\Omega}{\omega})\ [\psi(r) + 4\,\omega\,\Omega] \ \textit{with} \ \ \psi(r) = 2\frac{V}{r}(\frac{dV}{dr} + \frac{V}{r}) \qquad (10)$$

where the term $\psi(r)$ is the discriminant associated with the flow curvature. In the absence of curvature, the term $V/r = 0$ and $\Phi(r) = 2\Omega(2\Omega + dV/dr)$ which is nothing else than the Rayleigh discriminant for rotation induced perturbations [5]. Hence, explicitly, the generalized Rayleigh discriminant contains the sum of two terms expressing the stability condition of perturbations driven by either Coriolis force or by centrifugal force, and it contains an additional term which accounts for the coupling between rotation and curvature induced effects. The instability will occur only for negative stratification of kinetic momentum :

$$(1+\frac{\Omega}{\omega})\ [\psi(r) + 4\,\omega\,\Omega] < 0$$

This gives rise to two different cases :

a) $(1+\frac{\Omega}{\omega}) < 0$, $\psi(r) + 4\,\omega\,\Omega > 0$, then $\frac{\Omega}{\omega} < -1$: the angular velocity of solid rotation of fluid particles is smaller than and antiparallel to the rotation of the reference frame ($\boldsymbol{\omega} \uparrow\downarrow \boldsymbol{\Omega}$), while the centrifugal discriminant is positive : $\Psi(r) > - 4\,\omega\,\Omega > 0$. The flow is stable to centrifugally driven instability but unstable to rotationally driven instability.

b) $(1+\frac{\Omega}{\omega}) > 0$, $\psi(r) < - 4\,\omega\,\Omega$. Three different possibilities may be realized :

1) $-1 < \frac{\Omega}{\omega} < 0$, $0 < \psi(r) < -4\,\omega\,\Omega$: no centrifugal instability and only Coriolis force is destabilizing

2) $-1 < \frac{\Omega}{\omega} < 0$, $\psi(r) < 0$: both centrifugal force and Coriolis force are destabilizing

3) $\frac{\Omega}{\omega} > 0$, $\psi(r) < -4\omega\,\Omega < 0$: no Coriolis instability and only centrifugal force is destabilizing.

All these cases are schematically represented in the diagram on the figure 3.

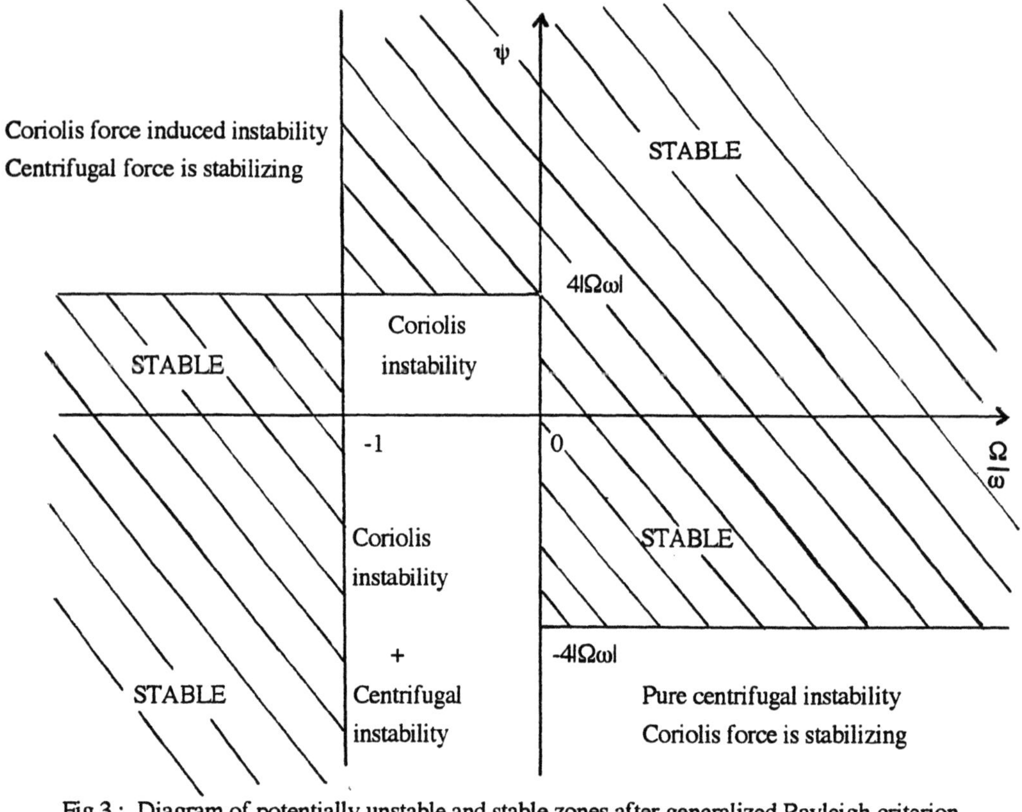

Fig.3 : Diagram of potentially unstable and stable zones after generalized Rayleigh criterion

The structure of the generalized Rayleigh discriminant is noticeable by the existence of the coupling term between the solid rotation of fluid particle induced by the curvature and the rotation of the frame of reference. Such term does not exist in the similar problem of unidirectional flow of electrically conducting fluids in the presence of a uniform magnetic field. For example, let us consider the stability of non-dissipative Couette flow in electrically conducting fluids, when a stationary distribution of current flows parallel to the axis [1]. The magnetic field does not affect the structure of the base flow velocity but it modifies the radial pressure gradient and the structure of the Rayleigh discriminant. In fact, a magnetic field in the azimuthal direction prevails together with a stationary velocity distribution :

$$\mathbf{H} = H(r)\, \mathbf{e}_\theta \qquad \mathbf{v} = V(r)\, \mathbf{e}_\theta \qquad \frac{dP}{dr} = \frac{V^2}{r} + \frac{\mu}{4\pi\rho}\frac{H^2}{r} \tag{11}$$

where μ is the magnetic permeability. It is possible to introduce a Rayleigh circulation criterion in order to describe the stability of the flow to axisymmetric perturbations [1] :

$$\Psi(r) = 2\frac{V}{r}\left(\frac{dV}{dr} + \frac{V}{r}\right) - \frac{\mu}{4\pi\rho}\, r\frac{d}{dr}\left[\frac{H}{r}\right]^2 \tag{12}$$

The flow will be potentially unstable to axisymmetric perturbations if $\Psi(r) < 0$. Nonetherless, there is no coupling between curvature effects and magnetic field, the linear stability system of equations is more complex than that corresponding to the curvature-rotation induced instability [1].

3.2. LINEAR STABILITY ANALYSIS IN THE SMALL GAP APPROXIMATION

In the small gap approximation, after elimination of the pressure and the axial velocity component from the equations (7-a,b), the linear stability equations may be written in the form :

$$[(s+i\, mTa\, V)K - i\, mTa\, D^2V\, E - A + Ta\, B]\Psi = 0 \tag{13}$$

where the control parameter $Ta = Re\, \delta^{1/2} = (V_o d/\nu)\, \delta^{1/2}$ is sometimes called the *Taylor number* and the linear operators K, A , E are defined as follows :

$$K = \begin{pmatrix} D^2-q^2 & 0 \\ 0 & 1 \end{pmatrix} \quad A = \begin{pmatrix} (D^2-q^2)^2 & 0 \\ 0 & D^2-q^2 \end{pmatrix} \quad E = \begin{pmatrix} 1 & 0 \\ 0 & 0 \end{pmatrix} \tag{14}$$

the structure vector $\Psi(x) = \begin{bmatrix} u(x) \\ v(x) \end{bmatrix}$, the *mean flow operator* B has different forms depending on the values of the product Ro :

a) For the small radius of curvature and small values of the rotation ratio (Ro<<1), we have

$$B = \begin{pmatrix} 0 & 2q^2(V+\delta^{-1}Ro) \\ DV & 0 \end{pmatrix} \tag{15}$$

The quantity $V_e = V + \delta^{-1}Ro$ is the *equivalent velocity profile*. In that case the rotation modifies the velocity profile and renders the flow potentially unstable or stable to centrifugal or Coriolis force induced perturbations. The competition between destabilizing mechanisms in the Coriolis or centrifugally unstable zones may lead to critical oscillatory modes [15]. In that sense, it represents the same physical behavior as the Taylor-Dean system investigated previously [16].

b) For small radius of curvature and higher rotation values such that $Ro \approx 1$, the matrix B has the following form :

$$B = \begin{pmatrix} 0 & 2q^2\,\delta^{-1}Ro \\ DV+2Ro & 0 \end{pmatrix} \tag{16}$$

This operator is similar to that of the Rayleigh-Bénard convection for DV = const and to that of the Taylor-Couette problem when DV is a linear function of x.

Figure 4 shows the diagram of states obtained by numerical solution of equations (13) together with the homogeneous boundary conditions at x = 0 and x = 1.The rotation inhibits the onset of the instability for large values of Ro and for $Ro \in [-1,-\delta] \cup [0,\delta]$. For $Ro < -\delta$, the instability branch corresponds to Coriolis driven perturbations moderated by curvature effects, while for $Ro > 0$, the instability branch is associated with centrifugally driven perturbations, moderated by rotation effects. We retrieve the qualitative behavior from the inviscid model (Fig.3). The dependence $q_c(Ro)$ shows that in general, the wavenumber increases or decreases with Ro, i.e. the rotation or the curvature compresses or elongates the cells occcuring at onset in the absence of rotation. For higher rotation numbers, the critical wavelength vanishes rapidly and the flow remains bidimensional in the Taylor-Proudman regime : rotation completely restabilizes the flow.

We should mention that similar results have been recently obtained using an integral method of Aihara for the stability of a boundary layer flow over a rotating concave wall [17]. It has been shown that counterotation stabilizes while corotation destabilizes the boundary layer flow over a rotating concave wall.

If we set $u = \delta^{1/2}$ u and remind that $Ta = Re\,\delta^{1/2}$, the equation (13) will describe the evolution of axisymmetric perturbations (m = 0) of plane Couette flow or the plane Poiseuille channel rotating about its own axis as derived by Lezius and Johnston [5].

The linear stability of flow in the plane Poiseuille channel flow without rotation is described by the linear 4th order Orr-Sommerfeld equation [2] :

$$(D^2+\alpha^2-i\omega)(D^2+\alpha^2)F = -\alpha\, Re\,[F\,(D^2\,V - V(D^2+\alpha^2)F] \tag{17}$$

where F(x) is the structure factor of the streamfunction, ω is the frequency of the perturbation, α is the perturbation wavenumber in the parallel to wall direction, V(x) is the base flow velocity profile. This equation has solutions represented by Tollmien-Schlichting waves propagating in the y-direction with a celerity $c = \omega / \alpha$ .The linear onset of Tollmien-Schlichting waves in straight channel is $Re_c \approx 5772$ [2].

The rotation or curvature couples radial and azimuthal velocity components of the perturbations, transforms the 4th order Orr-Sommerfeld equation into a linear 6th order system equation, introduces the axial wavenumber and considerably modifies the onset of the instability. This explains the origin of the dynamic similarity of the rotating plane channel flow with the Taylor-Couette flow. In fact, the fluid motion in a rotating straight channel and in circular Couette flow are analogous since their linear stability is governed by equations of similar form and their solutions have the same form and are represented physically by longitudinal rolls.

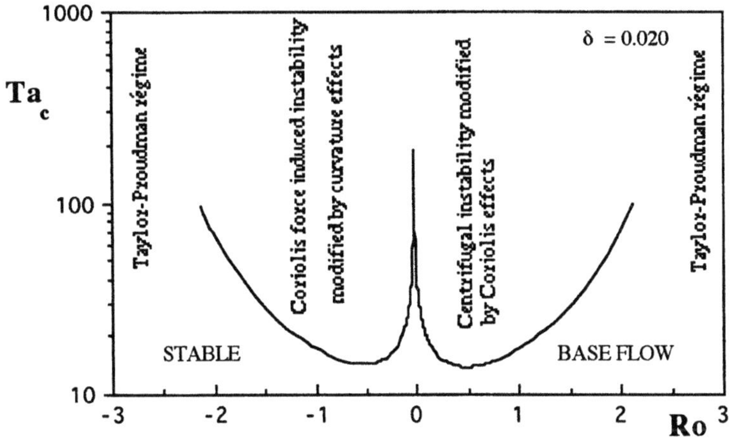

Fig. 4 : Different states in the parameter space (Ro, Ta_c)

3.3. GAP SIZE INDUCED ASYMMETRY

We have pointed out that the Rayleigh discriminant is not invariant under the rotation inversion, this property is recovered in the asymmetry of the mean flow operator : $B(-Ro) \neq \pm B(Ro)$. This asymmetry is more pronounced for large values of the curvature δ. Numerical solution of the linear stability equations shows that the finite gap size induces an asymmetry in the critical parameters values for positive and negative rotation. This asymmetry can be measured by the quantities :

$$\Delta Ta = Ta_c(Ro) - Ta_c(-Ro) \qquad \text{or} \qquad \Delta q = q_c(Ro) - q_c(-Ro) \tag{18}$$

Figure 5 shows the asymmetry parameter ΔTa as function the finite gap size δ computed at the value of Ro^* where the restabilization begins. The gap size increases the asymmetry between the positive and negative rotation, and makes the distinction between positive and negative rotation more pronounced.

3.4. EXPERIMENTAL STATE OF ART

While for the curved flow configurations, a lot of experimental studies are available, the experimental investigation of rotating systems has drawn less attention than considerable numerical work done the last five years. To the best knowledge of the authors, in configuration under consideration, recent experiments have been performed by the Swedish group [15,18]. They have shown that straight or curved channel flow rotating about its axis exhibit longitudinal rolls like those observed in the curved channel flow.These observations are in a qualitatively good agreement with theoretical predictions. However, while linear stability analysis predicted oscillatory rolls near the value $Ro = -\delta$, these rolls were not observed in the experiment. Furthermore, some of the observed rolls properties have not yet found an exhaustive explanation : the mechanisms underlying the occurrence of higher modes of instability (ondulation, twisting, chaotic states,...), the spatial growth of Dean rolls, etc.

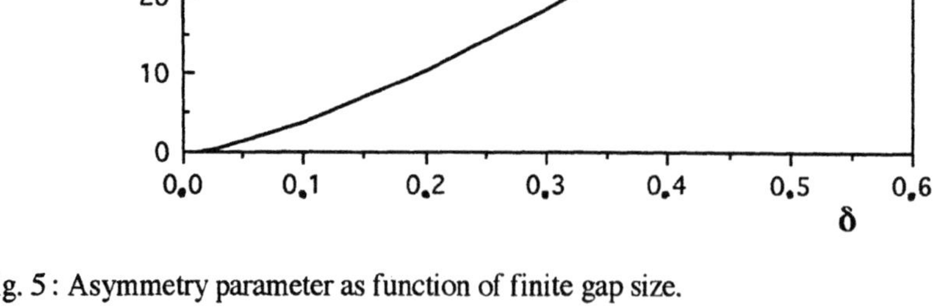

Fig. 5 : Asymmetry parameter as function of finite gap size.

4. Analogy between rotation and density stratification effects

A flow is said to be stratified when the density of the fluid has a spatial variation $d\rho/dx_i$ in the i- direction, stratified flows are of great importance in geophysics. Rotation and stratification can each play a stabilizing role if they exert constraints which inhibit the amplification of certain types of perturbations. The analogy betwen rotation and density stratification effects has been first investigated by Rayleigh [1] and after him a lot of studies have addressed this question [19]. We use a simplified version of the recent formulation of Vladimirov [20] for an incompressible viscous fluid homogeneous with respect to density in a frame of reference rotating at constant velocity $\mathbf{\Omega} = \Omega\mathbf{e}_3$. The Navier-Stokes equations in rotating frame of reference (1) have a class of solutions

$$\{\mathbf{v}(t,\mathbf{r}) = \mathbf{v}(t,x,z), \qquad \Pi(t,\mathbf{r}) = \Pi_0(t,x,z) - q(t)y\} \tag{19}$$

which satisfy the following equations :

$$D\rho = q \tag{20-a}$$

$$Du = -\frac{\partial \Pi}{\partial x} + \rho\, g \tag{20-b}$$

$$Dw = -\frac{\partial \Pi}{\partial z} \tag{20-c}$$

$$\frac{\partial u}{\partial x} + \frac{\partial w}{\partial z} = 0 \tag{20-d}$$

where the operator $D \equiv \frac{\partial}{\partial t} + u\frac{\partial}{\partial x} + w\frac{\partial}{\partial z} - \nu(\frac{\partial^2}{\partial x^2} + \frac{\partial^2}{\partial z^2})$, the function $\rho \equiv v(x,z) + \Omega x$ may be considered as a "density", $\mathbf{g} = g\mathbf{e}_1 = -2\Omega\mathbf{e}_1$ is the "gravitational field" oriented along the direction $\mathbf{e}_1$. A particular solution corresponds to stationary unidirectional flow $\mathbf{v} = (0, V(x), 0)$ for which

$$\frac{d\Pi}{dx} = -\rho(x)g \qquad \rho(x) = V(x) + 2\Omega\, x, \qquad \frac{d^2\rho}{dx^2} = -\frac{q}{\nu} \tag{21}$$

The stability of this "state of hydrostatic equilibrium " with density $\rho(x)$ is given by the sign of a quantity that measures the "density stratification" in the x-direction :

$$\Phi(x) = -g\frac{d\rho}{dx} = 2\Omega\,[\frac{dV}{dx} + 2\Omega\, x] \tag{22}$$

The flow is stable if $\Phi > 0$ and unstable if $\Phi < 0$. For positive value of the gravitational field due to the Coriolis force, the condition of stability corresponds to a negative stratification of density, while the instability is associated with a positive stratification of the density. The quantity Φ is nothing else than the Rayleigh discriminant for rotation and in dimensionless form, it becomes : Φ = 2Ro (DV + 2Ro). The combination $\Phi(x)(dV/dx)^{-2}$ defines a dimensionless parameter called the *Richardson-Bradshaw* number

$$Ri = -g\frac{d\rho}{dx} / [\frac{dV}{dx}]^2 = 2\Omega\,(2\Omega + \frac{dV}{dx}) / [\frac{dV}{dx}]^2 \qquad (23)$$

which plays the same role as the Richardson number for stratified flows [19]. The Richardson-Bradshaw number may be represented as Ri = S(S+1) where $S \equiv 2\Omega/(dV/dx)$.

For the plane Poiseuille channel flow of witdh d, the gravitational field induced by rotation is oriented along the x-direction and the corresponding "density stratification " is given by

$$\rho(x) = -\frac{q}{2\nu}\,x\,(x-d) + 2\Omega\,x \qquad (24)$$

where q is the driving pressure gradient. Equations (20) may also have the following natural thermal interpretation : identifying ρ with the temperature T, one sees that they correspond to the flow evolution in the presence of heat sources that change temperature and density without changing the expansion or the compression of the fluid.

5. Recapitulation and concluding remarks

We have shown that the instabilities induced by Coriolis force bear a strong analogy with centrifugal instabilities, which can be summarized in the following table 1.

For reasonable curvature and low rotation rate, the centrifugal or rotation induced instabilities occur for Reynolds number (Re $\approx$ 100) very lower than the Tollmien-Schlichting waves in form of longitudinal rolls superimposed to the base flow. The rotation has an effect of either decreasing the threshold of the instability in curved geometries or suppressing the instability for higher rotation number values. The occurrence of the longitudinal vortices aligned along the base flow modifies consequently the transfer properties of the flow. In particular, the rotation effect of inhibiting the onset of instability can play an important role in the heat transfer enhancement of the turbo-machinery. For example, the understanding of flow and heat transfer in the inner cooling of turbine blades is a permanent challenge of engine designers.

The rotation, the curvature and the density stratification have stabilizing or destabilizing effect on flow depending on the coupling with shear vorticity of the base flow. Flows encoutered in the reality combine the effects of rotation, curvature and density stratification, investigation of role played by each factor or by two combined factors represents a first step towards the understanding of the complex flows and of their technical applications such as transfert properties. A tentative unification scheme of both the rotation-induced and centrifugal instabilities have been performed using the Rayleigh discriminant and the similarity of the linear stability equations.

Table 1 : Comparative properties of centrifugal and rotation-induced instabilities

Kinematic parameters	Curvature	Rotation (plane channel)
Equilibrium conditions	$\frac{\partial \Pi}{\partial r} = \frac{V^2}{r}$	$\frac{\partial \Pi}{\partial x} = 2\Omega V$
Rayleigh discriminant	$\Phi = 2\frac{V}{r}(\frac{dV}{dr} + \frac{V}{r})$	$\Phi = 2\Omega (\frac{dV}{dx} + 2\Omega)$
Linear stability (small gap approximation)	$(sK - A + Ta\, B)\, \Psi = 0$	$(sK - A + Re\, B)\, \Psi = 0$
Bifurcation solutions	Longitudinal rolls	Longitudinal rolls

Acknowledgement

The authors would like to thank C. Normand, A. Aouidef, V.A. Vladimirov for many useful discussions on this project.

References

[1] S. Chandrasekhar, *Hydrodynamic and Hydromagnetic Stability*, Oxford Uiversity Press, 1961

[2] P.G. Drazin and W.H. Reid, *Hydrodynamic Stability*, Cambridge University Press, 1981

[3] H. Görtler, 'On the three-dimensional instability of laminar boundary layers on concave walls', *NACA Technical Memorandum* **1375** (1954)

[4] I. Mutabazi and J. E. Wesfreid, 'Spatio-temporal properties of centrifugal instabilities', in E.Tirapegui and W. Zeller (eds.), *Instabilities and Nonequilibrium Structures **III***, 201-216, Kluwer Academic Publishers, Dordrecht (1991)

[5] D.K.Lezius and J.P. Johnston, 'Roll cell instabilities in rotating laminar and turbulent channel flows', *J. Fluid Mech.* **77**, 153 (1976)

[6] R. Wolkind and R.C. DiPrima, 'Effects of a Coriolis force on the stability of plane Poiseuille flow', *Phys. Fluids* **16**, 2045 (1973)

[7] R.J. Wienner, P.W. Hammer, C.E. Swanson and R.J. Donnelly, 'Stability of Tayor-Couette flow subjected to external rotation', *Phys. Rev. Lett.* **64**, 1115 (1990)

[8] L. Ning, M. Tveirtereid, G. Ahlers and D.S. Cannel, 'Taylor Couette flow subjected to external rotation', *Phys. Rev. A* **44**, 2505 (1991)

[9] L.Landau and E.Lifshitz, *Mécanique des Fluides,* éditions MIR, Moscou (1990)

[10] M. Lesieur, *Turbulence in Fluids,* Kluwer Academic Publisher, Dordrecht (1990)

[11] I. Mutabazi, C.Normand and J.E. Wesfreid, 'Gap size effects on centrifugally and rotationally driven instabilities', *Phys. Fluids A* **4**, 1199 (1992)

[12] C.D. Andereck, S.S. Liu and H.L. Swinney, 'Flow regimes in circular Couette system with independently rotating cylinders', *J. Fluid Mech.* **164**, 155 (1986)

[13] I. Mutabazi, J.J. Hegseth, C.D. Andereck and J.E. Wesfreid, 'Pattern formation in the flow between two horizontal coaxial cylinders with a partially fille gap', *Phys. Rev. A* **38**, 4752 (1988)

[14] B.J. Bayly, 'Three-dimensional centrifugal-type instabilities in inviscid two-dimensional flows', *Phys. Fluids* **31**(1), 56-64 (1988)

[15] O.J. E. Matsson and P.H. Alfredsson, 'Curvature and rotation-induced instabilities in channel flow', *J. Fluid Mech.* **210**, 537 (1990)

[16] I. Mutabazi, C.Normand, H.Peerhossaini and J.E. Wesfreid, 'Oscillatory modes in the flow between two horizontal coaxial cylinders with a partially filled gap', *Phys. Rev. A* **39**, 763 (1989)

[17] A. Aouidef, J.E. Wesfreid and I. Mutabazi, 'Coriolis effects on Taylor-Görtler instability', Accepted for publication in *AIAA Journal* (1992)

[18] P.H. Alfredsson and H. Persson, 'Instabilities in channel flow with system rotation', *J. Fluid Mech.* **202**, 543 (1989)

[19] D.J. Tritton and P.A. Davis, 'Instabilities in geophysical fluid dynamics', in H.L. Swinney and J.P. Gollub (eds.), *Hydrodynamic Instabilities and the Transition to Turbulence,* 229, Springer Verlag, Berlin (1981),

[20] V. A. Vadimirov, 'Analogy between density stratification and rotation effects', *J. of Applied Mech. Tech.Phys.* (in Russian) **3**, 58 (1985)

AMPLITUDE EQUATIONS FOR VISCOELASTIC CONVECTIVE FLUIDS

J. Martínez-Mardones, R. Tiemann and W. Zeller
Instituto de Física
Universidad Católica de Valparaíso
Casilla 4059, Valparaíso, Chile.

C. Pérez-García
Departamento de Física y Matemáticas Aplicadas
Universidad de Navarra
31080 Pamplona, Navarra, Spain.

ABSTRACT. The possible bifurcations of convective instabilities in viscoelastic polymeric Oldroyd type fluids are studied. We analyse stationary and oscillatory convection. The treatment of the amplitude equations in the scope of weak nonlinear analysis shows that the convective solutions depend mainly on the viscoelastic parameters and the constitutive equation used to describe the fluid.

INTRODUCTION

The viscoelastic properties of polymeric fluids can lead to very rich behaviours.[1,2] The constitutive equation of highly concentrated solutions of polymers in a fluid is normally given by the Maxwell model. But such a model does not satisfy for many polymeric solutions. They can be better modeled by combining the properties of a Maxwell fluid for the polymer and a Newtonian fluid for the solvent. In such a way, specially for very low concentrations, the more complex Oldroyd model arises.[3] When the concentration grows, this model evolves to a Maxwellian fluid. In general, the dependence of the stress tensor on the rate of the strain tensor, expressed by the constitutive relation, is the main purpose of rheology. Moreover, dispite the polymeric solutions are multifase mixtures, the governing equations will be only focused on the viscoelastic properties.

E. Tirapegui and W. Zeller (eds.), Instabilities and Nonequilibrium Structures IV, 317–324.

GOVERNING EQUATIONS OF VISCOELASTIC CONVECTION

In the present work, a shallow viscoelastic fluid layer of depth d is heated from below and remains at rest until a critical temperature gradient is reached.[4] Under Boussinesq approximation, in which the fluid is taken as incompressible, with exception of the bouyancy term register in Navier-Stokes equation, this viscoelastic Rayleigh-Bénard problem can be described as

$$\nabla\cdot\mathbf{v} = 0 \tag{1}$$

$$\rho_o[\partial_t\mathbf{v} + (\mathbf{v}\cdot\nabla)\mathbf{v}] = -\nabla p + \nabla\cdot\tau + \rho o[1 - \alpha(T-To)]\mathbf{g} \tag{2}$$

$$\partial_t T + (\mathbf{v}\cdot\nabla)T = \kappa\nabla^2 T, \tag{3}$$

where $\mathbf{v}$ is the velocity field, p the pressure, τ the extra-stress tensor, T the temperature, ρ_o the density at the temperature T_o of the bottom, α the thermal expansion coefficient, κ the thermal difusivity and $\mathbf{g}$ the gravity. It is important to notice that the stress tensor τ is given by the Oldroyd constitutive equation

$$[1+\lambda_1 D_t]\tau = 2\eta_o[1+\lambda_2 D_t]\mathbf{D}(\mathbf{v}) \tag{4}$$

where η_o is the viscosity, λ_1 the relaxation time and λ_2 the retardation time. The coefficients η_o, λ_1 and λ_2 are assumed to be constant. The convective derivative D is defined as

$$D_t\tau = \partial_t\tau + (\mathbf{v}\cdot\nabla)\tau - \mathbf{W}\cdot\tau + \tau\cdot\mathbf{W} + a(\mathbf{D}\cdot\tau + \tau\cdot\mathbf{D}), \tag{5}$$

being rate of deformation $\mathbf{D}(\mathbf{v}) = 1/2\,[\nabla\mathbf{v}+(\nabla\mathbf{v})^T]$, the vorticity $\mathbf{W}(\mathbf{v}) = 1/2\,[\nabla\mathbf{v}-(\nabla\mathbf{v})^T]$, and the constant a $(-1\leq a\leq +1)$. For $a=0$, D_t is the corrotational or Jaumann derivative, for $a=+1$, the so called upper (contravariant) time convective derivative, and for $a=-1$, the lower (covariant) one. On linear analysis, D_t becomes the normal partial time derivative ($D_t\to\partial_t$) and we arrive at Jeffreys equation

$$[1+\lambda_1\partial_t]\tau = 2\eta_o[1+\lambda_2\partial_t]\mathbf{D}(\mathbf{v}) \tag{6}$$

The viscosity of the viscoelastic fluid, $\eta_o=\eta_P+\eta_S$, is the result of contributions of the polymer and the solvent viscosities, η_P and η_S, respectively. The outcoming stress tensor, $\tau = \tau_P+\tau_S$, is formed by the addition of the Newtonian stress of the solvent $\tau_N = 2\eta_S\mathbf{D}(\mathbf{v})$, and the Maxwellian polymer stress $\tau_P = 2\eta_P/\lambda_1\int_{-\infty}^{t}\exp[-(t-t)/\lambda_1]\mathbf{D}[\mathbf{v}(t)]dt$. Of course, if concentration grows, the retardation time λ_2 vanishes and the constitutive equation (6) corresponds to the Maxwell model. It is important to notice that Oldroyd constitutive equation (4), specially in the case of the upper convective derivative utilized to specify Oldroyd (B) fluids, represents fairly well the properties of a low concentrated mixture of a mainly elastic polymer on a single viscous fluid.

The perturbed adimensional governing equations of the polymeric fluid initially at rest, conform the following system

$$\partial_t \nabla^2 \psi = - PR\partial_x \Theta + P\Delta^2 \tau_{xz} + P\partial_{xz} S - J(\psi, \nabla^2 \psi) \tag{7}$$

$$\partial_t \Theta = - \partial_x \psi + \nabla^2 \Theta - J(\psi, \Theta) \tag{8}$$

$$\begin{aligned} \Gamma\partial_t \tau_{xz} - \Gamma\Lambda\partial_t \Delta^2 \psi = {} & \Delta^2 \psi - \tau_{xz} - \Gamma[J(\psi, \tau_{xz}) + 1/2(\nabla^2\psi)S - 1/2a(\Delta^2\psi)U] \\ & + \Gamma\Lambda[J(\psi, \Delta^2\psi) + 2a(\partial_{xz}\psi)(\Delta^2\psi)] \end{aligned} \tag{9}$$

$$\begin{aligned} \Gamma\partial_t S - 4\Gamma\Lambda\partial_t \partial^2_{xz} \psi = {} & 4\partial^2_{xz}\psi - S - \Gamma[J(\psi, S) - 2(\nabla^2\psi)\tau_{xz} - 2a(\partial^2_{xz}\psi)U] \\ & + \Gamma\Lambda[4J(\psi, \partial^2_{xz}\psi) - 2(\nabla^2\psi)(\Delta^2\psi)] \end{aligned} \tag{10}$$

$$\begin{aligned} \Gamma\partial_t U = {} & - U - \Gamma[J(\psi, U) - 2a(\Delta^2\psi)\tau_{xz} - 2a(\partial^2_{xz}\psi)S] \\ & - 2\Gamma\Lambda[(2\partial^2_{xz}\psi)^2 + (\Delta^2\psi)^2] , \end{aligned} \tag{11}$$

where $J(f,g)=\partial_z f \partial_x g - \partial_x f \partial_z g$ denotes the Jacobian, $\Delta = \partial^2_{zz} - \partial^2_{xx}$ corresponds to a differential operator, and the two dimensional Laplacian is $\nabla^2 = \partial^2_{xx} + \partial^2_{zz}$. The perturbations are represented by the stream function Ψ, the temperature Θ, the in-plane shear tensor τ_{xz}, the normal stress difference $S=\tau_{xx}-\tau_{zz}$ and the trace $U=\tau_{xx}+\tau_{zz}$. The Rayleigh number $R=\rho_o g\alpha\Delta T d^3/\eta_o\kappa$, the Prandtl number $P=\eta_o/\rho_o\kappa$, the nondimensional stress relaxation time $\Gamma=\lambda_1\kappa/d^2$, and the ratio between retardation and relaxation times $\Lambda=\lambda_1/\lambda_2=\eta_S/\eta_o$ have also been introduced.

The boundary conditions for free and perfectly conducting surfaces on the bottom and on the top limiting infinite horizontal layers, are given by $\Psi=\partial^2_{zz}\Psi=0$ and $\Theta=0$ at $z=0,1$. As a consequence, at these levels we found that $\tau_{xz}=\partial_z S=U=0$. These boundary conditions allow us to obtain simple analytical solutions of eqs. (7) - (11). Furthermore, by defining the vector $\mathbf{v}(x,z,t) = [\psi,\Theta,\tau_{xz},S,U]^T$, these equations may be written in the compact form

$$\partial_t \mathbf{Lu} = \mathbf{Mu} + \mathbf{N(u,u)} , \tag{12}$$

The perturbation fields matrices **L** and **M** are linear and **N** takes into account the nonlinearities.[5]

LINEAR THEORY

Assuming that the perturbation **u** is small enough, the nonlinear term on the left handside in eq. (12) is negligible. The perturbations may be expanded in normal modes of the form $\mathbf{u}(x,z,t) = \mathbf{U}_o(z)\exp(ikx + st)$, where $s=\sigma+i\omega$ is a complex parameter (σ is the growth factor of a perturbation and ω its frequency) and the function $\mathbf{U}_o(z)$ is subject to the boundary conditions pointed out earlier. Introducing the normal modes in the linearized form of eq. (12), an eigenvalue problem arises from

$$\det(M_{mn} - sL_{mn}) = s^j + \sum_{i=o}^{j-1} a_i s^i = 0 . \qquad (m,n\geq 1) \qquad (13)$$

For the most unstable modes, when m=n=1, two roots are always negative and have stable eigenvectors. Therefore, the eigenvalue problem is restricted to the cubic characteristic polynomial $P(s) = s^3+a_2s^3+a_1s+a_o = 0$, with $a_2 = \Gamma^{-1} + q^2(1+\Lambda P)$, $a_1=P(k/q)^2[q^6\Lambda/k^2+ q^4(1+P)/k^2P\Gamma-R]$, $a_o=(P/\Gamma)(k/q)^2[q^6/k^2-R]$ and $q^2=k^2+\pi^2$. The complicate dependence of the coefficients on (R,P,Γ,Λ,k) do not permit to visualize physical relevant implications without going through a detailed examination. The possible roots of the critical polynomial P(s) give the different bifurcations that can appear in the problem.

The stationary bifurcation for s=0 occurs if $a_o=0$. This correponds to $R_S=q^6/k^2$, giving the critical Rayleigh and wave number values for stationary convection $R_{cs} = 27\pi^4/4$ and $k_{cs}=\pi/\sqrt{2}$, respectively. It should be noted that the solution for stationary instability is identical to the one corresponding to a Newtonian fluid.

The oscillatory bifurcation is expected for $s=i\omega$, when $a_o=a_1a_2$ and $a_1=\omega^2>0$. Since the Rayleigh number is

$$R_o = q^6/k^2 - (q\omega/k)^2[q^2\Gamma(P^{-1}+\Lambda) + P^{-1}] \qquad (14)$$

where $\omega^2 = [q^2\Gamma(1-\Lambda)-(1+P^{-1})]/\Gamma^2(P^{-1}+\Lambda)$, both critical values for oscillatory instability, R and k, must be obtained numerically.[6-9] Moreover, the direction of the already presented inequality $\Gamma(1-\Lambda) > (1+P)/q^2P$ changes.

The solution of the characteristic polynomial is reduced to solve $s^2(s+a_2)=0$, when $a_o=a_1=0$. Then it has double zero solutions $s^2=0^2$ and also

$s=-a_2$. These leads to the critical Rayleigh number $R_{ct}=q^6/k^2$ and $\Gamma(1-\Lambda) = (1+P)/q^2P$. The marginal curve coincides with the curve that arises on the stationary state.

The critical Rayleigh number R_c for the onset of overstability is presented graphically in Fig. (1) as function of Γ for Maxwell and Jeffreys models, for highly concentrated solutions (Λ=0.0,P=1000.0) and low concentrations of the polymer in the fluid (Λ=0.75,P=100.0). The pointed line shows the stationary instability. The oscillatory curves are illustrated only for critical Rayleigh numbers R_c lower than the stationary Rayleigh number R_{cs}. For values of Γ smaller than the minimun represented in each curve, the oscillatory instability cannot appear spontaneously.

NONLINEAR ANALYSIS

The linear analysis allows us to specify the dynamical situations when the system becomes unstable. However, a nonlinear analysis must be carried out in order to determine the convective motions observed in experiments. The useful weakly nonlinear Stuart-Landau method is based on the amplitude equation, assuming basically that the constant amplitudes of the linear analysis are variables in space and time above the threshold. But these amplitudes depend on space and time scales different from the usual linear ones.

The solutions of the nonlinear problem at the nearthreshold behavior of convection, for small values of the reduced Rayleigh number $\varepsilon=(R-R_c)/R_c$ and small amplitudes may be expressed as

$$\mathbf{u}(x,z,t) = \varepsilon^{1/2}\mathbf{u}_o + \varepsilon\mathbf{u}_1 + \varepsilon^{3/2}\mathbf{u}_2 + \dots \tag{15}$$

Here, $\mathbf{u}_o$ corresponds to the equivalent linear problem solution

$$\mathbf{u}_o(x,z,t) = \mathbf{U}_o(z)A(T)\exp[i(k_c x+\omega_c t)]\ \text{c.c.} \tag{16}$$

where c.c. refers to the complex conjugate of the first term. The time differentiation has to be transformed as $\partial_t \to \partial_t + \varepsilon\partial_T$, with the slow time variable $T=\varepsilon t$. This multiscale approach has not taken into account the gauge of the spatial coordinates and differentiation operators, which would lead to a broader Ginzburg-Landau problem.

Equation (12) is shortened by using only one linear matrix operator $\mathfrak{L}=\partial_t\mathbf{L}-\mathbf{M}$, which can be expanded as $\mathfrak{L}=\mathfrak{L}_o+\varepsilon^2\mathfrak{L}_1+\dots$ In addition, the nonlinear term on (12) may be develop as

$$N(u,u) = \varepsilon^2 N_{oo}(u_o,u_o) + \varepsilon^3[N_{o1}(u_o,u_1) + N_{1o}(u_1,u_o)] + \ldots \tag{17}$$

By gathering terms of the same order in ε, the hierarchy of equations below is obtained

$$\begin{aligned} \mathfrak{L}_o u_o &= 0 \\ \mathfrak{L}_o u_1 &= N_{oo}(u_o,u_o) \\ \mathfrak{L}_o u_2 &= -\mathfrak{L}_1 u_o + N_{o1}(u_o,u_1) + N_{1o}(u_1,u_o) \end{aligned} \tag{18}$$

The first equation is equivalent to the linear problem treated before. Once u_o is found, u_1 can be determined and, having both, the last equation allows us to obtain the amplitude equation.

Several studies have been done in aim of Landau nonlinear theory.[10-12] In this context, the amplitude equation for stationary bifurcation, with $\omega_c=0$ and $R_c=R_{cs}$, has the form

$$d_T A = \mu A - KA^3 \tag{19}$$

where the coefficients μ and K are functions of the parameters a, P, Γ, Λ, k_{cs} and R_{cs}. The bifurcation is always supercritical for $a=\pm1$, since K is strictly real and positive. Of course, this is also true in the particular case of Maxwell fluid (Λ=0.0), disagreeing with some authors[12] that show the possibility of having the presence of a tricritical point for a subcritical transition. In Fig. (2a) K is represented as function of Γ for different Maxwell, and upper and lower convective Oldroyd fluids ($a=\pm1$).

It must be emphasized that the value of a ($|a|<0.825$) is crucial in order to find subcritical bifurcations and polycritical points. For instance, it may be observed on the corrotational Jeffreys model ($a=0$) (fig. 2b). As function of Γ, the coefficient K runs from positive to negative values for viscoelastic fluids whose concentrations vary from the Maxwell (Λ=0.0) to a high dilute polymeric fluid (Λ=0.75).

The complex amplitude equation for oscillatory bifurcation is

$$d_T Z = \mu_o Z - |Z|^2 Z \tag{20}$$

where μ_o and $K_o=K_{or}+iK_{oi}$ are complex coefficients. The real part of K_o determines if a supercritical [$\mathrm{Re}(K_{or})>0$] or a subcritical [$\mathrm{Re}(K_{or})<0$] bifurcation is going to be found. In figure (3) $\mathrm{Re}(\kappa_o)$ is presented as function of Γ. Figure (3a) shows that the convective Maxwell model is always supercritical, but the convective Oldroyd fluids are subcritical for low values of Γ and they change to supercritical when Γ is over a certain value ($\log\Gamma > 1.20$). And, figure (3b) illustrates that the corrotational Maxwell fluid has subcritical bifurcations for $\log\Gamma < 1.26$ and it changes to supercritical branching for higher values of $\log\Gamma$.

For the corrotational Jeffreys model (with P=10.0) it can be pointed out a supercritical and stable Hopf bifurcation can be pointed out, with the exception of values of Λ between 0.1 and 0.5 (fig. 5).

CONCLUSIONS

We have shown how convection can be affected by the viscoelastic parameters and by the type of the constitutive relation used. The linear stability for Jeffreys model has been analysed, which in particular includes the Maxwell fluid. Accordingly, the convective solution is steady in certain parameter range and periodic in others.

The amplitude equations for the bifurcations studies are also obtained. From these equations we can determine if the bifurcation is subcritical or supercritical and the existence of polycritical points. In the case of a stationary bifurcation these characteristics depend on the value of the parameter a taken in the constitutive equation. A similar conclusion can be inferred from the results obtained for oscillatory convection. The parameter a is characteristic of the fluid and its value cannot be determined from symmetry arguments but it must be deduced from experimental results. The present study suggests that convection can be a useful tool to determine this property. In fact the precision of convective experiments allows to infer some rheological properties of polymers.

ACKNOWLEDGMENT

We greatly acknowledge many helpful comments and remarks from Professor Dr. Daniel Walgraef. This work was partially supported by FONDECYT.

REFERENCES

1. R.B.Bird et al., Dynamics of Polymeric Fluids, Vol. 1. Wiley and Sons, New York (1977).
2. D.D.Joseph, Fluid Dynamics of Viscoelastic Liquids. Springer-Verlag, Heidelberg (1990).
3. J.G.Oldroyd, Proc. R. Soc. A 200, 523 (1950).

4. S.Chandrasekhar, Hydrodynamics and Hydromagnetic Stability, Dover, New York (1981).
5. J.Martínez-Mardones, Convección en Fluidos Viscoelásticos. Doctoral Thesis, Universidad Autónoma de Barcelona, (1991).
6. C.M.Vest and A.Arpaci, J.Fluid Mech. 36, 613 (1969).
7. M.Sokolov and R.I.Tanner, Phys. Fluids 15, 534 (1972).
8. R.W.Kolkka and G.R.Ierly, J. Non-Newtonian Fluid Mech. 25, 209 (1987).
9. J.Martínez-Mardones and C.Pérez-García. J. Phys.: Condens. Matter 2, 1281 (1990).
10. I.A.Eltayeb, Proc. R. Soc. A 356, 161 (1977).
11. B.J.A. Zielinska, D. Mukamel and V. Steinberg, Phys. Rev. A 33, 1454 (1986).
12. M.Renardy and Y.Renardy, Z. angew Math. Phys. 43, 154 (1992).

Fig. 1

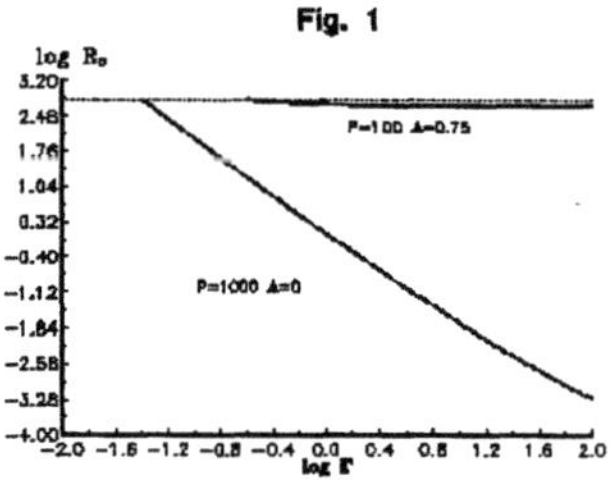

Figure 1: Critical Rayleigh number for the onset of overstability on the Maxwell (Λ=0) and Jeffreys (Λ=0.75) fluid . The pointed line represent them for the onset of stationary convection.

Figure 2: The cubic coefficient in the stationary amplitud equation $\mathcal{K}$ as function of Γ . Fig.(2a) for covariant and contravariant Oldroy model and Fig.(2b) for corrotational Oldroy fluid.

Fig. 2a

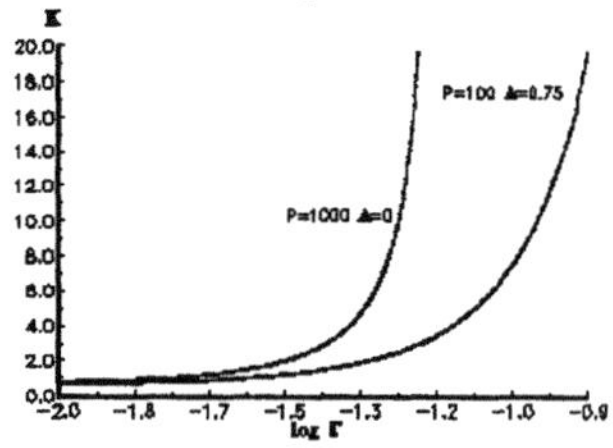

Fig. 2b

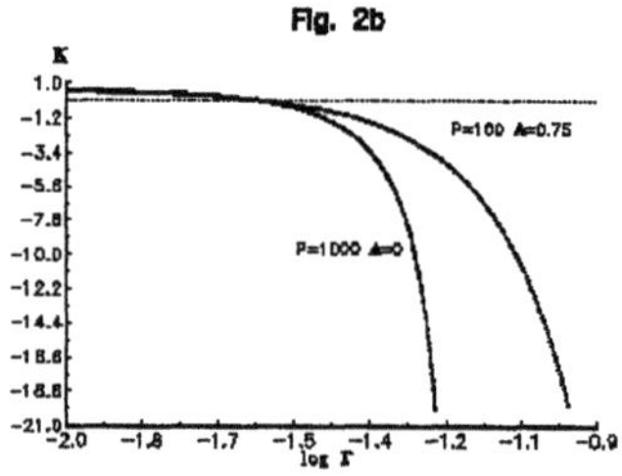

Fig. 3a

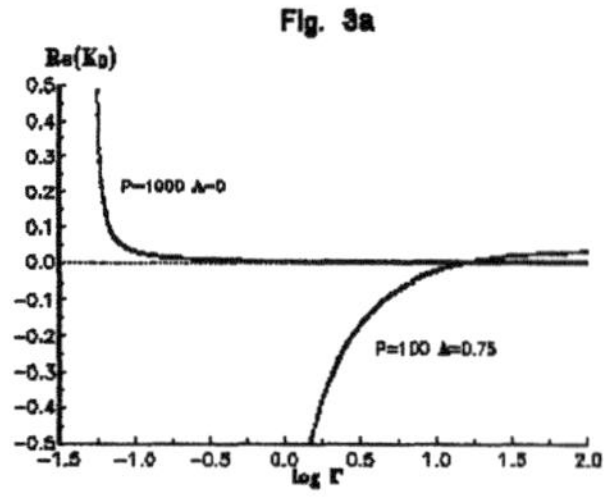

Fig. 3b

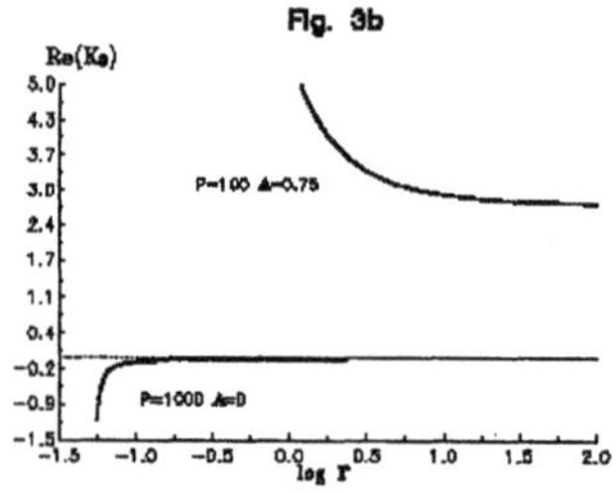

Figure 3: The values of the real part of the cubic coefficient in the oscillatory amplitud equation Re($\mathcal{K}_0$) as function of Γ. Fig.(3a) for covariant and contravariant Oldroy model and Fig.(3b) for corrotational Oldroy fluid.

INSTABILITIES MEDIATED BY LINE DEFECTS IN THREE DIMENSIONS WITHOUT UNBINDING

ALEJANDRO F. RAMIREZ
Courant Institute of Mathematical Sciences
251 Mercer St., N.Y., N.Y., 10012, U.S.A.

ANDREAS REISENEGGER
Physics Department, California Institute of Technology
Pasadena, CA 91125, U.S.A.

FERNANDO LUND
Facultad de Ciencias Físicas y Matemáticas
Universidad de Chile, Casilla 487-3, Santiago, Chile

ABSTRACT. Vortex loops in a superfluid and dislocation loops in an elastic solid drive an instability in three dimensions that closely resembles the lambda transition in He4 and shear melting, respectively. The mechanism is much the same as Kosterlitz-Thouless in two dimensions. However, contrary to what happens in that case, the transition is achieved without the sudden proliferation of unbound loops.

1. Introduction

The physical idea to be exploited is as follows (Figure 1): there are systems that may present one dimensional defects (alias strings), for instance vortex or dislocation loops, that can be thermally excited. At low but nonzero temperature there will be a finite population of such strings. Since the (finite) energy needed for their creation goes roughly like their size, as the temperature is increased both their number and their size will increase and at a certain point loops of sizes spaning many different scales will coexist. This is the essential point. Consider a "large" loop: to use a magnetostatics analogy, it will polarize the medium making it easier for smaller loops to appear. The "small" loops, on the other hand, will screen the self-energy of the "large" one, making it easier for it to increase its size. There is clearly here a potential mechanism for an instability. The questions to be answered are then whether this instability will occur at a finite temperature, whether it is possible to estimate this temperature on the basis of some model, whether the value so estimated is reasonable, and what is the behaviour of the system near the instability.

E. Tirapegui and W. Zeller (eds.), Instabilities and Nonequilibrium Structures IV, 325–331.

The system whose statistical mechanics is to be studied is the set of all interacting strings in three dimensions. This phase space is far too complicated and in order to make progress its infinitenumber of dimensions will be slashed down to six dimensions by considering only circular loops. This is a severe and uncontroled (in the sense of being able to estimate what is being neglected) approximation. However, any closed curve can be thought of as being made up of many circles of possibly widely different scales. The interactions among loops will be approximated by way of a scale dependent mean field approximation: the self energy of one loop will depend on linear response functions (superfluid density or elastic constants) that depend on scale, determined by the loop radius R. The macroscopic value of these response functions will be read off from the behaviour for very large scales, $R \to \infty$.

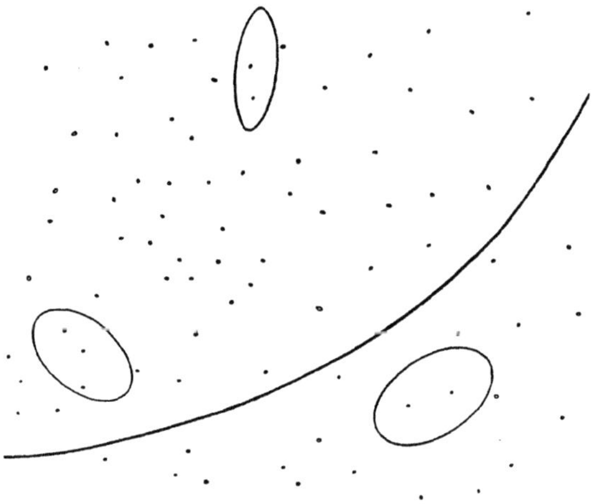

Figure 1: The physical system that is considered in this article is that of many, thermally excited, circular strings with sizes over many length scales. The small loops screen the self energy of the large loops, which in turn 'polarize' the medium so that it becomes easier to create small loops. This process drives an instability occuring at finite temperature.

2. Vortex Loops

The energy associated with an isolated vortex ring of radius R and circulation Γ, both in classical and quantum fluids, is

$$U_0(R) = \frac{\rho\Gamma^2 R}{2}(\ln(R/\tau) + C) \tag{1}$$

where ρ is the fluid density and C a constant whose exact value depends on what goes on inside the vortex core, of radius τ. For a superfluid, the value $C = 0.464$ has been proposed[1], and expression (1) is valid for a thin vortex ring, that is, for $R \gg \tau$. As we are interested in the critical behaviour of the system, what goes on at a microscopic scale should not matter and the precise values of τ and C should be irrelevant as far as the universal quantities are concerned. A system of noninteracting vortex rings will clearly not have a phase transition, as the energy that it costs to get one grows like $R \ln R$ for large R [4]. Interactions have to be taken into account, and now we follow the ideas of Kosterlitz–Thouless[2], as explained in particular by Young[3], whereby this interaction is approximately considered by way of a scale dependent screening. This makes sense, since a vortex ring behaves in many respects as a magnetic dipole, and we expect that the self–interaction of a large ring will be screened by smaller ones, whose creation will be facilitated by the presence of the larger one, and one might expect that as in two dimensions, it will cost a finite energy to form infinitely large rings. It has also been noted[4] that for a hypothetical system of vortices whose energy was proportional to length, allowing for arbitrarily shaped vortices would mean that at a finite temperature the gain in entropy would dominate over the cost in internal energy allowing for infinitely large vortices to occur.

The scaling differential equations describing this state of affairs are as follows, with the renormalized superfluid density ρ being given by $\rho = \varepsilon^{-1}$ [5] :

$$\frac{d\varepsilon}{dR} = \frac{4\pi^3}{3}\beta\rho\Gamma^2\left(\frac{R}{\tau}\right)^6 e^{-\beta U(R)} \tag{2}$$

$$\frac{dU}{dR} = \frac{\rho\Gamma^2(\ln(R/\tau)+C+1)}{2\varepsilon(R)} \tag{3}$$

with initial conditions $U(R=\tau)=\mu$ and $\varepsilon(R=\tau)=1$. The first initial condition is the energy needed to create a vortex of minimum size, and the second is the statement that for a vortex of minimum size there are no smaller vortices, and hence no screening. The critical behaviour of (3) should not depend on the precise values of the parameters or on the precise values of the initial conditions. For a numerical solution of (3) we choose $\mu = 1.39\rho\Gamma^2\tau$ as given by Jones and Roberts[6] on the basis of a Landau–Ginzburg theory for an isolated vortex.

In terms of the scaled variables

$$\tilde{\varepsilon} = \frac{2\varepsilon}{\beta\rho\Gamma^2\tau} \tag{4}$$

$$\tilde{U} = \beta U \tag{5}$$

$$r = \frac{R}{\tau} \tag{6}$$

equations (3) become

$$\frac{d\tilde{\varepsilon}}{dr} = \frac{8\pi^3}{3}r^6 e^{-\tilde{U}} \tag{7}$$

$$\frac{d\tilde{U}}{dr} = \frac{1}{\tilde{\varepsilon}}(\ln r + C + 1). \tag{8}$$

Different initial conditions now correspond to different temperatures and what one does is to solve these equations nuerically for different initial conditions and look for a sudden change in the asymptotic behaviour of the solutions This was done in[5] and it was shown that there is a transition at

$$K_c = 0.54222,$$

where $K = \beta\rho\Gamma^2\tau(C+1)/12$, corresponding to a temperature $\sim$ 2K above which the superfluid density $\rho = \varepsilon^{-1}(\infty)$ vanishes. Moreover, it was established that, in terms of reduced temperature t, the superfluid density behaves as a power law

$$\rho \sim t^{\nu}$$

with $\nu \sim 0.57$, in qualitative accord with experimental facts[7]. The value of the critical temperature is of the right order of magnitude for He4 but depends on the adjustable parameters τ and C. The value of ν, on the other hand, is universal in that it does not depend on those parameters.

Below the critical temperature $\tilde{\varepsilon} \rightarrow \tilde{\varepsilon}_{\infty}$, a finite value, as $r \rightarrow \infty$. Substitution into (8) leads to the following expression for the number dn of rings per unit volume with radius between R and $R + dR$:

$$dn(R) = \frac{4\pi R^2}{\tau^6} e^{-R(\ln(R/\tau)+C)/\tau\tilde{\varepsilon}_{\infty}} dR.$$

Thus, this number decays faster than exponentially at large R.

Let us now see what happens at the critical temperature, where we know that $\tilde{\varepsilon}(r) \rightarrow \infty$ as $r \rightarrow \infty$. We try the following ansatz for the asymptotic behaviour:

$$\tilde{\varepsilon} = Ar(\ln r + B)$$

where A and B are constants to be determined. Substitution into (8) leads to

$$\tilde{\varepsilon}(r) = \frac{1}{6} r \ln r + O(r) \tag{9}$$

$$\tilde{U}(r) = 6 \ln r - \ln(\ln r) + O(1). \tag{10}$$

Thus we see that, although the renormalized self-energy of a vortex loop still grows (weakly) with R, the string "tension" dU/dR decays like $1/R$, showing that indeed at criticality it is much easier to have large vortex loops. The number dn of loops per unit volume with radius between R and $R + dR$ is

$$dn \sim \frac{1}{4\pi^2 R^4} \ln \frac{R}{\tau} dR. \tag{11}$$

That is, the nuber of vortices of radius R decays algebraically at the critical temperature rather than (faster than) exponentially below the critical tempertuare. From (11) we see that both $< R >$ and $< R^2 >$ are *finite* at criticality, and that the vanishing of the superfluid density is achieved without the sudden proliferation of unbound loops. This is a surprising result, at variance with the expectation mentioned at the beginning of this secton, that the transition would be characterized by the sudden proliferation of infinitely large rings.

3. Dislocation Loops

The energy of an isolated, circular, dislocation loop of radius R and core radius τ in an isotropic elastic solid is, for $R \gg \tau$[8]

$$W_s = \frac{R\mu}{2(1-\nu)} \left(\frac{b_{\parallel}^2(2-\nu)}{2} \left(\ln \frac{R}{\tau} + C_1 \right) + b_{\perp}^2 \left(\ln \frac{R}{\tau} + C_2 \right) \right), \tag{12}$$

where $b_{\|}$ (resp. $b_{\perp}$) is the component of the Burgers vector b_i along (resp. perpendicular to) the plane of the dislocation loop, μ is the shear modulus, ν is Poisson's ratio, and C_1, C_2 are numbers of order one that depend on what happens inside the dislocation core. Physically, W_s is the energy needed to increase the loop radius from τ to R. How is this changed when the elastic constants depend on R because of the screening due to the presence of dislocation loops of many different size scales? The answer to this question proceeds in two steps. First, just as in two dimensions, the response of an elastic solid to an external stress will be modified ("renormalized") by the presence of dislocation loops[9]. This modification will be the response of a single loop times the number of loops, which will be given essentially by a Boltzmann factor. Second, the energy it takes to *change* the size of a loop, say from radius R to radius $R+dR$ depends on the renormalized response functions $\mu(R)$ and $\nu(R)$. With these ingredients it is possible to write down scaling equations whose solutions are the renormalized response functions as a function of temperature, specified by initial conditions. What one is now looking for is a sudden change in the asymptotic behaviour of solutions. This would signal an instability, a phase transition. The algebra leading to the equations has been given elsewhere[10], so that only the equations themselves will be written down here. Let

$$\begin{aligned} A &= \frac{15(3G+4F)}{2(3G-F)(G+3F)} \\ B &= \frac{15(2G+F)}{2(3G-F)(G+3F)}, \end{aligned} \tag{13}$$

and $A \equiv \mu(2-\nu)/(1-\nu)$ and $B \equiv \mu/(1-\nu)$. The equations are

$$\frac{dF}{dr} = 6b^2\beta\pi^3(\frac{r}{\tau})^6 \int_0^\pi d\theta \sin\theta \cos^2\theta E(r,\theta) \tag{14}$$

$$\frac{dG}{dr} = 3b^2\beta\pi^3(\frac{r}{\tau})^6 \int_0^\pi d\theta \sin\theta(\cos^2\theta + 1)E(r,\theta) \tag{15}$$

$$E(r,\theta) = e^{-\beta(V(r)\cos^2\theta + W(r)\sin^2\theta)}$$

$$\frac{dV}{dr} = \frac{b^2 A}{4}(\ln\frac{r}{\tau} + C_1 + 1) \tag{16}$$

$$\frac{dW}{dr} = \frac{b^2 B}{2}(\ln\frac{r}{\tau} + C_2 + 2). \tag{17}$$

They have to be solved with initial conditions

$$V(\tau) = b^2 A_0 \tau C_1/4 \tag{18}$$

$$W(\tau) = b^2 B_0 \tau C_2/2, \tag{19}$$

and A, B, F, G at $r = \tau$ given by the bare values of the elastic constants. This can be done numerically and it is found that, for $C_1 = C_2 = 1$, there is a transition at

$$\tilde{\mu}^{-1} = 0.124602$$

where $\tilde{\mu} = \tau\beta b^2\mu$ is a dimensionless shear modulus, above which the renoralized shear modulus vanishes, signaling shear melting. For $\mu \sim 10^{10}$ erg/cm^3, $b \sim 10^{-8}$ cm, $\tau \sim 10^{-8}$

cm, this means a transition temperature $T_s \sim 10^3 K$, which is the right order of magnitude for the melting temperature of a metal. The behaviour of the shear modulus as a function of temperature was found in [10], where it was shown that, below T_s, it can be fit to a very good approximation by a power law, $\mu \sim t^\gamma$, $\gamma \sim 0.5$, where t is the reduced temperature. What is truly remarkable is that the exponent γ is independent of the constants C_1 and C_2, as well as of the values of the bare parameters, suggesting that near the mechanical melting point the shear modulus of overheated solids should behave universally: it should be the same for all solids whose long wavelength properties can be modelled by continuum, isotropic elasticity. It should be possible to test this striking prediction with superheated solids[11].

Nelson and Toner[12] have shown that a system of unbound dislocation loops, while not having long range translational order, does have long range orientational order and for this reason does not behave as a liquid. The gas of circular dislocation loops that we are considering here is able to renormalize the elastic constants of a solid to zero while most of them remain bound. This is in marked contrast with the situation in two dimensions where the transition is marked by the dissociation of dipoles[9], but agrees with the analogous results for vortex loops presented in Section 2. This is seen in Figure 2 where the mean dislocation radius is plotted as a function of temperature. It is seen that although the qualitative behaviour of this function does change across the transition point, the mean radius remains finite.

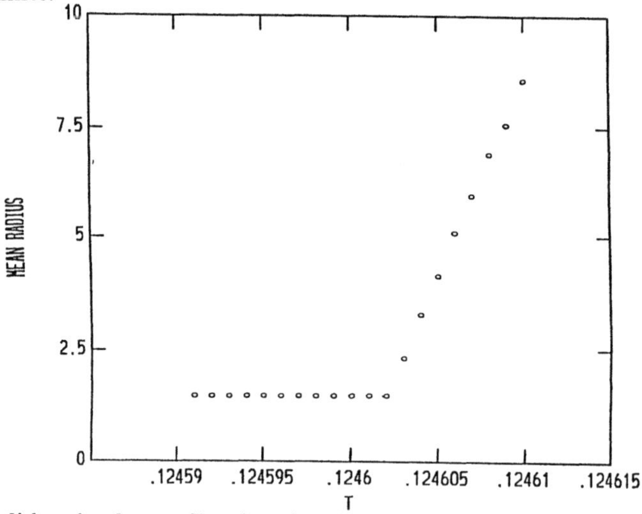

Figure 2: Mean dislocation loop radius, in units of core radius, as a function of temperature in dimensionless units across the transition point. There is a qualitative change in the behaviour of the mean radius as a function of temperature but it remains finite across the transition.

Acknowledgements

This work has been supported by DTI grant E-2854-9244 and FONDECYT grant 1265-91.

References

[1] P. H Roberts and J. Grant *J. Phys.* **A4**, 55 (1971).

[2] J. M. Kosterlitz and D. J. Thouless, *J. Phys.* **C6**, 1181 (1973); J. M. Kosterlitz, *J. Phys.* **C7**, 1046 (1974); A. P. Young, *J. Phys.* **C11**, L453 (1978); D. R. Nelson and J. M. Kosterlitz, *Phys. Rev. Lett.* **39**, 1201 (1977); J. V. José, L. P. Kadanoff, S. Kirkpatrick and D. R. Nelson, *Phys. Rev.* **B16**, 1217 (1977). For a review see P. Minnhagen, *Rev. Mod. Phys.* **59**, 1001 (1987).

[3] A. P. Young, *Phys. Rev.* **B19**, (1979).

[4] F. W. Wiegel, *Physica* **65**, 321 (1973).

[5] F. Lund, A. Reisenegger and C. Utreras, *Phys. Rev. B*41, 155 (1990); A. F. Ramírez, F. Lund and R. Hernández, *Phys. Rev. B*, to appear.

[6] C. A. Jones and P. H. Roberts, *J. Phys.* **A15**, 2599 (1982).

[7] J. C. Le Gillou and J. Zinn–Justin, *Phys. Rev.* **B21**, 3976 (1980); P. Hohenberg, *Physica* **109 & 110 B+C**, 1436 (1982), and references therein; J. A. Lipa and T. C. P. Chui *Phys. Rev. Lett.* **51**, 2291 (1985).

[8] J. P. Hirth and J. Lothe, *Theory of Dislocations*, 2nd edition, Wiley, 1982.

[9] A. P. Young, *Phys. Rev. B* **19**, 1855 (1979); D. R. Nelson and B. I. Halperin, *Phys. Rev. B* **21**, 5312 (1980); *Phys. Rev. B* **19**, 2457 (1979); J. M. Kosterlitz and D. J. Thouless *J. Phys. C* **6**, 1181 (1973).

[10] F. Lund,*Phys. Rev. Lett.* **69**, 3084 (1992).

[11] J. J. Métois and J. C. Heyraud, *J. Physique France* **50**, 3175 (1989); J. Daeges, H. Gleiter and J. H. Perepezko, *Phys. Lett. A* **119**, 79 (1986); C. J. Rossouw and S. E. Donnelly, *Phys. Rev. Lett.* **55**, 2960 (1985).

[12] D. R. Nelson and J. Toner, *Phys. Rev. B* **24**, 363 (1981).

GENERATION OF SIDE JETS IN FORCED AXISYMMETRIC JETS

M. Abid and M. E. Brachet
CNRS, Laboratoire de Physique Statistique
Ecole Normale Supérieure
24 rue Lhomond 75231 Paris Cedex 05
FRANCE

ABSTRACT. The generation of side jets from axisymmetric jets is studied by direct numerical integration of the Navier-Stokes equations. After vortex rings are formed via the primary Kelvin-Helmholtz instability, pairs of axially counter-rotating vorticity filaments appear which eject fluid out of the jet. We describe the three-dimensional topology of the rings, filaments, and side jets.

In certain experimental configurations, the dynamics of spreading, entrainment, and mixing in axisymmetric jets occurs primarily via what are called *side jets* [1]. These structures have been observed both in hot jets and also in acoustically forced jets of uniform density. In the case of hot jets, linear theory predicts a transition from convective to absolute instability; this leads to a self-oscillating ("preferred") mode for the jet [2]. In the second case, the jet is forced acoustically at the most convectively unstable frequency. In both cases, these mechanisms lead to the formation of high-intensity vortex rings.

Based on his experimental results, Monkewitz has suggested [1] that the formation of side jets is related to the instability of vortex rings observed by Widnall. More recently, he suggested that the side jets observed *between* the vortex rings were produced by pairs of counter-rotating longitudinal vortex filaments [3]. These filaments would be analogous to the streamwise vortices observed in mixing layers.

In this note, we use direct simulation of the three-dimensional Navier -Stokes equations:

$$\partial_t \mathbf{u} + (\mathbf{u} \cdot \nabla)\mathbf{u} = -\nabla p + \nu \Delta \mathbf{u} \ , \tag{1}$$

$$\nabla \cdot \mathbf{u} = 0 \ , \tag{2}$$

to characterize the mechanism responsible for the generation of side jets. We also solve the equation of a passive scalar:

$$\partial_t \rho + (\mathbf{u} \cdot \nabla)\rho = \nu \Delta \rho \ . \tag{3}$$

This passive scalar, which has no effect on the velocity field, is used only as a numerical tracer to visualize the mixing between the interior and exterior of the jet.

E. Tirapegui and W. Zeller (eds.), Instabilities and Nonequilibrium Structures IV, 333–338.

To integrate numerically (1), (2) and (3), we use a standard pseudo-spectral method [4]. The flow is taken to be periodic along the axis of the jet, and is expanded in sine or cosine functions in the lateral directions. The use of Fourier transforms adapted to this geometry reduces the computation time and storage by a factor of four relative to a general periodic code. Time stepping is carried out by a second-order-accurate Adams-Bashforth/Crank-Nicolson scheme. The code uses multitasking, and one time-step at a resolution of $32 \times 80 \times 80$ takes 2 CPU seconds on a CRAY 2 computer.

The initial condition is a superposition of the basic profile of the jet and several unstable eigenmodes from linear stability theory. The basic profile is that studied by Michalke [5] :

$$u(r) = (U_0/2)[1 + \tanh((R - r)/(2\theta))] \ , \tag{4}$$

where U_0 is the centerline speed, θ is the momentum thickness, r is the distance to the axis and R is the radius of the jet. The unstable modes are obtained by solving the Rayleigh equation by a shooting method and interpolating the solutions onto the grid-points of the pseudo-spectral code. The passive scalar is initialized with the same profile as (4).

We checked the three-dimensional code by reproducing the growth rates predicted by linear theory for sufficiently small amplitudes of the unstable mode. We then followed the nonlinear temporal evolution of the dynamics of the jet. For mixing layers, it has been established that this type of calculation captures the essential features of the vorticity dynamics [6]. The calculations presented here used a resolution of $64 \times 160 \times 160$.

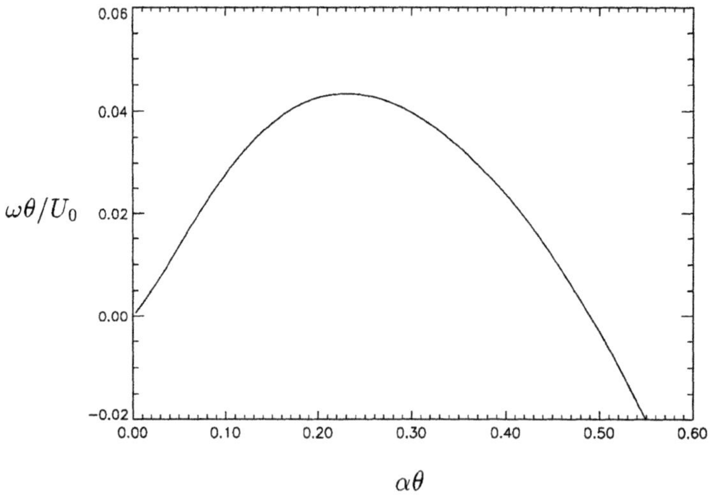

Figure 1 Non-dimensionalized growth rate $\sigma\theta/U_0$ of the linear instability of the axisymmetric mode as a function of the non-dimensionalized wavenumber $\alpha\theta$. For $\theta/R = 0.08$, the maximum growth rate is obtained at $\alpha\theta = 0.233$.

Figure 1 shows the results of linear stability analysis for the axisymmetric profile given by (4). The temporal evolution of the most unstable axisymmetric mode is shown on figure 2. By $tU_0/R = 3.84$, the Kelvin-Helmholtz instability has saturated, and Figure 3(a) shows the tracer distribution in a plane containing the jet axis at that time. This distribution is typical of the Kelvin-Helmholtz instability.

After saturation of the Kelvin-Helmholtz instability, further evolution occurs via instability to non-axisymmetric perturbations. This evolution therefore depends on the non-axisymmetric perturbations present in the initial condition. Figure 3(b) is a 3D visualization of the vorticity field (in black) and of the gradient of the passive scalar (in gray). It shows a case in which an initial perturbation with azimuthal wavenumber $m = \pm 6$ evolves spontaneously into 6 pairs of counter-rotating filaments located between the vortex rings formed by the Kelvin-Helmholtz instability. The rings themselves display a waviness characteristic of Widnall's instability [7]. Note that the disconnected appearance of the rings in figure 3 (b) is due to the threshold applied by the graphics software used to visualize the vector field: only the 4000 largest vectors are shown.

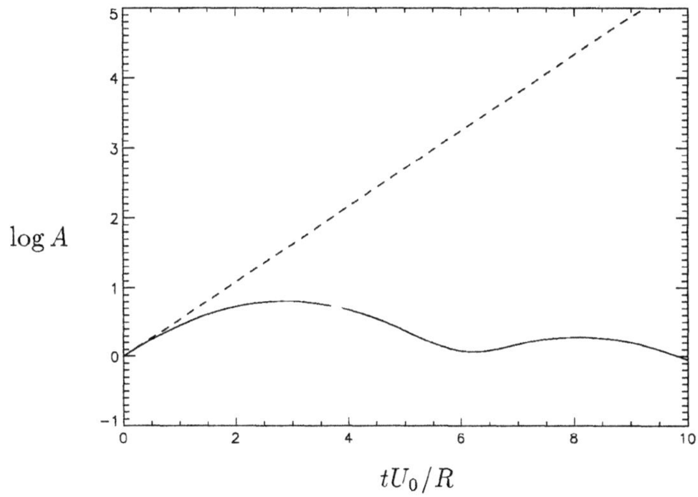

Figure 2 Saturation of the primary instability. Logarithm of the amplitude of the unstable axisymmetric mode as a function of time. The straight line is the extrapolation from linear theory. The initial amplitude of the unstable mode was chosen so that saturation would occur approximately at $tU_0/R = 3.5$.

The gradient of the tracer is seen on figure 3 (b) to be localized between the rings, showing that the basic mechanism for the generation of side jets, as suggested by Monkewitz [3], is the expulsion of fluid by the counter-rotating axial filament pairs. Each axial filament originates outside a vortex ring and terminates inside the following ring along the jet. Detailed examination of the 3D visualizations shows

Figure 3

(a)

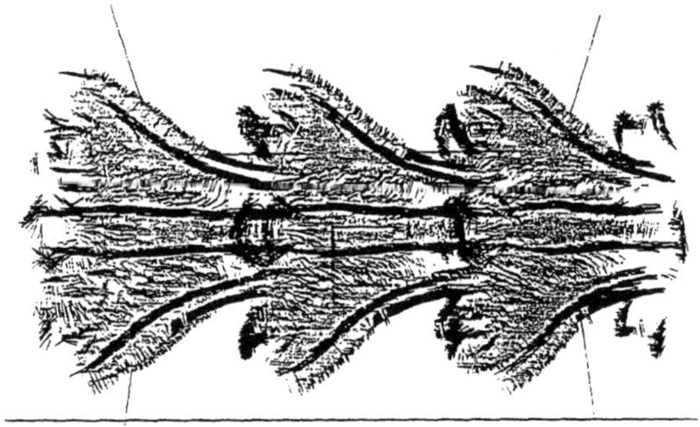

(b)

(c)

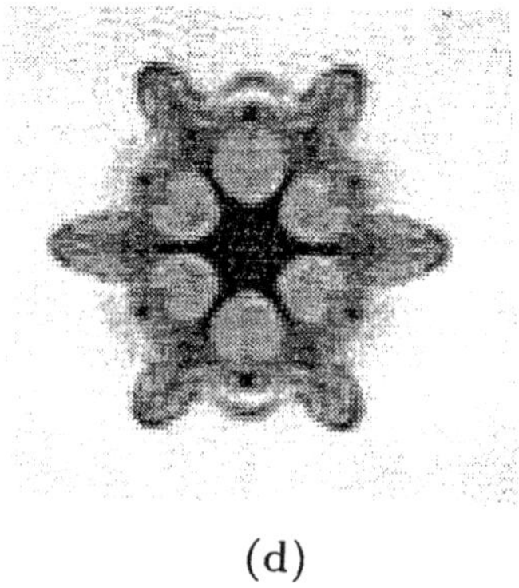

(d)

that the filaments exchange partners as they enter a vortex ring. Here, we witness the tracer being drawn into the jet. These results are confirmed by studying cross-sections perpendicular to the axis of the jet (figures 3 (c) and 3 (d)). Both the side jets and the concentration of the tracer towards the interior of the jet are visible in figure 3 (c), a cross-section near a vortex ring. In figure 3 (d), a cross-section between the rings, the side jets are still present.

In conclusion, we have shown numerically that side jets are generated by the pairs of axial counter-rotating filaments that are in turn produced by the secondary instability of the braid and vortex-ring structure obtained by the saturation of the Kelvin-Helmholtz instability of the jet.

The filaments are thus generated by a mechanism analogous to that which generates axial vortices in mixing layers (1) via condensation of the vorticity ribbons connecting the primary vortices [8], [9] et [10]. The two sides of a mixing layer are, however, symmetric, in contrast to the inside and outside of a jet.

Finally, we suggest that it might be possible to observe these axial filaments experimentally by seeding the flow with micro-bubbles, as has already been done to study turbulence [11].

Acknowledgments

Our computations were carried out on the Cray-2 of the Centre de Calcul Vectoriel pour la Recherche. The 3D visualizations were performed using the VFFS software package developed at CERFACS. This research was supported by DRET contract 90-173. We thank P. Huerre and J. M. Chomaz for giving us a code solving the linear theory of mixing layers by shooting methods, and for useful scientific discussions.

References

-1. P.A MONKEWITZ et al, The spreading of self-excited hot jets by side jets, *Phys. Fluid*, A1, 1989, p. 446-448

-2. C. M. HO et P. HUERRE, Perturbed Free Shear Layers, *Ann. Rev. Fluid Mech.*, 16, 1984, p. 365-424.

-3. P.A MONKEWITZ et E. PFIZENMAIER, Mixing by "side jets" in strongly forced and self-excited round jets, *Phys. Fluid*, A3, 1991, p. 1356-1361.

-4. D. GOTTLIEB et S.A. ORSZAG, Numerical Analysis of Spectral Methods : Theory and Applications, 1977.

-5. A. MICHALKE, Survey on jet instability theory, *Prog. Aeorespace Sci.*, 21, 1984, p. 159-199.

-6. R. W. METCALFE, S. A. ORSZAG, M. E. BRACHET, S. MENON et J. J. RILEY, Secondary instability of a temporally growing mixing layer, *J. Fluid Mech.*, 184, 1987, p. 207-243.

-7. S. E. WIDNALL et al, The instability of short waves on a vortex ring, *J. Fluid Mech.*, 66, 1974, p. 35-47.

---

(1) See [6] (fig. 21 (a,b)) page 238, for the equivalent of side jets in mixing layers.

-8. J. C. NEU, The dynamics of stretched vortices, *J. Fluid Mech.*, 143, 1984, p. 253-276.

-9. J. C. NEU, The dynamics of a columnar vortex in an imposed strain, *Phys. Fluids*, 27, 1984, p. 2397-2402.

-10. S. J. LIN et J. M. CORCOS, The effect of plane strain on the dynamics of stream-wise vortices, *J. Fluid Mech.*, 141, 1984, p. 139-178.

-11. S. DOUADY, Y. COUDER et M. E. BRACHET, Direct observation of intermittency of intense vorticity filaments in turbulence, *Phys. Rev. Lett.*, 67, 1991, p. 983-986.

Parametric forcing of coupled pendula

F. Belin, P. Coullet, T. Frisch and J.C. Robinet

Institut Non-linéaire de Nice,

U.M.R C.N.R.S 129

Universite de Nice Sophia-Antipolis Parc Valrose, F-06034, Nice CEDEX, France

Patterns which arises as the result of the parametric forcing have been particularly well studied in the context of surface waves [1][2][6]. In this paper we study the parametric forcing of a chain of pendula. We report the spontaneous appearance of a structured standing wave, when the forcing is close to twice the natural frequency of the individual pendulum. Futhermore due to the existence of the cut-off frequency homogeneous oscillatory states are also observed. Those states are the analog of the Ising magnetic phase in anisotropic magnets. The transition from modulated patterns to Ising like phases is induced by the boundaries or the defects as usual in first order transition [5].

As a simple model of a medium characterized by a finite cut-off frequency let us consider a chain of pendulas parametrically pumped. In the continuum approximation such a system is described by the equation

$$\Theta_{tt} + \delta\Theta_t + (1 + \varepsilon \sin 2(1 - \nu) t) \sin\Theta = \Theta_{XX} \qquad (1)$$

where δ, ε and ν represents respectively the damping, the amplitude of the forcing and the detuning parameter. The forcing can be easily achieved by a periodic modulation of the gravity. The unforced, undamped system can sustain waves whose dispersion relation is given by

$$\omega = \sqrt{1 + k^2} \qquad (2)$$

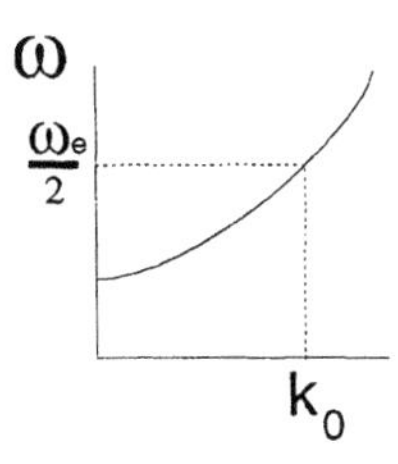

Figure 1 : Graph of the dispersion relation.

E. Tirapegui and W. Zeller (eds.), Instabilities and Nonequilibrium Structures IV, 339–342.

We consider in this paper the case of weak dissipation and weak forcing. In this parameter regime one can look for a solution of Eq. (3) under the form

$$\Theta = Ae^{i(1-v)t} + \bar{A}e^{-i(1-v)t} + \ldots \quad (3)$$

where the dots involve corrections due to higher order harmonics. At the leading order the equation for the amplitude reads[3]

$$A_t = ivA - \mu A + i\alpha|A|^2 A + i\beta A_{xx} + \gamma\bar{A} \quad (4)$$

where $\alpha=-1/4$, $\beta=-1/2$, $\mu=\delta/2$ and $\gamma=-\varepsilon/4$. The stability analysis of the rest state of the chain is straigthforward. Let us write the perturbation of the rest equilibrium solution as $A = X_k e^{\sigma t} e^{ikx} + iY_k e^{\sigma t} e^{ikx}$

The growth rate of such a perturbation then reads

$$\sigma = -\mu \pm \sqrt{\gamma^2 - (\nu - \beta k^2)^2} \quad (5)$$

For negative detuning the instability of the rest state sets in when $\gamma = \mu$. It is characterized by a finite wave number given by the dispersion relation $(\nu - \beta k^2) = 0$.

For positive detuning a zero wave-length instability sets in when $\gamma^2 = \mu^2 + \nu^2$. The homogeneous locked state solution of Eq. (3) reads

$$X_L = -\eta Y_L$$

$$X_L^2 = \frac{(\gamma - \mu)\eta - \nu}{\alpha(1+\eta^2)} \quad (6)$$

where $\eta = \sqrt{(\gamma+\mu)/(\gamma-\mu)}$

Another solution in which $= -\sqrt{(\gamma+\mu)/(\gamma-\mu)}$ exists for some parameter range but is always unstable. An analysis of Eq. (4) shows that the transition to modulated phases is supercritical but the saturation term comes in at fifth order.

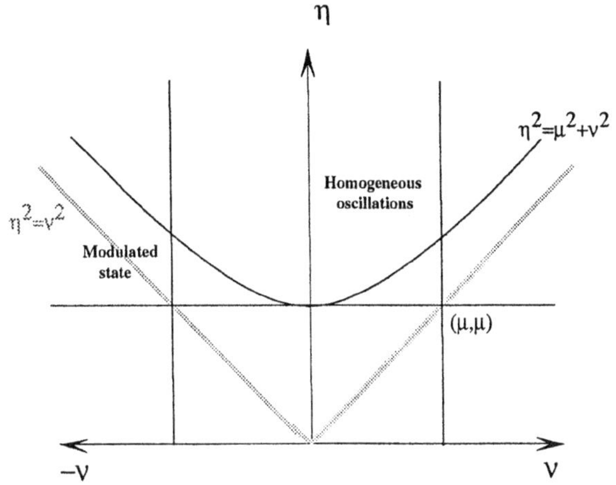

Figure 2 : Stability diagram

The border of the stability domain of this state corresponds to a sub-critical transition towards modulated states.

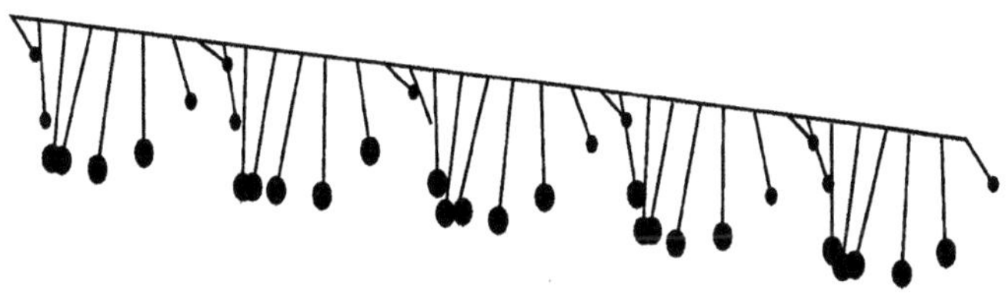

Figure 3 : Modulated State.

When the locked solution is stable, domains in phase opposition are observed. They are reminiscent of the magnetic domains observed in anisotropic magnets. As usual, defects play an important role in the transition. Close to the instability boundaries of the homogeneous locked state the core of the defect consists in a localized standing wave. In the presence of defects the transition from homogeneous oscillations to standing waves occurs as a nucleation of the modulated phases into the homogeneous states.

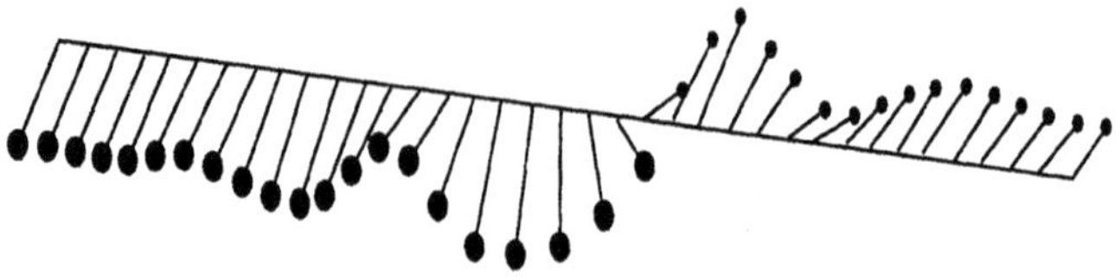

Figure 4. Kink State.

Photosensitive chemical reactions[4], below their oscillation's threshold could be the ideal candidate to observe the phenomenon described in this paper. Other experiment in which our considerations could have some relevance are transverse pattern formation in optics and parametrically driven array of Josephson junctions. Detailed calculation will appear elsewhere.

Acknowledgment: This work has been sponsored by the EEC grant SC1*CT91 083 and SC1*CT92 0006.

Notes and references.

[1] J. Miles and D. Henderson, Ann. Rev. Fluid Mech. 143 (1990).

[2] S. Douady and S. Fauve, Euro. Phys. Lett. **6**, 221 (1988).

[3] A.C. Newell, Lect. Appl. Math. **15** (1974) 157.

[4] S. Muller private communication.

[5] S. Cilliberto, P. Coullet, J. Lega, E. Pampaloni, C. Perez-Garcia, Phys. Rev. Lett. **65**, 2370 (1990).

[6] M. Faraday, Phil. Trans. R. Soc. London, **52**, 319 (1831).

TRACER DISPERSION IN THE TAYLOR-COUETTE INSTABILITY WITH AXIAL FLOW

PIVA, M., CALVO, A., BARRANTES ,A., GABANELLI, S., ROSEN, M.
Grupo de Medios Porosos
Facultad de Ingenieria- Univ. de Buenos Aires
Paseo Colon 850 - 1063 Buenos Aires, Argentine.
IPPOLITO, I. and WESFREID, J.E.
Laboratoire d'Hydrodynamique et Mécanique Physique
Ecole Supérieure de Physique et Chimie de Paris
10, rue Vauquelin - 75231 Paris Cedex 05 - France

1. Introduction

We present experimental results on the mass tracer dispersion in an organized cellular flow , with stationary and propagating cells, arising from the centrifugal Taylor-Couette (TC) hydrodynamic instability [1]. An axial flow is superimposed to the annular cells and a pulse of passive tracer particles is injected in the middle of the cylindrical TC column.
The tracer concentration is measured at the exit of the apparatus and records of concentration as a function of time are obtained. These records are compared with the classical Gaussian profiles obtained when tracers homogeneously disperse in flows. The deviations with respect to Gaussian profiles (anomalous diffusion) are correlated with the observations of the cellular structure obtained by flow visualization methods.
Quantitative results of the dispersion coefficients are obtained fitting the experimental curves with models usually employed for studying tracer dispersion in porous media. So, we can compare analogous behaviors: the tracer trapping in recirculating trajectories, as occurs inside the pores of porous media, and the complex trajectory of tracer particles across the TC cells.
Many interesting physical problems are involved in this experiment:
i) Anomalous diffusion behavior where molecular diffusion near cellular flow separatrices (low velocity regions) produces jumps between neighboring cells [2-5]
ii) The axial flow across the cell induces rich spatio - temporal patterns in the flow as was extensively observed [6-10].
iii)The trajectory of the particles established across the cells and around the separatrices lines, quickly pushing a fraction of the mass towards the ends of the system in a forced way. This advection phenomenon, enhanced by the presence of waves and/or axial flow, is known as "Levy's flight" [11].

2. Experimental Set up

The experimental set up is shown in Figure 1. The inner cylinder rotates with angular velocity Ω, while the outer one and end plates are at rest. The radius of the inner cylinder is $R_1 = 4.5$ cm and the gap is $d = 1.1$ cm, resulting a radius ratio $\eta = R_1/R_2 = 0.8$. The axial aspect ratio L/d = 32 cm/1.1 cm = 29.
A mass flux is imposed in the axial direction, through the gap, by means of an external circulation imposed by a constant pressure head. The flow rate Q is expressed by the dimensionless Reynolds number, $Re = Q / \pi \nu (2R_1 + d)$, where *n* is the kinematic viscosity of the fluid. The inlet of the flux is at the left end and the outlet at the right one.

E. Tirapegui and W. Zeller (eds.), Instabilities and Nonequilibrium Structures IV, 343–349.

During flow visualization experiments the system is filled with a 2 % solution of Kalliroscope tracers in distilled water ($\nu = 1$ cm^2/sec) and the same solution is used for the axial flow.

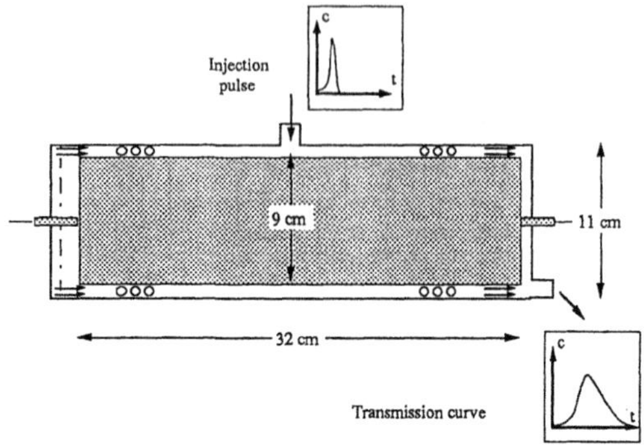

Figure 1: Experimental setup

For the dispersion experiments, the system is filled with distilled water, an axial flow of the same fluid is imposed and a pulse of tracer (solution of fluoresceine in water) is injected at the mid-length of the system. The evolution with time of the concentration of tracer in the outflowing fluid is measured by means of an optical detector located in the outlet edge. In this way, and using a previous calibration performed with cells filled with fluoresceine-water solutions of known concentrations, transmission curves are obtained.

3. Experimental results

3.1 FLOW VISUALIZATION

A set of experiments of flow visualization was performed by filling the system with a solution of Kalliroscope in distilled water. At a given supercritical Taylor number: $Ta = \{d (R_1+R_2)/2\}^{1/2} \Omega d/\nu$, different axial flow rates of the same solution were imposed and the evolution with time of the structure was recorded, at each flow rate, with a video camera. In Figure 2 spatio-temporal diagrams for Ta = 65 and for different Reynolds numbers *Re* are shown. These diagrams, obtained with image processing, represent the temporal evolution of a central line along the axis.

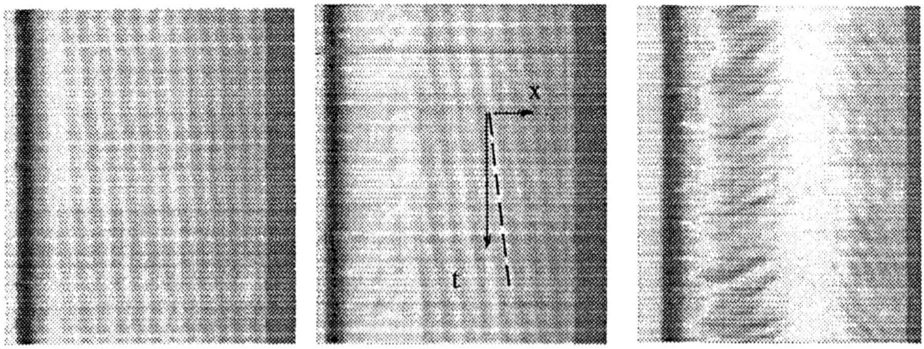

Figure 2 : Spatio-Temporal evolution of a central line : a) Re = 0.5; b) Re = 3.3 and geometric measurement of the phase velocity from the spatio-temporal diagram; c) Re = 9.3

The spatio temporal diagrams also give informations about the advancing vortex velocity: the slope of the lines, computed as shown in Figure 2 b, gives a geometrical measurement of the vortex phase velocity *c*. These results show a linear variation of the velocity with the mean axial velocity U with c ≈ 1.1 U.

It can be seen that the axial flow imposes an amplitude modulation on the velocity field, resulting in a typical healing distance *l*, measured from the inlet edge, in which the structure seems to be destroyed. The experimental variation of *l* with *Re* is shown in Figure 3 where it can be seen that *l* increases with *Re* [9]. In the numerical simulations of Müller et al. [10], a diverging *l* is predicted when the Taylor number approaches the onset Ta_c (Figure 4).

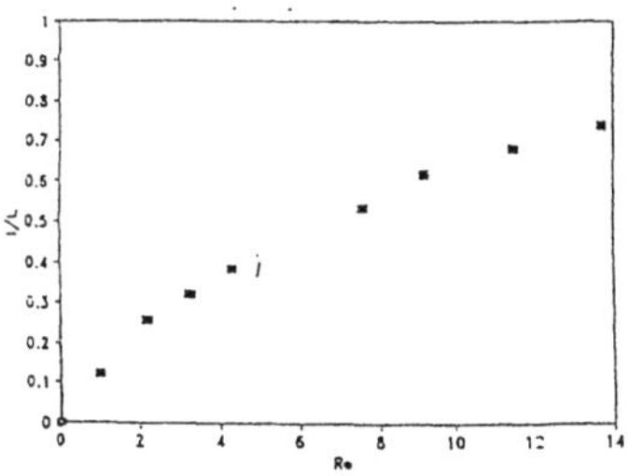

Figure 3 : Healing length in function of the Reynolds number evaluated from the visualizations

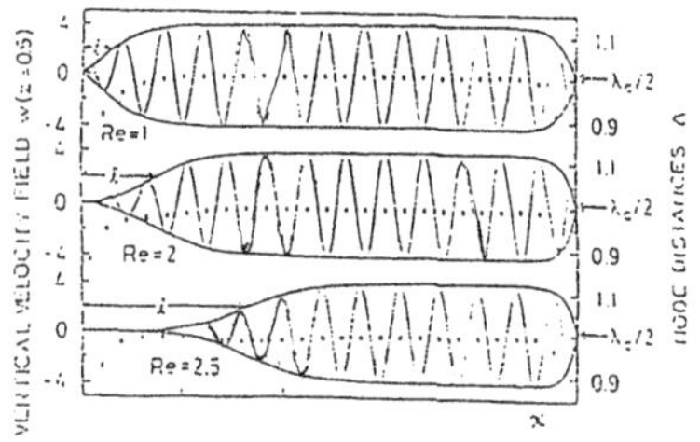

Figure 4 : Numerical simulations of the envelope, after Ref [10]

3.2 DISPERSION EXPERIMENTS

This set of experiments was performed in order to study the dispersion of a tracer through the travelling cellular structure.

As has already been mentioned, the system is filled with distilled water and an axial flow of the same fluid is imposed. A pulse of 0.02 cm^3 of tracer (solution of fluoresceine in water) is injected at the middle length of the structure and close to the inner cylinder. The injection time is very short (~ 5 sec) compared with the tracer mean residence time.

The system is cuter with a sheet of UV light and this unables a direct visualization of the tracer evolution inside the cells.

In Figure 5 a photographic sequence of that evolution, for Re = 0.5, is shown. It can be seen that, in the initial stage of the process, the fluoresceine runs over the vortex boundaries towards the ends of the system, showing a large scale current and producing an hyperdifussive regime (Levy's flight phenomenon).

On the other hand, part of the tracer remains trapped in the injection cell and, as time goes by, invades, by molecular diffusion, the neighbor cells.

At the same time, the structure as a whole moves towards the outlet of the system with the velocity *c* .

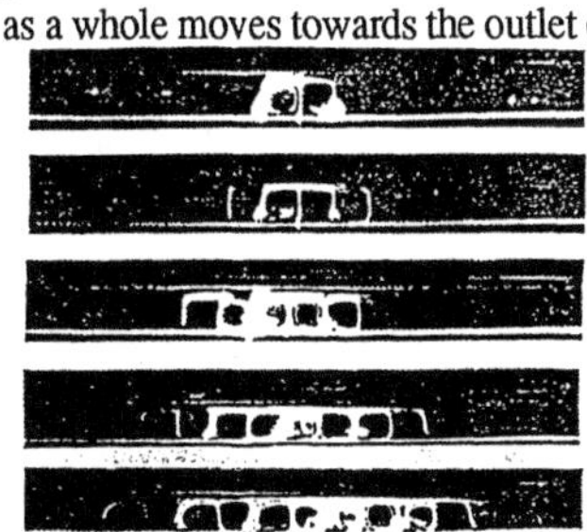

Figure 5: Photographic sequence of the tracer invasion in the rolls

Simultaneously with the direct visualization of the tracer evolution, the tracer concentration is measured at the outlet of the system. By means of an optical detector the tracer concentration of the emerging fluid is recorded as a function of time.
A family of typical transmission curves is shown in Figure 6. It can be seen that, as expected, the tracer transit time decreases as *Re* increases. However, the concentration growth begins nearly at the same time for all *Re*, showing the existence of a short time scale related to a large-scale current. This "arrival time" seems to be weakly influenced by the Reynolds number.

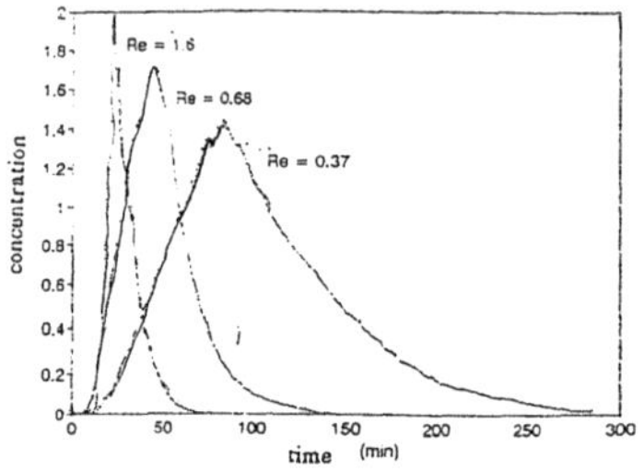

Figure 6: Transmission curves using fluoresceine tracer

4. Analysis of the results

The transmission data were analyzed assimilating the cell structure studied here with a porous medium with stagnation zones. In this way, the present experiment can be compared with a typical porous media dispersion experiment: a pulse of tracer is injected in a saturated porous medium with a continual axial flow imposed, and the tracer concentration of the emerging fluid is measured [12] .
Under this point of view, the transmission curves obtained can be analyzed using two different fitting procedures.
In the first one, it is assumed that the tracer concentration *C* at the outlet corresponds to a solution of a Gaussian convection-difussion equation of the type:

$$\partial C/\partial t + U\ \partial C/\partial x = D\ \partial^2 C/\partial x^2$$

The characteristic parameters in this model are the longitudinal dispersion coefficient *D* and the tracer mean transit time T_o through the porous sample associated with the mean velocity *U*.
This fitting, shown in Figure 7 with large dots, gives an estimation of T_o, computed from the first moment of the distribution curve, resulting in $T_0 \sim 1/\mathrm{Re}$. (Figure 8a)

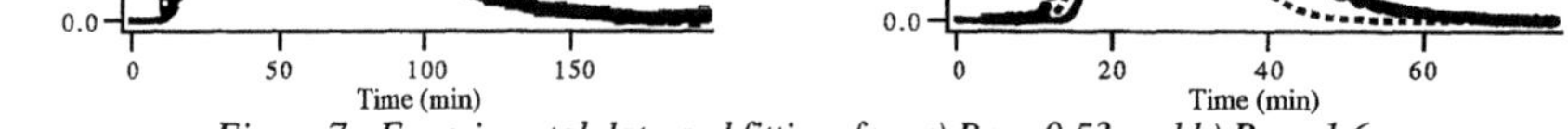

Figure 7 : Experimental data and fitting, for a) Re = 0.53 and b) Re = 1.6

With these values a rough estimation of the volume visited by the tracer, $V_0 = Q\ T_0$, can be made. In the limit $\mathrm{Re} \to 0$, we observe by visualization, than the tracer goes upstream to the left end of the system and then downstream to the opposite end, so $T_0 \to \infty$ and V_0 takes in this case its maximum value: $V_t/2+V_t$. On the other hand, when Re $\rightarrow \infty$, the tracer is quickly advected by the flow and only visits the volume $V_t/2$ included between the injection point and the outlet of the system without backflow motion.

It can be seen in Figure 8 b that, as expected, the values tend asymptotically to 0.5 for Re >> 1 and to 1.5 for Re $\rightarrow$ 0, while unphysical values are obtained with this fitting for Re ~ 0.5

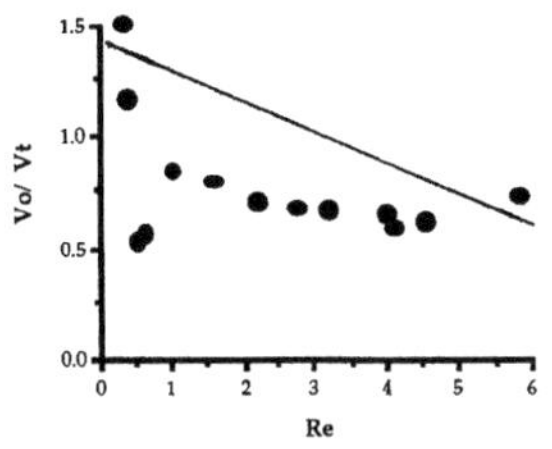

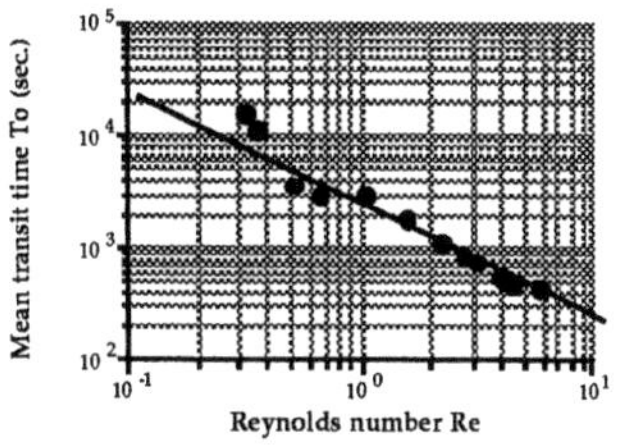

Figure 8 : a) Ratio between $V_o = T_o Q$ and V_t as a function of the Reynolds number and the maxima values V_{max}/V_t b) Mean transit time T_o as a function of the Reynolds number.

The maxima values of the relative volume occupied by rolls V_{max}/V_t, that could be visited by the tracer, was computed with the healing length l (Re) mentioned above, as

$$V_{max}/V_t = \{L/2 + 2\,(L/2 - l)\} / L = 1.5 - 2\,l/L.$$

4.1. ANALOGY WITH POROUS MEDIA WITH TRAPPING ZONES

The presence of a cellular structure (TC rolls), implies the existence of recirculating zones where the tracer can be trapped for certain time. We can assimilate our system to a porous medium with stagnant zones and in consequence, is possible to analyze our transmission curves using mathematical models similar to those used for analyze data obtained in porous media dispersion experiments. As can be seen in Figure 7b, our experimental curves are asymmetric, indicating a non-gaussian or anormal diffusion behavior.

Coats and Smith [13] have proposed a model for dispersion in presence of stagnant zones. This model includes a modified convection diffusion equation and a coupling between the concentration C in the moving and the concentration C^* in the stagnant zones characterized by the mean volume fraction $(1-f)$ occupied by the stagnant zones and the time of exchange $T_r = (1-f)\,T_f$. The coupled equations are:

$$f\,\partial C/\partial t + (1-f)\,\partial C^*/\partial t + U\partial C/\partial x = D_{c\text{-}s}\,\partial^2 C/\partial x^2$$

$$\partial C^*/\partial t = \{(1-f)\,T_f\}^{-1}\,(C - C^*)$$

An additional parameters is the effective dispersion coefficient $D_{c\text{-}s}$ corresponding to the moving region. From this last three parameters, one can compute, in the case of very long samples [14], an effective "asymptotic" dispersion coefficient D_{as}, related to the second moment of the experimental curve, and representing the total dispersion effects: those associated to the moving volume and those associated to the exchange between moving and stagnant zones. In dimensionless notation, where U and L are the characteristic velocity and length, this can be written as:

$$D_{as} / UL = D_{c\text{-}s} / UL + (1-f)\,T_r/T_o$$

This dispersion coefficient, can be understood as the dispersion coefficient that would have a gaussian curve of the same T_o and variance as the experimental curve.

The experimental transmission curves were adjusted with this model and the results are shown in Figure 7 with small dots, for 2 different values of the Reynolds number.

It can be seen that for large *Re* the experimental curves are far from the gaussian fitting, showing long tails and a good agreement with the Coats & Smith model.
In Figure 9 we have plotted D_{as} / UL and we observe than this value is approximately constant for Re > 1.
In Figure 10, the values of the trapping volume fraction *(1-f)* are plotted as a function of the Reynolds number. For Re > 1, the convected fraction of the volume becomes constant.

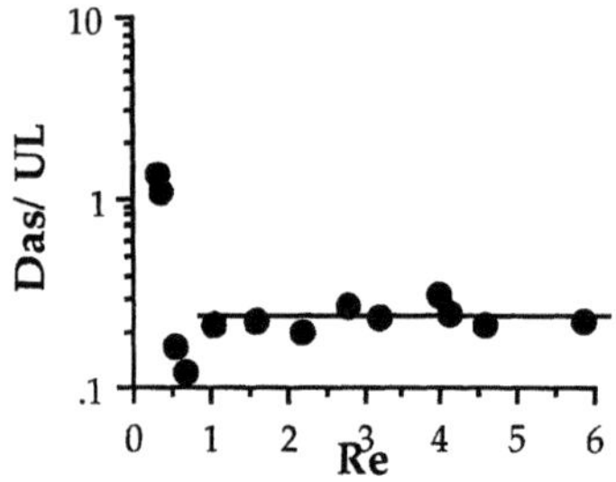

Figure 9 : Dimensionless effective dispersion coefficient Das / UL as a function of the Reynolds number

The trapping typical time, $T_r = (1\text{-f})\ T_f$, decays as $T_r \sim 1/\text{Re}$ (Figure 10b), showing the advective regime for the tracer. For the other side Tr is greater for low values of the axial flow, showing the way as the tracer is trapped in the rolls as occurs in recirculating regions.

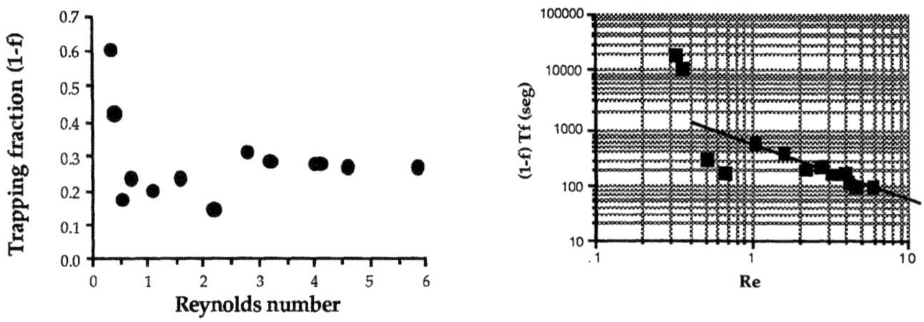

Figure 10: a) Trapping volume fraction and b) Trapping time as a function of the Reynolds number.

5. Conclusions

In this paper we have presented a first experimental work about passive tracer displacement pushed by an axial flow across a cellular structure (TC rolls) at constant Taylor number.
From the results of the Figure 9 we observe than D_{as} grows linearly with the mean velocity. In porous media dispersion experiments, this type of dependence reveals a geometrical dispersion regime, with dispersion coefficients $D = U\ \zeta$, where ζ is the typical length in the fluid trajectory visited by the tracer during the random exchange between recirculating zones (rolls) and the axial flow across the separatrices in the gap [12][14]. We estimated this value as $\zeta = 0.24L = 7.7\text{cm} \sim 3\ \lambda$, where λ is the wavelength of the rolls.
We also note that the increasing of the coefficient Das with the Reynolds number, shows the strong

capacity of effective mixing of the axial flow when Re >1.
The asymmetric shape of the transmission curves, show in the Figure 7, is due to the fact that around 70% of the mass (value of *f*) is pushed through the sinuous separatrices way, while the remaining 30% is transitory trapped in the rolls contributing to the long tails of the non gaussian transmission curves. This anomalous behavior may be quantified by means of the exponent b in the general law $<x - x_0>^2 = D\,t^{1+\beta}$, where β is a measure of the deviation from normal behavior. Analysis of our results in order to estimate β are in progress using the third moment of the transmission curves and also fractional diffusion laws [15-16].

ACKNOWLEDGMENT

This work has been supported by the cooperation program PICS between CNRS (France) and CONICET (Argentine) We are grateful to J.P.Hulin for stimulating discussions and help.

6. References

[1] Drazin P.G. and Reid W.H. *Hydrodynamical Instabilities* Cambridge Univ.Press (1981)
[2] Pomeau Y. C R Acad. Sc. Paris **I 301**, 1323 (1985)
[3] Solomon T. and Gollub J. P. Phys. of Fluids **31**,1373 (1988).
[4] Cardoso O. and Tabeling P. Eur. J. Mech. B / Fluids **8**, 459 (1989)
[5] Barrantes, A., Calvo A., Rosen, M. and Wesfreid, J.E. in *Instabilities and Nonequilibrium Structures III* ed. by Tirapegui, E. and Zeller, W. Kluwer Ac. Pub. p. 233 (1991)
[6] Bühler, K. and Polifke, N. in *Nonlinear Evolution of Spatio-Temporal Structures in Dissipative Continous Systems* ed. by Busse, F.H. and Kramer, L.,Plenum p. 21 (1990)
[7] Tsameret A. and Steinberg, V. Europhys. Letters **14**, 331 (1991)
[8] Takeuchi, D.I. and Jankowski, D.F. J. Fluid Mech. **102**, 101 (1981)
[9] Babcock K.L. , Cannell D.S. and Ahlers G. Phys. Rev. Lett. **67**, 3388 (1991)
[10] Müller H.W., Lücke M. and Kamps M. Europhys. Letters **10**, 451 (1989)
[11] Levy P. *Processus Stochastiques et Mouvement Brownien* Gauthier - Villars (1948)
[12] Bacri J.C., Bouchaud J.P., Georges A., Guyon E., Hulin J.P., Rakotomalala N. and Salin D. in *Hydrodynamics of dispersed media* ed. by Hulin,J.P., Cazabat, A.M., Carmona F. and Guyon,E. North-Holland p. 249 (1990)
[13] Coats K.H. and Smith B.D. Soc. Pet. Eng. J. Trans. AIME **231**, 73 (1964)
[14] Charlaix E., Hulin J.P. and Plona T.J. Phys. of Fluids **30**, 1690 (1987)
[15] Mastrangelo V. private communication
[16] Lenormand R. Report 39541 Inst. Français du Pétrole (1992)

Critical Velocities and Nucleation of Vortices in a Model of Superflow

Sergio Rica* Yves Pomeau†

Abstract

We study the superfluid flow around an obstacle and observe a transition to dissipation which is explained by the nucleation of quantized vortices. This transition between a potential flow state, dominated by sounds waves (phonons), toward a vortical state occurs at a well determined velocity threshold. We study the dependence of the critical velocity in different geometries in 2 or 3 spaces dimension. We also report a new mechanism of nucleation of a pair of vortices (vortex-antivortex) from the collision between (very short wavelength) sound waves.

1 Introduction

The theory of superfluidity in He II of Landau introduces the concept of "critical velocity" [1]. At sufficiently small temperatures this liquid can flow without any energy dissipation if its velocity is small. The usual picture assumes an energy spectrum of excitations of the quantum liquid $\varepsilon(p)$, $\vec{p}$ linear momentum of this excitation. As a consequence of the energy and momentum conservation laws for an excitation interacting with a moving body, the superfluidity of the system disappears if the velocity v of the body is bigger than a critical velocity v_c defined by

$$v_c = min\, \frac{\varepsilon(p)}{p}, \qquad (1)$$

in other words this critical velocity is such that the phase and group velocity are equal.

*Institut Non Linéaire de Nice, Nice, France.
†Laboratoire de Physique Statistique, Ecole Normal Supérieure, Paris, France.

E. Tirapegui and W. Zeller (eds.), Instabilities and Nonequilibrium Structures IV, 351–364.

If the only posible excitations are those of the Landau's spectrum[1], the critical velocity would be the threshold for roton nucleation.

$$v_c^{roton} = min\ \frac{\varepsilon(p)}{p} \approx 60\ m/sec. \tag{2}$$

However experiments in superfluid helium show the existence of dissipation of energy for velocities down to 1 cm/s.

Another picture show that the superfluidity breaks due to the creation of quantized vortices[2]. The classical experience is the problem of a superflow that flows by an orifice out of a tank. In this case we have that near the wedges there are big accelerations which permit the creation of vorticity, like a macroscopical ring which radius is of order of the length of the orifice. The calculus of the critical velocity for generate a vortical state give us (Feynman, 1955)

$$v_c^{vortex} = \frac{\hbar}{md}\log(d/a) \tag{3}$$

where d is the size of the orifice and a is the core size of the vortex. For $d = 10^{-5}\ cm,\ a = 10^{-8}\ cm,\ v_c^{vortex} \approx 100 cm/sec\ (\hbar/m = 1.5\ 10^{-4} cm^2/sec)$[3].

We will study the superfluid flow around an obstacle. We show that there is a criterium "a la Landau" for the critical velocity without any quantum fluid properties of the liquid, as assumed in the Landau theory of superfluidity [1]. Also we present a mechanism to reduce the critical velocities. Later we consider the big acceleration induced when the fluids flow near pointed objects. If we submerge an obstacle, with arbitrary form, the speed of the fluid on the surface can be several order of magnitudes bigger than the velocity of the current, then we believe that little pointed obstacles increase the local velocity[4].

[1]The form of such a spectrum was suggested by Landau in 1947 [2] from the data of the thermodynamical properties of He II. Feynman proposed a qualitatively theory os such spectrum [3]. This spectrum was verified experimentally via neutron scattering [4].

[2]Quantized vortices were suggested by Onsager (as a footnote in an article on classical theory of vortices [5]) in 1949 and Feynman in 1955 independently [3].

[3]We note that the expression for the critical velocity follows from the values of the energy and momentum of a ring of vorticity of radius R, $E(R) \approx R\log R$, $p \approx R^2$ but $R \approx d$ since the nucleated vortex has a maximun size of the order of the channel size.

[4]"Little" means that their characteristic size is of the order of magnitude of the coherence length, which is the characteristic size of a quantized vortex core ($\approx 1\AA$). This object may be crystals of water or nitrogen which a size of a hundred of Amstrongs.

2 The Non-Linear Schrödinger equation (NLS): A model of Superflow

2.1 The weakly ideal Bose gas

We consider the "Bogoliubov's superflow" which is a quasi-ideal bosonic gas introduced for the first time in 1947 to explain in a macroscopic way the phenomenon of superfluidity [6]. The interaction between the particles of the gas is repulsive[5], at zero temperature a big percentage of the particles are in the ground state, which is considered homogeneous by Bogoliubov. If we suppose that this fundamental state $\hat{a}_0$ is described by complex numbers[6] we compute the perturbations (often called excitations) to the ground state, the result is an ideal gas of quasiparticles[7] which energy spectrum coincides with Landau's for a long wavelength perturbations and these perturbations cannot be interpreted as atoms. A rederivation of Bogoliubov's ideas was made on the basis of a non-homogeneous fundamental state by Gross and Pitaevskii [8, 9]. Of course this picture of a superflow can represent in detail neither the He II nor the famous Landau' spectrum. We hope however and claim that certain robust phenomena in neutral superflows are easier to treat in this framework and in certain cases the details of the calculus can be done to the end.

2.2 Properties of NLS

The Gross-Pitaevskii equation becomes the Non-Linear Schrödinger equation when the interactions between bosons are localized, and can be written after proper rescaling[8] :

$$i\partial_t\psi(\vec{x},t) = \frac{1}{2}\nabla^2\psi - |\psi|^2\psi; \tag{4}$$

ψ is the condensate wave function and $|\psi|^2$ means it density[9].

NLS has the following properties:

[5] More precisely the mean of the interaction.

[6] In fact the fundamental state conmute with its respective creation operator $\hat{a}_0^\dagger$ at order $1/N$, N the number of particles.

[7] Ideal meens that we neglected the interaction betwen three quasiparticles, which in a quantum picture drive the system to a termodynamic equilibrium [6].

[8] The Gross-Pitaevskii equation reeds in the correct units: $i\hbar\partial_t\psi = \frac{-\hbar^2}{2m}\nabla^2\psi + V|\psi|^2\psi$, V meens the mean of the interparticle interaction and m the He^4 mass.

[9] N.B. The condensate appears at the same time than the superfluid and the velocity of the condensate is the same as the one of the superfluid. However the density of the condensate is not the one of the superfluid.

-conserved quatities $N = \int |\psi|^2 d\vec{x}$, $H = \frac{1}{2}\int(|\nabla\psi|^2 + |\psi|^4)d\vec{x}$

-reversible in time $\psi(\vec{x},t) \rightarrow \psi^*(\vec{x},-t)$.

-Galilean invariance $\psi(\vec{x},t) \rightarrow \psi(\vec{x}-\vec{v}t,t)e^{i(\vec{v}\cdot\vec{x}-\frac{1}{2}v^2t)}$

-if $\psi = \rho^{1/2}e^{i\phi}$, NLS can be trasformed in a set of hydrodynamical type equations, $\rho = |\psi|^2$ being the fluid density and the phase ϕ of ψ the velocity potential, i.e. $\vec{v} = -\vec{\nabla}\phi$, we obtain the continuity and Bernoulli equations,

$$\partial_t\rho = \vec{\nabla}\cdot(\rho\vec{\nabla}\phi); \tag{5}$$

$$\partial_t\phi = \rho + \frac{1}{2}(\vec{\nabla}\phi)^2 - \frac{\nabla^2\rho^{1/2}}{2\rho^{1/2}}. \tag{6}$$

-NLS is invariant in a global phase change $\psi \rightarrow \psi e^{i\alpha}$ and scale $\psi(\vec{x},t) \rightarrow \lambda\psi(\lambda\vec{x},\lambda^2t)$, α,λ real. This implies that the long wavelength perturbations of phase and modulus decay weakly. In particular we have phonons which satisfy the Bogoliubov's relation of dispersion ($\omega_k^2 = c_s^2k^2 + \frac{1}{4}k^4$). The perturbation of phase and modulus around a uniform solution $\psi_0 = \rho_0^{1/2}e^{i\rho_0t}$ satisfies waves equations, eg. for ϕ we have

$$\frac{\partial^2\phi}{\partial t^2} = c_s^2\nabla^2\phi; \tag{7}$$

with $c_s^2 = \rho_0$.

-One of the great success of this model is the existence of spatially extended quantized vortices [7, 8, 9]. The solutions describing topological defects of NLS are given by,

$$\psi = R(r)e^{i(m\varphi+\rho_0t)} \tag{8}$$

with m an integer, R satisfy an ordinary differential equation of second order with boundary conditions: $R(0) = 0$, $R(\infty) = \sqrt{\rho_0}$, r and φ the polar coordinates centered at the core of the vortex.

3 Drag on objects in NLS and Critical Velocity

We realized the following numerical experiment: a disc is "submerged" in a "superfluid bath" which satisfies NLS and moves with a speed v toward the disc. It is observed that the drag on the disc is zero till a critical velocity v_c, then the drag grows linearly in $v - v_c$. The threshold of dissipation is simultaneous with the nucleation of vortices in the perimeter of the disc. In an ordinary fluid we can see that near the critical velocity a shock wave begins to form, but the existence of such shock waves is forbidden because there is not explicit dissipation in NLS then the shocks waves are replaced by vortex emission. Now the fluid is in a vortical and dissipative state [10]. Below the threshold for vortex nucleation we have that the velocity potential ϕ and the density ρ are steady solutions, then the drag is null by the d'Alembert paradox for perfect fluids.

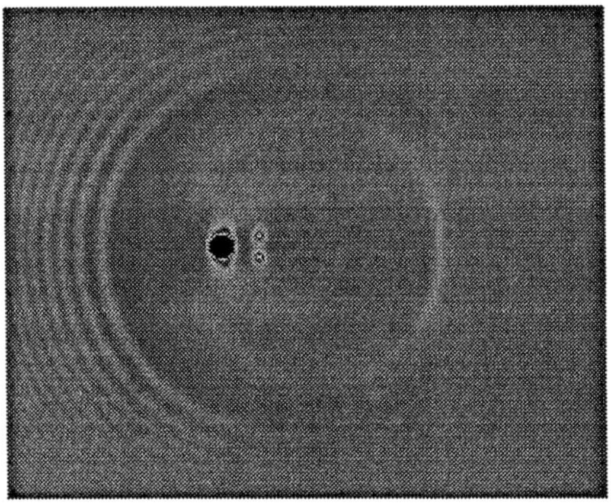

Figure 1: Numerical solution of NLS of a disc in a flow from the left. One represents the modulus of ψ. The disc is black and the boundary conditions is $\psi = 0$. The vortices are the white dots on the right of the disc and they are convected by the flow. The sound waves are only transients and they are not relevant for the long time evolution.

This transition from an irrotational to rotational fluid can be undertood in the following way:

Below the critical velocity the fluid is in a steady state, if we neglect the compressibility effects (variations of ρ) the phase satisfies the Laplace equation ($\nabla^2\phi = 0$) which solution is $\phi = v_\infty r \cos\varphi(1 - \frac{d^2}{r^2})$, v_∞ the velocity of the flow at infinity, d the radius of the disc,

$(r,\ \varphi)$ the usual polar coordinates centered at the center of the disc. It is easy to see that the local speed has a maximum value at the extremum superior and inferior of the disc and this value is twice the velocity of the fluid at infinity, then if the flow velocity is $0.5c_s$ we will have a shock wave at the extremum of the disc.

In the figure 1 we see a pair of vortices emitted which are advected by the current, while another pair begins to form at the points of maximun velocity which is of the order of the sound speed[10]. The field velocity generated by the emitted pair decreases the local velocity along the perimeter of the disc, then for a certain time the totality of the fluid is subsonic[11]. Only when the distance between the pair and the disc increases sufficiently the local velocity becomes supersonic and at the perimeter we will have nucleation of a new vortex pair. Qualitatively calculations give us that if τ is the period of vortex emission, then $\tau \approx \frac{1}{v-v_c}$, moreover if F is the drag on the disc $F \approx (v - v_c)$ [10].

The fact that the frequency of vortex nucleation is proportional to the bifurcation parameter $v - v_c$ suggests us that the transition from a potential fluid to a rotational ones follows via a Hopf bifurcation

$$\partial_t A = (v - v_c)(1 + i\nu)A - (1 + i\alpha)|A|^2 A; \tag{9}$$

where A is the order parameter and $\alpha \& \nu$ two real numbers. A solution of 9 is $A(t) = \sqrt{v - v_c} e^{i(v-v_c)(\nu-\alpha)t}$. then we see that the frequency of vortex emission is $\omega = (v - v_c)(\nu - \alpha)$. The resistence $F \approx |A|^2$.

4 Nucleation of vortices as a fustrated shock wave formation.

The validity of the Laplace equation for the potential flow ϕ is restricted to stationary solutions only and when we neglect the variations of the density ρ. Of course this is possible only if the Mach number is small. When the velocities are comparable with the sound speed c_s it is necessary to take into account the phenomena with density variations (shock waves).

We neglect the time scales of evolution in comparison to the spatial phenomena also we have that the long wave variations of the density are smaller than long wave variations of the phase, then we neglect the "quantum pressure" $(\frac{\nabla^2 \rho^{1/2}}{2\rho^{1/2}})$ in the hydrodynamical

[10] We remark that the velocity at the center of a vortex is of the order of sound speed c_s.

[11] The only exception is in the vortex core.

equations (5,6). Time derivatives will be $\partial_t \rho \approx 0$ and $\partial_t \phi = \rho_0 + \frac{1}{2}v_\infty^2$, where ρ_0 and v_∞^2 are the density and speed of the superflow at infinity. This approximation leads to:

$$\vec{\nabla} \cdot (\rho \vec{\nabla} \phi) = 0 \tag{10}$$

and the density ρ is given by:

$$\rho = \rho_0 + \frac{1}{2}v_\infty^2 - \frac{1}{2}(\vec{\nabla}\phi)^2. \tag{11}$$

Now our problem reduces to solve the non-linear equation (10) with the adequate boundary condition. With the method of the hodograph transformation[12] we reduce the complicated non-linear equation (10) to a linear equation and show that the maximal velocity is reached on the boundary of the obstacle. Also if is easy to see that the partial differential equation (10) has a transition from an elliptical equation to a hyperbolic equation for a value of the speed $v = |\nabla \phi|$. It happens that this qualitative change in the PDE is well defined by the nucleation of vortices.

The equation (10) is elliptic if $\rho > 0$ and $\rho + v\rho' = d(\rho v)/dv > 0$. which gives the condition

$$\rho_0 + \frac{1}{2}v_\infty^2 - \frac{3}{2}v^2 > 0. \tag{12}$$

The boundary condition in the hodograph plane is complicated, and nonlinear, but we obtain as a general and essential result that the velocity attains its maximun at the perimeter of the obstacle [10, 11].

If we approximate the maximum value of the speed on the perimeter of the disc by twice its value at infinity $v = 2v_\infty$, we obtain from (12) that the equation (10) is elliptic for all v if the mean flow speed v_∞ satisfies: $\rho_0 + \frac{1}{2}v_\infty^2 - \frac{12}{2}v_\infty^2 > 0$, $\rho_0 = c_s^2$. Simple algebra gives us $v_\infty/c_s < \left(\frac{2}{11}\right)^{1/2} = 0.426$, which is less than the previous value 0.5 found in the incompressible approximation. The value $v_c = 0.426\ c_s$ agrees very well with the numerical simulations [10]. We note that the critical velocity does not depend on the radius of the disc d, this independence is expected to be valid if $d >> a$ (a size of a vortex core) only. Numerical experiments on a three dimensional tapered cylinder are in agreement with the analytical estimations.

[12] See Ref. [11] and [12] for details.

5 Critical velocity for a pointed object

As the speed in the superflow is bigger on the obstacle it is of general interest to find the form of the obstacle which causing the biggest velocity. We will study here wedges obstacles in two space dimensions and pointed obstacles in three space dimensions.

5.1 Case 2-d

Here we study the problem of the critical velocity where the superfluid flows around an edge. Let v_∞ flow at infinity and there is a point O where the plane changes its orientation. Let α the external angle between the two planes, the flow potential for the incompressible case, i.e. when ϕ satisfies the Laplace equation, was calculated for the first time by Maxwell, and has a well known solution via the Schwarz-Christoffel transformation, near the point O the value of ϕ is

$$\phi = v_\infty d \left(\frac{r}{d}\right)^{\frac{\pi}{\alpha}} \cos(\frac{\pi}{\alpha}\varphi); \tag{13}$$

where d is a characteristic length of the geometry, (r, φ) the polar coordinates centered in O. Therefore the velocity near the point O ($r = 0$) is

$$v \approx v_\infty \left(\frac{r}{d}\right)^{\frac{\pi}{\alpha}-1}; \tag{14}$$

then if $\alpha > \pi$ the velocity diverges for $r \to 0$. Then the velocity near the point O will be very large[13]. When the velocity is of order c_s the fluid nucleates vortices. the condition (12) gives the following relation[14]

$$v_c \approx c_s \left(\frac{a}{d}\right)^{1-\frac{\pi}{\alpha}}. \tag{15}$$

If $\alpha = \frac{3\pi}{2}$, i.e. for instance the Feynman's problem in 2-d, which we discused at the introduction, we have[15]

$$v_c \approx c_s \left(\frac{a}{d}\right)^{\frac{1}{3}}. \tag{16}$$

[13]The velocity does not diverge since the approximation of incompressible flow breaks down for $r \approx \mathcal{O}(a)$.

[14]We take $r \approx \mathcal{O}(a)$ how we have said above.

[15]This velocity is different of the critical speed for nucleation of a vortex out of a circular hole in 3d, which-as explained in footnote 3, should be $v_c' = c_s \frac{a}{d} \log(d/a)$, according to Feynman. However, it must be noticed that the two velocities (v_c and v_c') have a different meaning, often overlooked in the litterature. The velocity v_c is the one beyond which a vortex is nucleated because of the loss of a stationary solution. On the contrary, this velocity v_c' is the one beyond which it is energetically favorable to create a fixe vortex ring somewhere in the fluid. In general v_c' will not be the actual velocity of nucleation of a vortex, because of a very large energy barrier to be crossed to reach the state of lowest energy.

The case of a barrier ($\alpha = 2\pi$)

$$v_c \approx c_s \left(\frac{a}{d}\right)^{\frac{1}{2}}. \tag{17}$$

This law for the critical velocity (17) was verified in a direct simulation of NLS on a Connection Machine (see figure 2).

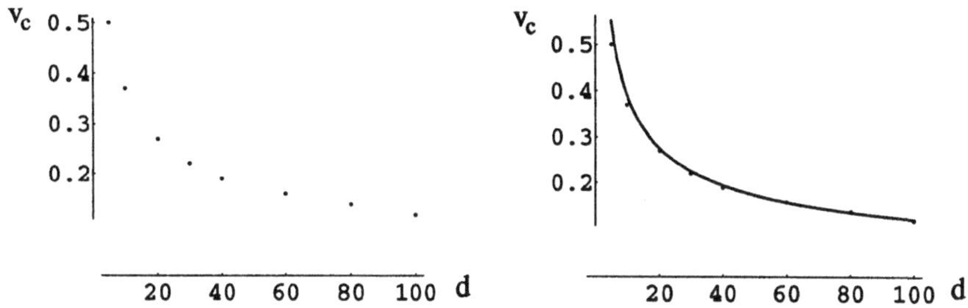

Figure 2: Dependence of the critical velocity as a function of the height of the barrier. The dots are the data from a numerical experience nad the continuos line are the fitting $v_c = (0.66\ d)^{-\frac{1}{2}}$ with $c_s = \sqrt{2}$.

It is interesting to note that in figure 3 we observe in front of the barrier the formation of dipoles of two quantized vortices by the scattering of very short wavelength phonons and they are stable due to a kind of turbulence.

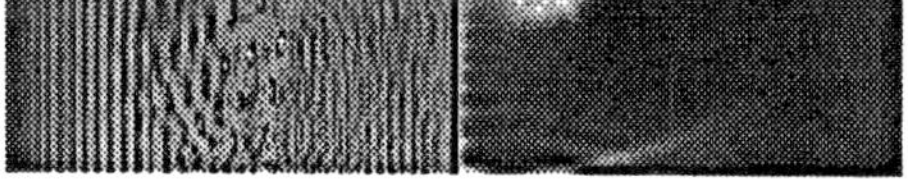

Figure 3: The flow (from left to right) through a barrier (vertical line in black), the vortices are nucleated and convected away.

5.2 Case 3-d

For potential problems in three space dimensions we cannot apply methods of complex analysis, then it is necessary to solve the partial differential equations directly. As an example we will determine the formula (13) via a solution of a fundamental eigenvalue problem. As an example of the method, let us solve the 2d edge problem without complex variables. Let us solve the Laplace equation with Neumann boundary condition in $\varphi = 0$ and $\varphi = \alpha$. Due to the symmetries we separe the variables and try a solution $\phi = r^{\ell} y(\varphi)$. The problem is reduced to

$$\frac{d^2y}{d\theta^2} = -\ell^2 y, \quad y'(0) = 0, \; y'(\alpha) = 0 \tag{18}$$

an elementary calculation gives $y = \cos(\ell\theta)$ and $\ell = \frac{\pi}{\alpha}$, then ϕ agree with (13), when one imposes the boundary condition at infinity.

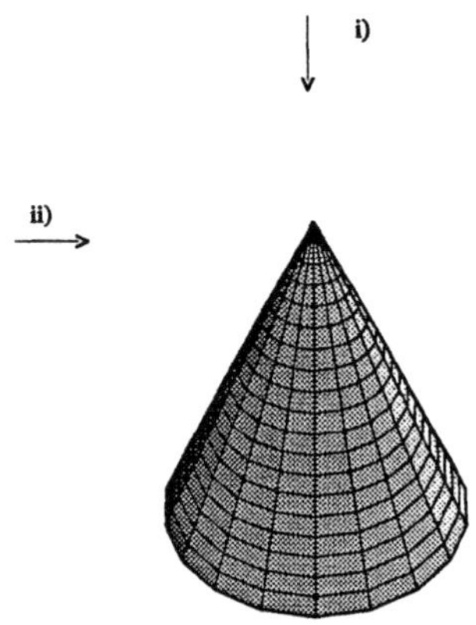

Figure 4:

In the case of singularities in 3-d we will use a similar method. We will restrict our selves tosurfaces with conical points, i.e. there is a point O who generates a curve in the solid angle (the unitary sphere), the pointed obstacle is the surface generated between O and the curve on the sphere. The Laplace equation is separable in the same way: $\phi = r^{\ell} Y(\theta, \varphi)$, where θ, φ are the polar and azimuthal angles, usual in spherical coordinates, respectively.

The spherical harmonics $Y(\theta,\varphi)$ satisfy the equation:

$$L^2Y = \ell(\ell+1)Y, \quad L^2Y = -\frac{1}{\sin\theta}\frac{\partial}{\partial\theta}\left(\sin\theta\frac{\partial Y}{\partial\theta}\right) - \frac{1}{\sin^2\theta}\frac{\partial^2 Y}{\partial\varphi^2} \tag{19}$$

and a Neumann boundary condition on the surface of the obstacle.

Two cases have a simple solution:

i) Flow parallel to the polar axis.

By symmetry Y does not depend on the variable φ, then Y has to satisfy:

$$\frac{1}{\sin\theta}\frac{\partial}{\partial\theta}\left(\sin\theta\frac{\partial Y_\ell}{\partial\theta}\right) + \ell(\ell+1)Y_\ell = 0, \tag{20}$$

the solution being a generalized Legendre polynomial for ℓ real. The regular one in $\theta = 0$ is

$$Y_{\ell\,0}(\theta) = {}_2F_1(-\ell,\ell+1;1;(1-\cos\theta)/2); \tag{21}$$

where ${}_2F_1$ is a hypergeometric function, and ℓ is fixed by the boundary condition

$$\frac{dY_{\ell,\,0}}{d\theta}(\theta=\alpha) = 0; \tag{22}$$

which gives $\ell = \ell(\alpha)$. Using some properties of hypergeometric functions the boundary condition can be written as

$${}_2F_1(1-\ell,\ell+2;2;\sin^2\alpha/2) = 0. \tag{23}$$

It is possible to solve asymptotically this equation for α near zero and near π

$$\ell = \frac{x_1}{2\sin\alpha/2} - \frac{1}{2}, \quad \alpha \to 0; \tag{24}$$

$$\ell = 1 + \frac{1}{4}(\alpha-\pi)^2, \quad \alpha \to \pi; \tag{25}$$

$x_1 = 3.8317$ is the first zero of the Bessel function J_1. Also it is easy to see that $\ell = 2$ for $\alpha = \pi/2$.

Via Mathematica one can compute the function $\ell(\alpha)$. Since $v \approx r^{\ell-1}$ with $\ell > 1$ we conclude that the velocity field never becomes singular[16].

ii) Flow perpendicular to the polar axis (See figure 4, case ii).

[16] In the case of electrostatic fields we have a Dirichlet boundary condition which changes completely this situation in fact we have an infinite electric field at the point of a perfect conductor.

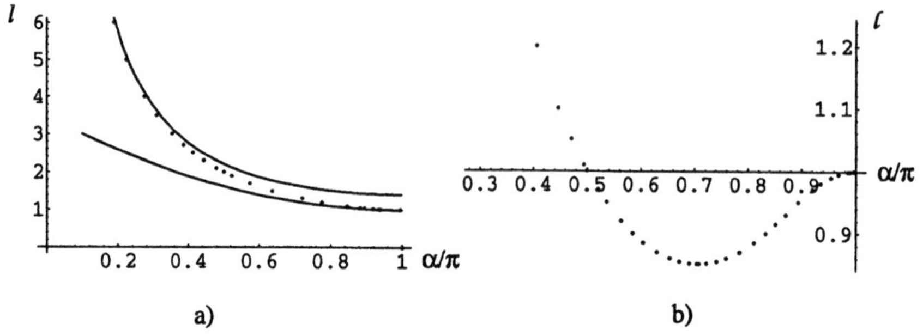

Figure 5: a) The function $\ell(\alpha)$ for the case i). The dots are the exact values and the other two lines are the asymptotic expression (24) & (25). b) Ibid. but the case ii).

From the problem of a flow around a disc in two dimensions we expect that the function Y will be proportional to $\cos\varphi$. Then we find a spherical harmonic with $m = 1$

$$Y = Y_{\ell,\,1}(\theta,\varphi) = (\cos\theta)\ {}_2F_1(1-\ell,\ell+2;2;(1-\cos\theta)/2)\cos\varphi; \tag{26}$$

and the boundary condition is written as

$$\frac{\partial}{\partial\theta}\left(\cos\theta\ {}_2F_1(1-\ell,\ell+2;2;(1-\cos\theta)/2)\right)(\theta=\alpha) = 0. \tag{27}$$

With Mathematica we obtain a plot $\ell = \ell(\alpha)$ drawn in figure 5.b.

For $\theta > \pi/2$, $\ell < 1$ and since $v \approx r^{\ell-1}$ is singular for $\theta > \pi/2$ but the degree of singularity is not big since there is a $\ell_{min} = 0.851$ minimun which is for an angle $\alpha_{min} = 0.706\pi = 2.218$. Near $\alpha = \pi$ the asymptotic expansion

$$\ell = 1 - \frac{1}{4}(\alpha-\pi)^2, \quad \alpha \to \pi; \tag{28}$$

Therefore we conclude that the maximum singularity which is possible is of the form

$$v \approx v_\infty \left(\frac{r}{d}\right)^{-0.149}; \tag{29}$$

then the critical velocity will be

$$v_c \approx c_s \left(\frac{a}{d}\right)^{0.149} \tag{30}$$

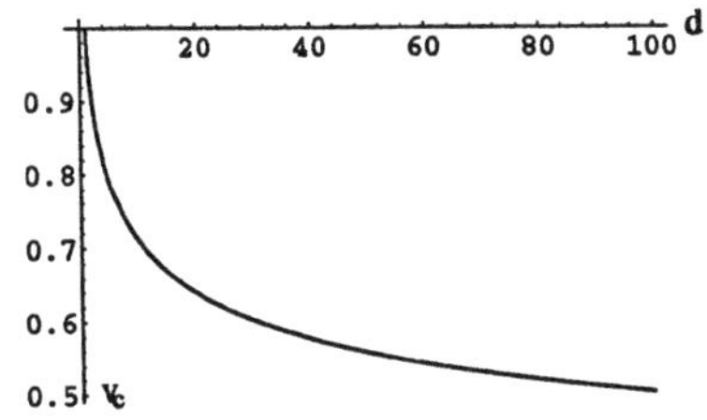

Figure 6: The minimum critical velocity of a pointed cone obstacle.

We can see in figure 6 for a and c_s equal to unity that there is not a great lowering of v_c, for instance for $d \approx 100$ the critical velocity takes the value $v_c \approx 0.5$ which is of the order of the critical velocity for a sphere[17].

Finally in three dimensions the effect is not very important. Numerical simulations on NLS in 3-d show qualitatively this result. For instance in the flow arround a conic point one could imagine the nucleation of a little ring of a quantized vortex line which radius is order the coherence length a, in analogy to the bidimensional case, however we have a line of vorticity generated on the surface of the cone (see figure 7). This line is not closed, which is required by topological reasons, but ends on the walls. It develops then as a consequence of the advection or self-interaction. Nucleation of this kind of lines of vorticity was recently observed by E. Varoquaux *et al.* at Saclay (France) in experiments of flows through holes [14].

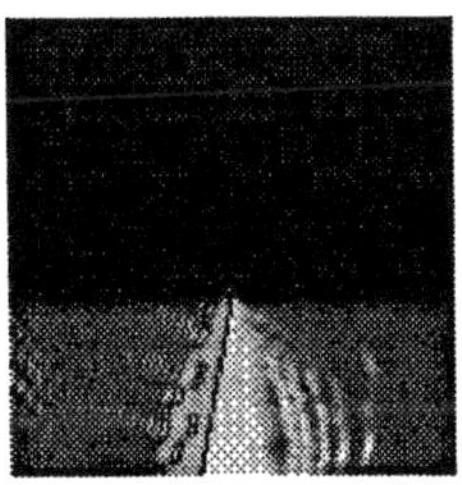

Figure 7: Cut of the three dimensional cone, we observe the vortex lines at the right of the cone. We can also the spontaneous nucleation of a ring of vorticity by the scattering of sound, in the front of the cone.

[17] The solution of the Laplace equation for a flow around a sphere show that the velocity increases by a factor 3/2 (instead of 2 for a cylinder) which implies that the critical velocity is order $\frac{2}{3}c_s$.

The numerical simulations were done on the Connection Machine of the "Centre Regional de Calcul PACA,antenne INRIA-Sophia-Antipolis" through the R3T2 network.

The authors thanks Professor Alan C. Newell for useful discussion and the *Department of Applied Mathematics of the University of Arizona (Tucson)* where part of this work was done. We thank T. Frisch for his interest and collaboration at the beginning (chapter 3). Also we thank M.E. Brachet, P. Coullet, K. Mallick, C. Nore and E. Varoquaux for their comments and constants interest in the subject.

References

[1] L.D. Landau, J. Phys. (U.S.R.R) **5**, 71 (1941).

[2] L.D. Landau, J. Phys. (U.S.R.R) **11**, 91 (1947).

[3] R.P. Feynman, *Progress in Low Temperature Physics*, Vol I, Chap II (1955).

[4] D.G. Henshaw & A.D.B. Woods, Phys. Rev. **121**, 1266 (1961).

[5] L. Onsager, Nuovo Cimento Suppl. **6**,2, 249 (1949).

[6] N.N. Bogoliubov, J. Phys. (U.S.R.R) **11**, 23 (1947).

[7] V.L. Ginzburg & L.P. Pitaevskii, JETP **34**, 1240 (1958).

[8] L.P. Pitaevskii, JETP **40**, 646 (1961).

[9] E.P. Gross, J. Math. Phys. **4**, 195 (1963).

[10] T. Frisch, Y. Pomeau & S. Rica, Phys. Rev. Lett. **69**, 1644 (1992).

[11] L.D. Landau & E.M. Lifshitz, *Fluid Mechanics* Vol. 6.

[12] Y. Pomeau, *Remarques sur les écoulements externes compressibles*, (Masson Éditeurs 1992).

[13] J.C. Maxwell, *A Treatise on Electricity & Magnetism*, Art. 191, Dover (1954).

[14] G.C. Ihas, O. Avenel, R. Aarts, R. Salmelin & E. Varoquaux, Phys. Rev. Lett. **69**, 327 (1992).

SUBJECT INDEX

A

B

C

G

H

I

J

K

L

M

N

O

P

Q

R

S

T

U

V

W

Z

Other *Mathematics and Its Applications* titles of interest:

F. Langouche, D. Roekaerts and E. Tirapegui: *Functional Integration and Semiclassical Expansions.* 1982, 328 pp. ISBN 90-277-1472-X

N.E. Hurt: *Geometric Quantization in Action. Applications of Harmonic Analysis in Quantum Statistical Mechanics and Quantum Field Theory.* 1982, 352 pp. ISBN 90-277-1426-6

C.P. Bruter, A. Aragnal and A. Lichnerowicz (eds.): *Bifurcation Theory, Mechanics and Physics. Mathematical Developments and Applications.* 1983, 400 pp. *out of print,* ISBN 90-277-1631-5

N.H. Ibragimov: *Transformation Groups Applied to Mathematical Physics.* 1984, 414 pp. ISBN 90-277-1847-4

V. Komkov: *Variational Principles of Continuum Mechanics with Engineering Applications. Volume 1: Critical Points Theory.* 1986, 398 pp. ISBN 90-277-2157-2

R.L. Dobrushin (ed.): *Mathematical Problems of Statistical Mechanics and Dynamics. A collection of Surveys.* 1986, 276 pp. ISBN 90-277-2183-1

P. Kree and C. Soize: *Mathematics of Random Phenomena. Random Vibrations of Mechanical Structures.* 1986, 456 pp. ISBN 90-277-2355-9

H. Triebel: *Analysis and Mathematical Physics.* 1987, 484 pp. ISBN 90-277-2077-0

P. Libermann and Ch.M. Marle: *Sympletic Geometry and Analytical Mechanics.* 1987, 544 pp. ISBN 90-277-2438-5 (hb), ISBN 90-277-2439-3 (pb).

E. Tirapegui and D. Villarroel (eds.): *Instabilities and Nonequilibrium Structures.* 1987, 352 pp. ISBN 90-277-2420-2

W.I. Fushchich and A.G. Nikitin: *Symmetries of Maxwell's Equations.* 1987, 228 pp. ISBN 90-277-2320-6

V. Komkov: *Variational Principles of Continuum Mechanics with Engineering Applications. Volume 2: Introduction to Optimal Design Theory.* 1988, 288 pp. ISBN 90-277-2639-6

M.J. Vishik and A.V. Fursikov: *Mathematical Problems in Statistical Hydromechanics.* 1988, 588 pp. ISBN 90-277-2336-2

V.I. Fabrikant: *Applications of Potential Theory in Mechanics. A Selection of New Results.* 1989, 484 pp. ISBN 0-7923-0173-0

E. Tirapegui and D. Villarroel (eds.): *Instabilities and Nonequilibrium Structures (II). Dynamical Systems and Instabilities.* 1989, 328 pp. ISBN 0-7923-0144-7

V.G. Bagrov and D.M. Gitman: *Exact Solutions of Relativistic Wave Equations.* 1990, 344 pp. ISBN 0-7923-0215-X

Other *Mathematics and Its Applications* titles of interest:

O.I. Zavialov: *Renormalized Quantum Field Theory.* 1990, 560 pp. ISBN 90-277-2758-9

S.S. Horuzhy: *Introduction to Algebraic Quantum Field Theory.* 1990, 360 pp. ISBN 90-277-2722-8

V.G. Makhankov: *Soliton Phenomenology.* 1990, 461 pp. ISBN 90-277-2830-5

S. Albeverio, PH. Blanchard and L. Streit: *Stochastic Processes and their Applications in Mathematics and Physics.* 1990, 416 pp. ISBN 0-9023-0894-8

V.A. Malyshev and R.A. Minlos: *Gibbs Random Fields. Cluster Expansions.* 1991, 248 pp. ISBN 0-7923-0232-X

E. Tirapegui and W. Zeller (eds.): *Proceedings of the 3rd Workshop on Instabilities and Nonequilibrium Structures.* 1991, 370 pp. ISBN 0-7923-1153-1

C.A. Marinov and P. Neittaanmaki: *Mathematical Models in Electrical Circuits. Theory and Applications.* 1991, 160 pp. ISBN 0-7923-1155-8

C. Bartocci, U. Bruzzo and D. Hernandez-Ruiperez: *The Geometry of Supermanifolds.* 1991, 242 pp. ISBN 0-7923-1440-9

E. Goles and S. Martinez: *Statistical Physics, Automata Networks and Dynamical Systems.* 1992, 208 pp. ISBN 0-7923-1595-2

A. van der Burgh and J. Simonis (eds.): *Topics in Engineering Mathematics.* 1992, 266 pp. ISBN 0-7923-2005-3

L. Aizenberg: *Carleman's Formulas in Complex Analysis.* 1993, 294 pp. ISBN 0-7923-2121-9

E. Tirapegui and W. Zeller (eds.): *Instabilities and Nonequilibrium Structures IV.* 1993, 371 pp. ISBN 0-7923-2503-6

GPSR Compliance

The European Union's (EU) General Product Safety Regulation (GPSR) is a set of rules that requires consumer products to be safe and our obligations to ensure this.

If you have any concerns about our products, you can contact us on ProductSafety@springernature.com

In case Publisher is established outside the EU, the EU authorized representative is:

Springer Nature Customer Service Center GmbH
Europaplatz 3
69115 Heidelberg, Germany

Batch number: 10584568

Printed by Printforce, the Netherlands